AF711196

to Yvonne and Al Martens -

with warmest wishes

Judith Taylor

THE GLOBAL MIGRATIONS OF ORNAMENTAL PLANTS

How the World Got into Your Garden

THE GLOBAL MIGRATIONS OF ORNAMENTAL PLANTS

How the World Got into Your Garden

BY

JUDITH M. TAYLOR

WITH A FOREWORD BY PETER H. RAVEN

Missouri Botanical Garden

ISBN 978-1-930723-69-6
Library of Congress Control Number 2008934783

Scientific Editor and Head: Victoria C. Hollowell
Managing Editor: Beth Parada
Associate Editor: Allison Brock
Editorial Assistant: Tammy M. Charron
Press Coordinator: Cirri R. Moran

Design and typesetting by Adrianna Sutton.
www.adriannasutton.com

Copyright © 2009 by Missouri Botanical Garden Press
P.O. Box 299
St. Louis, Missouri 63166-0299, U.S.A.
www.mbgpress.org
All rights reserved. Printed in the U.S.A.

To Virgil, Mabel, Cosmo, Noam, Ezra & Gavriel

CONTENTS

PART THREE **GLOBAL TRANSCENDANCE**

FOREWORD

WE FIRST KNOW of ornamental plants from an ancient tale told of the Egyptian Queen Hatshepshut, who imported 31 frankincense trees to Thebes to line the path to her memorial and also for commerce. Our cultivation of the singular Christmas or Lenten rose (*Helleborus niger*) can be traced back to Roman times, as can that of the true roses and many other ornamentals shown, for example, on the wall murals of Pompeii. The earliest account of the care of ornamental plants, for *Rosa damascena*, was presented by Virgil in the 1st century C.E.

From such early beginnings, in *The Global Migrations of Ornamental Plants,* Judith Taylor explores the origins of the modern garden landscape. Ornamentals were added progressively to European gardens from northern Africa and across Europe during the first sixteen centuries of the Common Era. Now a welcome spring symbol of rebirth, ornamental tulips were selected from native species that grow in central and western Asia. Public mania for the portable wealth that these bulbs represented built to such extent that the excesses led to the Dutch financial crisis of 1637. Highly fashionable ornamental plants from the Cape of Good Hope beautified the gardens of Europe subsequently. The discovery of the New World was followed in the seventeenth century by the horticultural excitement of so many new American plants traversing the Atlantic pond. Observed in the Georgia wilds by the American botanists John and William Bartram in 1765, the Franklin tree (*Franklinia alatamaha*) would disappear from the American landscape well before the passenger pigeon and the Carolina parakeet. *Franklinia* served as a potent symbol of the New World and would be cultivated in the great gardens of Europe.

Asian plants, such as new kinds of roses, camellias, forsythias, and many others, began to reach English and American gardens in large numbers in the nineteenth century. Asia has subsequently been a major source of ornamental plants, introductions having continued down to the present day. At about the same time, ornamentals from the Pacific regions of North America appeared in gardens, accompanied by new finds from Central and South America, Australia, and New Zealand. Hartweg and Graham's explorations in Mexico would bring into the world's garden many new annuals and perennials, including annual cut flowers such as zinnias and cosmos, as well as succulents such as *Agave, Dudleya, Nolina, Yucca,* and many cacti. Our modern hybrid petunias can claim ancestry back to the Argentinean plants sent back by James Tweedie to the Glasgow Botanical Garden in 1831, while poinsettias were introduced from Mexico at about the same time.

The first tropical orchid to be cultivated in Europe appears to have been *Brassavola nodosa* by the Dutch in 1698, and this was followed by the Bahamian orchid *Bletia verecunda* in England in 1731. From Rio de Janeiro, William Swainson sent back to England in 1818 the spectacular *Cattleya labiata*, named by John Lindley, who went on to issue the classic *Genera and Species of Orchidaceous Plants*, published in seven fascicles from 1830–1840. From Brazil also came Lindley's blue orchid (1853) as *Acacallis cyanea*. This elusive orchid was found first by the intrepid explorer Richard Spruce and later again by Richard Evans Schultes, one of many orchid discoveries by both naturalists. The world would become fascinated with the mystery and beauty of orchids, and their collection led to commercial endeavors, with results that were dramatic and sometimes devastating to the plants in nature.

As Taylor's narrative illustrates, horticultural exploration has been profitable, exciting, and sometimes dangerous. Often risking their lives, early plant explorers brought exotic plants into the common landscape of the modern world. By example, what we know as Boston ivy (*Parthenocissus tricuspidata*) came not at all from there, but originated from China. Now a popular houseplant, Norfolk Island pine was once restricted to this small island that lies between Australia and New Zealand. It is now established in the gardens of warmer climes across both the Northern and Southern hemispheres. Even our ubiquitous, diminutive Bermuda grass is, in fact, a tough turf grass from Eurasia, originally described from Bombay, India.

In total, Judith Taylor estimates that an impressive 6,459 species are grown as ornamental, non-native plants in the United States. Of these worldly introductions, Asia contributes most substantially with 29%, with more than 10% derived from China alone. In contrast, just 19% of the cultivated flora in the United States originated in Europe, with an additional 15% from Mexico and South America. *The Global Migrations of Ornamental Plants* explores first the classic history of these plant peregrinations, moving on to the history of introductions in modern horticulture. Appendix 1 presents an authoritative and invaluable listing of the accepted names of representative ornamental plants. Introducing us to an English cottage garden in her prologue, Taylor outlines the intertwined origins of the plants from every temperate region of the world. Judith Taylor masterfully demonstrates that our gardens are indeed global in origin. Her book is a green delight to read.

Peter H. Raven, *President, Missouri Botanical Garden*
Victoria C. Hollowell, *Scientific Editor and Head, Missouri Botanical Garden Press*

ACKNOWLEDGMENTS

THE AUTHOR IS GRATEFUL to many generous-spirited people who helped her in the research and preparation of this book.

Axel Borg, State Librarian for Viticulture, University of California at Davis
Gavin Bredon, Hunt Institute for Botanical Documentation, Pittsburgh, Pennsylvania
Kathleen Fisher, Librarian, Helen Crocker Russell Library, San Francisco Botanical Garden at Strybing Arboretum, San Francisco
John Flanagan, Director of Libraries, Royal Botanic Gardens, Kew
James Folsom, Director of Botany, Huntington Library and Botanical Gardens, San Marino, California
Victoria Hollowell, Missouri Botanical Garden Press
Darin Jensen, cartographer, for preparation of the maps in Chapter 1
Norma Kobzina, Senior Librarian, Life Sciences Library, University of California at Berkeley
John Mackenzie, Professor Emeritus of History, Lancaster University, United Kingdom
Serena Marner, curator, Fielding Druce Herbarium, University of Oxford
William McNamara, Director, Quarry Hill Arboretum, Glen Ellen, California
Gail Michaels, St. Louis
Dore Minotodani, Special Collections, University of Hawaii
David Mitchell, Curator, Scottish Plant Collectors Garden, Royal Botanic Garden, Edinburgh
Beth Parada, Missouri Botanical Garden Press
Barbara Pitschel, Director, Helen Crocker Russell Library, San Francisco Botanical Garden at Strybing Arboretum, San Francisco
Dr. Patricia Raven, Missouri Botanical Garden
Dr. Peter Raven, Director, Missouri Botanical Garden
Charlotte Tancin, Librarian, Hunt Institute for Botanical Documentation, Pittsburgh, Pennsyslvania
Angela Todd, Archivist, Hunt Institute for Botanical Documentation, Pittsburgh, Pennsylvania
Freek Vrugtman, International Lilac Registrar, Carlisle, Ontario, Canada
Bruce Williams, Librarian, East Asia Library, University of California at Berkeley

English cottage garden. Reproduced by courtesy of Margaret Hensel.

PROLOGUE

AH, THE ENGLISH COTTAGE GARDEN, famed artifact of legend and fable. Let us take a closer look. Here are the mainstays we see most often:

Hollyhocks came from western Asia.
Delphiniums came from Europe and Asia.
Roses came from China.
Marigolds came from Africa.
Wisteria came from China.
Hydrangea came from America and Japan.
Lilac came from Turkey and China.
Lavender came from the south of France.
Dahlias came from Mexico.
Begonias came from South America.
Tulips came from Turkey.
Nasturtiums came from South America.
Pinks came from Normandy.
Poppies came from Greece.
Wallflowers came from southern Europe.
Anemones came from the eastern Mediterranean.
Peonies came from southern Europe.
Crocus came from Russia and central Europe.
Lupins came from the eastern Mediterranean.
Ivy came from Japan.
Gladioli came from South Africa.
Lilies came from China.
Mock orange (*Pittosporum*) came from Australia.
Geraniums (*Pelargonium*) came from South Africa.
Chrysanthemums came from Japan.

Where is the "English" in the English cottage garden?

INTRODUCTION

AT ONE TIME NORTH AMERICA was the principal source for exotic plants in Great Britain and Europe. Today the tables are turned. Not only are European gardens alight with Asian and Southern plants, but North American gardens include more plants derived from Europe, Asia, and the Southern Hemisphere than native ones. Most of the ornamental plants we grow in the United States are descendants of imported species.

The American Horticultural Society A–Z Encyclopedia of Garden Plants lists the origin of the plants commonly grown in North America. The numbers in the table below are in descending order of frequency of origin.

Asia and the Southern Hemisphere have supplied more species for our American gardens than North America itself, 5,343 compared with 1,122. What is surprising is how many other plants came from Europe. The European imports were either indirectly from the colonial period or more recent, with the preponderance being more recent.

Mexico (together with Central America) and South America combined are the source of 979 species, almost as many as natives. Two hundred of these come from Brazil alone.

The Antipodes (Australia and New Zealand with their associated islands) add up to 488 species, a not insignificant number (see Appendix 2).

This migration of plants from across the globe took place not only in North America but also in Europe. The wholesale replacement of the cultivated floras of two continents was striking enough by itself, but the collateral development of

Origins of Garden Plants Grown in the United States at Present[1]

Asia (including China)	1,878 species	(China alone: 663 species)
Europe	1,222 species	
North America (including Canada)	1,122 species	(Canada alone: 7 species)
Africa (including South Africa)	770 species	(South Africa alone: 453 species)
South America	631 species	(Brazil alone: 215 species)
Mexico	348 species	
Australia	298 species	
New Zealand	190 species	

1. Calculated from Christopher Brickell and Judith D. Zuk, eds., 1997, *The American Horticultural Society A–Z Encyclopedia of Garden Plants*, New York: DK Publishing Company.

science and industry that was derived from this plant migration adds another, very rich dimension to the story. This text will examine the complex series of events that ended in this floral colonization and follow the fascinating twists and turns of the many stages.

At first, each player in the drama was intent on a single purpose. There was no overarching design. No one set out intentionally to make such sweeping changes. The changes resulted from the accumulation of one discovery after another. Finally they all coalesced. This could be called a chain of inadvertent causality, to use a voguish technical term.

It is doubtful whether any one of the plant explorers or the recipients of their bounty initially recognized what the ultimate effect would be. They were satisfying immediate interests.

From about the sixteenth to the nineteenth centuries, plant collection was still merely an aspect of capturing the natural history of new lands, coinciding with geographical exploration. At first, only an occasional scholar traveled abroad specifically in order to find new plants. Pierre Belon (1517–1564) was probably the first.[1]

A few of the travelers had specific agendas. The German physician Leonhard Rauwolf (1535–1596) went to the Levant in 1573 and 1575 because he wanted to identify the plants in the Bible. All he had to go on were the crude woodcuts in the old herbals, and he could not match these to the plants he knew in Germany. Rauwolf was forced to the conclusion that holy writ was fallible or, at the very least, incomplete. In the case of plants, the Bible was not the last word. This recognition led to the idea that may there were other cases in which men could think for themselves, on the basis of visible evidence and not authority.

The bizarre and unexpected attracted early adventurers. The first European to report back on western Australia in 1699 after the Dutch sea captain Abel Tasman (1603–1659) had discovered yet another new continent was a pirate "of exquisite mind," Captain William Dampier (1651–1715).[2]

The forces that drove so much exploration in the eighteenth century combined the effects of the Enlightenment with military and political needs. The Enlightenment generated intellectual freedom and released human thought from superstition and religious obscurantism. It was a heady and exhilarating age.

Some of the better-known explorers and plant collectors were Sir Joseph Banks (1743–1820), who went to Australia with Captain Cook (1728–1779) in 1768, and Archibald Menzies (1754–1842), who accompanied Captain George Vancouver (1757–1798) to the West Coast of North America in 1791. Meriwether Lewis of the Lewis and Clark expedition from 1804 to 1806 found dozens of new species for Thomas Jefferson, but this contribution tends to be overlooked in the richness of all their other discoveries. Alexander von Humboldt traveled in North and South America between 1799 and 1805, leaving a legacy of immense scientific importance.

Toward the end of the nineteenth century, the focus shifted to exploration for commercially viable plants. It paid nurserymen to send their own explorers to

South America or the East because there was an avid market for whatever they imported. We shall look at the activities of Messrs. Veitch and Son and their employees the Lobb brothers. This exploration coincided with the heyday of the British Empire and came to an end with World War I.

The economic goal superseded the more leisurely searches for scientific purposes, but there was considerable overlap. For every agent of an aggressive nursery, there were still the disinterested travelers such as Charles Sprague Sargent from the Arnold Arboretum in Massachusetts or Augustine Henry from Ireland.

The changes described in the chapters ahead evolved very slowly over time. The cultivated British landscape underwent a series of transformations over a period of about 1,500 years, from the time of the Roman invasions to the recent past. The last Ice Age many thousands of years before had left the British Isles with a somewhat poverty-stricken flora. Useful and ornamental plants that now appear to be native came from continental Europe.

Imported plants also transformed the landscapes of France, Belgium, Holland, Germany, Spain, and the other European countries. At first the movement was between neighboring provinces and territories. Once it became possible to travel longer distances, the new flora came from farther and farther away: Russia, the Balkans, the Near East, and the Middle East. Wars and conquests played a major role in these movements.

Modern North America, at first a series of European colonies, underwent the same type of transformation telescoped into a shorter historical period. The transformations of the North American flora were from a pure native state before colonization to a European horticultural period lasting about two centuries, from approximately 1600 to 1840, and finally to the one that is still current, the Asia–Southern Hemisphere period. In his recent book *1491* Charles Mann lays out the rich complexity of native culture and commerce before the European onslaught.[3]

Many searches for the native flora of North America were launched over the centuries after colonization. The results of these journeys fueled the commerce in American plants going back to Europe. The significant explorations will be considered in more detail in Chapter 7.

Plants traveled in both directions. The size of the American continent and its wide climatic variations and remarkable native flora prevented the American transformation from being as comprehensive as the European one, but the cultivated sections reflect it closely. Gardens in North American cities reflect the gardens in European cities more closely than they reflect their own native flora.

In places that are storehouses of botanical diversity, such as China, Japan, India, Mexico, South America, Australia, and South Africa, and from which the Western garden derives most of its modern plants, flowers from Western countries are also flourishing in reverse migrations. Part of this is a result of historical movements of settlers, but some is also because the lure of the foreign or different is very strong.

I propose to look at Europe, particularly Britain, and North America as the instigators and receivers while recognizing that there are other points of departure for studying the global migration of garden plants. This is necessary to extract the underlying sense of what happened. Just enumerating lists is only half the job. The lists are here and they are very important, in fact the bedrock of the story, but there is more to be learned.

The first section of this book will be a discussion of the Atlantic ascendancy, i.e., plants moving across the North Atlantic in both directions. Without a guide to the history, politics, and geography, even of the most superficial kind, that preceded this phase, making sense of the story would be difficult.

The second section will consider the Asian and Southern ascendancy. Stories about the amazing men who brought new plants from remote, often dangerous, places bring this history to life. The story of the orchid in nineteenth-century England illustrates how nurserymen seized the initiative as rich men competed with each other over who had the most gorgeous plants.

The third section will include some comments on today's global ascendancy, when all these things have become commonplace. Much of the credit goes to horticultural societies, which have provided the core knowledge and often instigated the searches themselves.

The ultimate meaning of all these separate events has to be sought in the forces that were at play when they took place. Looking for gold, silver, and other precious metals was one of the commonest reasons for exploration. Missionary work, the need to convert "savages" to Christianity, followed in the wake of treasure seeking. Queen Isabella was probably more concerned about this than about finding gold. For quite a time, new plants came back as secondary phenomena.

Once treasure seeking was played out, the colonial powers began to notice the other resources in their new possessions. Individual scientists, such as the early missionaries and other idealistic scholars, had always been interested in finding new flora and fauna, but exploitation still ruled organized expeditions. It was uppermost in the minds of business and political leaders of most nations. All they cared about was wealth.

Builders of empires considered plants in a strictly utilitarian manner. Having more beautiful gardens at home was a delightful by-product. Wealth through better crops and commercially useful products was the real goal. A plant's function was either to provide new medicines that would allow colonists and occupiers to remain in the subjugated territory more safely (e.g., quinine) or to make money in some cases (e.g., cotton, tea, rubber, and sugar) and to save money in others (e.g., breadfruit).* None of these essential functions would have been accomplished without the work of the plant explorers and collectors.

---

*It was typical of the raw capitalistic societies that they should want to feed their slaves and peasants on the cheapest possible diet compatible with staying alive and being able to work. The epic story of Captain Bligh and the mutiny on the *Bounty* occurred because he had been charged with transporting the breadfruit tree from Tahiti to the Caribbean islands to feed the slaves. Protecting one's investment and increasing productivity by improving the lot of the worker was a much later and more sophisticated notion.

Another practical function of living plants, seen only once, was for punitive purposes in India (see Chapter 8). The Raj, or East India Company, had a monopoly on salt. Salt was heavily taxed under the Raj but not in the native states. The company created an impenetrable hedge of thorn trees along the 2,000-mile-long border with the native states to prevent smuggling and avoidance of customs duty.

This analysis of early colonial exploitation takes account of the known facts in light of a fairly well-accepted theory, "botanical imperialism." It is not intended to be political commentary. The bad old days of stealing the resources of another country have been largely mitigated though they are possibly not gone entirely. International treaties prevent the wholesale theft of unique botanical treasures.

This hitherto unprecedented rearrangement of the world's floras hastened the development of botany and horticulture. The teaching and scientific functions of a botanical garden in Europe were useful adjuncts to trade. Colonial botanical gardens started as mere holding pens to foster the discoveries, and yet they had an unexpectedly subversive effect.

In spite of the four-square, business-like approach to managing these gardens, science and beauty crept in like water between the cracks. Hedonistic aristocratic rulers and elite groups dominated the sponsoring organizations in the home countries. There was great personal rivalry between wealthy men with immense estates, large hothouses, and wives with social ambition. Even men in the second rank, such as merchants and tradesmen, could indulge this passion if they had enough money. Enterprising nurserymen played a significant role in the distribution of the new material, catering to several levels of society in this way.

Studious vicars and university scholars benefited from the inflow of exotic plants. If all that the scientists had to study was the material within their own borders, there would have been less pressure to advance so quickly. With the deluge of new material, the all-purpose scholar and amateur quickly gave way to specialists.

OVERVIEW

THE RATE OF MOVEMENT of plants around the world over the past three centuries, markedly accelerating between the nineteenth and twentieth centuries, is quite remarkable. Movement itself is not, however, unprecedented.

Vascular plants are sedentary organisms. Their ability to move is confined to the rotation of terminal limbs or branches while the roots and stem remain fixed in the ground. Some seeds are equipped with organs allowing them to move passively on the wind or float in water. The coconut is a perfect example of this. It can cross the ocean, but most seeds remain close to the parent plant.

Taking a plant from its normal habitat and moving it somewhere else requires an external agent. In nature, this is done involuntarily by the wind and by birds and animals. The latter eat the fruit or seed, and the remains pass unchanged through their digestive tract. The intact core germinates in new territory.

Charles Darwin (1809–1882) devoted a good deal of time to these phenomena. He kept up a friendly argument with Sir Joseph Hooker over the unexpected resemblance of plant life on very distant continents. Hooker (1817–1911) believed, with Asa Gray (1810–1888), professor of botany at Harvard University, that the separation of Pangaea into continents and the subsequent isolation of landmasses were the cause. Sir Joseph, the son of Sir William Hooker, took over the administration of the Royal Botanic Gardens at Kew when his father died in 1865. He had spent many years collecting plants in the East and the Southern Hemisphere. His opinions were backed by strong empirical evidence.

Darwin agreed with that idea but also thought that seeds themselves might contribute to the end result. He carefully experimented with seeds that had soaked in seawater for months or that had been fed to birds and then recovered from their digestive tracts.

Hooker offered to help him. In May 1856, he wrote:

"My dear Darwin

Nymphæa seeds—we can get you plenty in Autumn if you will remind me."[4]

It made sense to use water lily seeds, as they were adapted to remain in water.

In each case the seeds germinated in sufficient numbers to maintain the population.[5] Darwin contended that plants could spread across the seas successfully in this fashion. It seems that both concepts are correct.

He also watched other ways in which plants move. He wondered about the twining of vines, how it was done, and for what purpose. We assume that the plants use twining to climb toward the sun. Darwin found that, although this was a useful side effect, the actual function of the twining was to propel the sap over the very long distance from the roots to the active growing point.[6] Darwin was like John Hunter, the great eighteenth-century surgeon and anatomist. He did not "suppose" but got down to cases and studied the phenomenon he wanted to understand the hard, slow way.

Intentional change in the location of plants has been the province of human beings. It began when farmers had to move to new areas as populations grew and settlements ran out of space. At first the move would be over a very short distance, but by the time of the Greeks, for example, colonists sailed many miles across the Mediterranean Sea.

They carried the seeds of their essential food plants such as grain with them as well as livestock. Inevitably, weed seeds were mixed in with the wheat and with the droppings of the animals. Both the desired plant and the undesirable one were redistributed in a new habitat. Human agency, though minute in comparison with evolutionary forces, nevertheless accomplished something new by causing plants to cross natural barriers.

Redistribution of plants continued in this episodic fashion, always involving cultivated plants and most often useful plants. The well-known story of the Egyptian Queen Hatshepshut sending emissaries to the Land of Punt (Somalia) to bring back 31 frankincense trees is one of the earliest on record (Figure 1). The trees were used to create an avenue for her memorial at Thebes across the Nile from Karnak and Luxor, but the aromatic extract was an important article of commerce.[7]

Queen Hatshepshut's nephew Thothmes III also imported plants. He created "Syrian gardens" in about 1500 B.C.E. Bas-reliefs of the *Oncocyclus* iris he presumably imported can be seen on the pillars of his temple at Karnak.[8]

The Egyptians cultivated several flowers still frequently used nowadays. Poppies, lilies, and crocus grew in the garden of one of Amenhotep III's officials. Knowledge about this garden has come down to modern times because a picture of it was painted on a tomb wall. Each of these flowers had some ritual or medicinal use, but it is likely that the Egyptians recognized their ornamental value. The Egyptian word for a flower has survived, via the Hebrew name Shoshana, in the form of the modern name Susan.

Cultivated plants were the first to be transferred or exchanged. Food was the fundamental requirement, and food plants took precedence over everything else. Flowers were collected and later grown for real or imagined therapeutic effects.

The capacity to recognize medicinal properties in plants is ancient. It has been handed down by nonliterate peoples through the generations. Richard Evans Schultes (1915–2001), professor of botany at Harvard and one of the great twentieth-century plant collectors and ethnobotanists, was in awe of the shamans and tribesmen he met in the forest around the Amazon River.

Each generation of native people learned about powerful psychoactive drugs from the previous one. These preparations were isolated from very inconspicuous sources. One liane looks very much like another, yet all the people Schultes met picked the correct ones unerringly. Modern pharmaceutical firms have sent explorers into these forests seeking sources of new and effective drugs.

Plant exchanges were widespread in the ancient world. Plants went between India and China; between China, Korea, and Japan; and between other civilizations. Robert Fortune found magnolias in China. He had no idea they had come from India hundreds of years before.

The idea of growing a plant purely for its appearance was not recorded for many centuries, but that does not mean it did not take place. Ancient myths, such as the story of Echo and Narcissus, attest to this. Narcissus wasted away admiring his own reflection in the lake, and his body was transformed into the flower now known as the narcissus. Romantic views of flowers flourished in all cultures. Color and fragrance appealed to men and women alike. Individual blossoms were widely used as adornments.

The thought of deliberately seeking out new, wild plants for ornamental purposes took a long time to develop. Danger was the principal deterrent. For a long time, wild things were suspect. They were the province of savages or barbarians. Children might play with wildflowers, and poor adults who had no other source picked them. The idea that they were beautiful or merited serious attention by the upper classes would only arise with increasing security and leisure.

Returning to the modern period, the ultimate crossover of the transplanted ornamental floras took place in the nineteenth century and was more or less complete by about 1870. The excitement of the "American" garden in Europe had ceased by then. Almost all the commonly grown plants, the annuals, the flowering shrubs, the vines, and many of the trees all now originated in Asia or the Southern Hemisphere.

On the other side of the Atlantic Ocean, the fashionable obsession with European style, and hence its ornamental plants, had also diminished. The same Asian and Southern Hemisphere imports held sway there, too. Their route had been circuitous. Some came via Europe as a way station. Others came into the United States directly from Japan and China.

Part One

ATLANTIC ASCENDANCY

CHAPTER 1

PLANTS BEGIN MOVING

The Onset of Floral Colonization

A GOOD PLACE TO START the examination of plant migration is with England. England stands for Britain as a whole in this context. Scotland underwent botanical transformation more or less simultaneously with England. Ireland and Wales were involved later. It is reasonable to refer to England as a proxy for Britain in the account that follows. Many serious advances in botany and horticulture were made in England, though by no means all of them.

There are several significant threads in the story and a fair amount of overlap. None of the changes was abrupt. In all, there were four major transformations of the English (British) landscape, dated arbitrarily for convenience only:

1. Roman period (55 B.C.E. to ca. 500 C.E.)
2. Medieval–Early Modern period: "Medieval" is a shorthand term used to describe complicated historical events. It is customarily considered to have started in the twelfth century C.E., but it may have been somewhat earlier in what are still known as the "Dark Ages." For our purposes it can be said to merge into the Early Modern period, which lasted until at least 1600. There were some periods between about 300 and 1100 C.E. when nothing was recorded about the movements of plants into Britain.
3. American period (1580s–1820s)
4. Asia–Southern Hemisphere period (1760s to present)

The earliest records indicate that the British Isles were not rich in native ornamental flowering plants compared with the countries across the Channel. Rackham notes that there were only about 50 or so native species of trees.[9] The islands suffered from the receding Ice Age for much longer than more southerly regions on the continent of Europe. Before the English Channel opened, plants from the continent were able to cross directly and took hold where conditions favored them.

Britain was said to be covered with forest for thousands of years. As far back as 11,000 B.C.E., birch, aspen, and sallow (goat willow, *Salix caprea*) predominated. By 8500 B.C.E., pine and hazelnut began to edge out the birch. Oak arrived soon after this phase. Later the pine trees receded and were to be found only in the northern reaches of the islands.

Recent archeological investigations suggest that there was considerable variation in the forest mantle.[10] The size and distribution of the oak trees have probably been exaggerated, for example, though oak became the dominant native tree. Rackham believes that elm was carried by humans and not by spontaneous natural processes.[9]

Rough heaths and wealds were widespread. Dartmoor was a menacing place into modern times; it became the site of a famous prison in 1809. Conan Doyle exploited this terror in *The Hound of the Baskervilles.* The Brontë children roamed the wild Yorkshire moors. In Thomas Hardy's *The Mayor of Casterbridge*, Egdon Heath was a significant feature of the landscape (and some would say a character in the book). Traces of the wealds close to London persist in local place names: Harrow and Wealdstone is one.

There were also grass-covered plains. Stonehenge was built on Salisbury Plain. Plains made agriculture possible.

Some of the native pine trees grew very tall, with the first branches not appearing for about 20 feet. This was the plant's physiological response to gaining light and food.

Acorns, nuts from other trees, and the forest's undergrowth supported populations of pigs. When beech arrived, its nuts, "beech mast," were an additional source of food for the pigs. The early inhabitants grazed sheep, horses, and cattle in meadows. Food from these and other sources, such as rye and oats, supported their population, but famine was always a possibility during the summer until the harvests began maturing,

The Britons kept up a flourishing trade with the Continent. Among other things, it was this rich trade that attracted the Romans. The islanders could not have reached such a level of civilization without a reasonably reliable source of nourishing food, but the standard foods now associated with rural Britain had not yet been imported. It is hard to imagine that there were no turnips, cabbage, carrots, potatoes, or apples.

Agriculture was more advanced than has been previously credited, according to Rackham.[9] The archeological record indicates that whole sections of the country were neatly divided up into fields by 1000 B.C.E.

Britain had wildflowers such as dog roses, violets, and primroses, and some of the tribes had discovered *Isatis tinctoria*, woad. They used this to paint their faces blue.

When Julius Caesar first tried to land in Britain in 55 B.C.E., crossing the Channel from what is now Boulogne in Gaul, he failed because of the weather. In the following year, he returned and established a bridgehead but still did not succeed in completely subduing the British tribesmen. Caesar never returned. It was not until 43 C.E., nearly a century later, that the emperor Claudius' legions truly conquered Britain. The Romans remained there for more than three centuries.

The aboriginal inhabitants are assumed to have migrated from the continent of Europe, but nothing is known of what plants they might have taken with them. The first transformation can be dated to the Roman invasions. We can't quite begin at the year one, but the relatively barren flora began to be enriched as valuable plants were taken there from western Europe, the Near East, and even North Africa, when the Romans settled in Britain in 43 C.E. The process was slow at first, but the pace picked up as communications improved.

With the arrival of the Romans, the number of edible and ornamental plants increased. The first steps had been taken. More would be taken over a period of about 1,500 years.

The Romans had access to numerous food and ornamental plants from their enormous empire. They included ancient and valuable foods from Greece and Asia Minor, such as wheat, olives, and grapes, that stretched back into antiquity. Roses were found in the lands that are now Turkey and Iran. By the time the Romans invaded Britain, they too assumed that everything that went with them was part of their heritage, not thinking whence so many valuable plants had come.

What follows is pure speculation, but one can imagine slaves in the Roman legions filling baskets in Gaul with foods their master enjoyed. Maybe they added a couple of roses, the master's favorite flower, to keep him in a good mood. Wiser ones also thought of taking vegetable seed.

Crossing the Channel did not take a long time unless the winds blew in the wrong direction. It was possible to keep foods and vegetable seeds in good condition.

The Romans probably made many voyages back and forth across the Channel over the years. At the outset, they would have taken only enough to sustain themselves in the short run. Armies always try to live off the land as they advance. Only when the Roman officers and officials built cities, estates, and private dwellings would ornamental plants even be considered.

The Medieval–Early Modern period saw further augmentation of the flora (Map 1). By the end of the fifteenth century, new plants coming from North and South America began to overlap very slowly with the movement of plants from the continent of Europe to England (Map 2). The second transformation was afoot.

This was the faint beginning of a major transformation, but for the time being a more robust expansion within the known "old" world was taking place. Broadened military and diplomatic contacts with Russia, Turkey, and eastern Europe led to more and more new species entering the British Isles in the sixteenth and seventeenth centuries (Map 3). This phase lasted until the early eighteenth century. The second transformation was complete.

Some of the flowers English emigrants later took to North America were these recent "foreigners." There was a constant flux of plants going in both directions for the next century, yet in the end more plants went to England than the other way about. The third transformation, to the American period, was in full swing (Map 4).

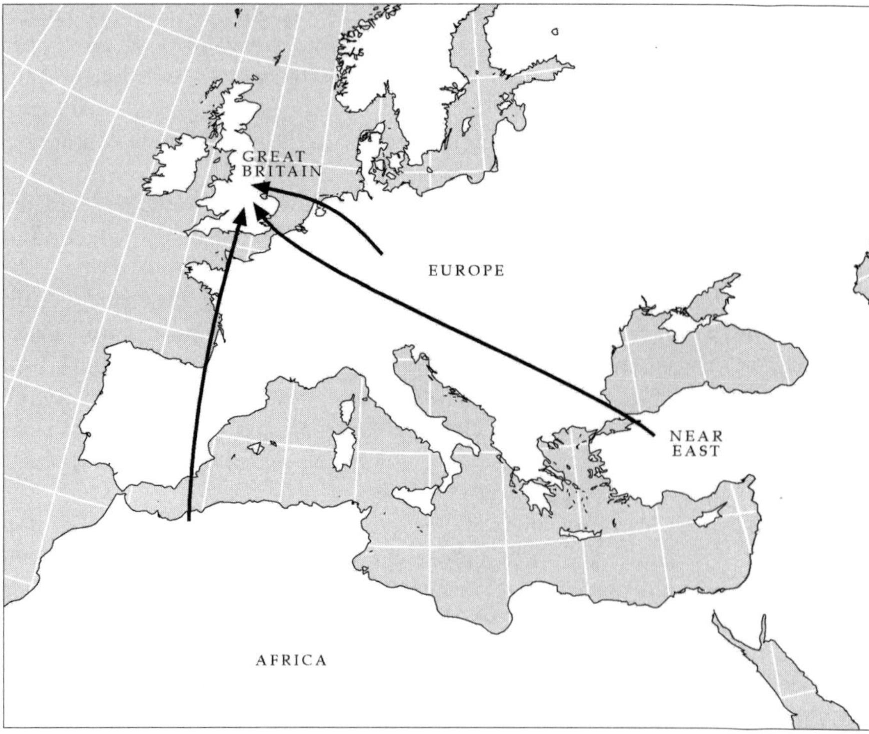

Map 1. Plants migrating into Britain before 1500.

Map 2. North and South American plants entering Britain ca. 1600.

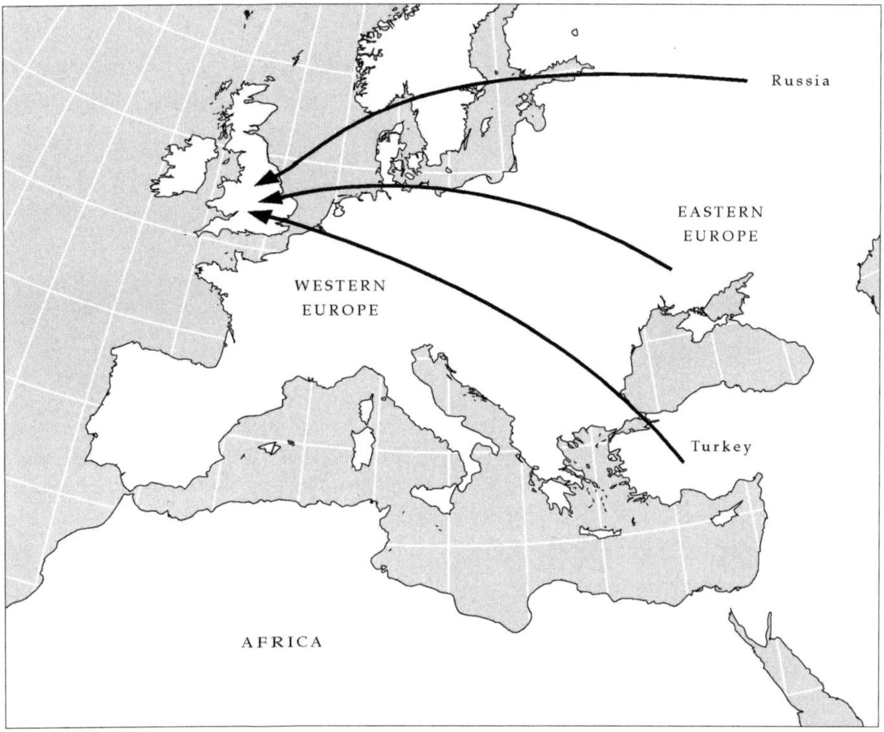

Map 3. Eastern European and western Asian plants entering England in the sixteenth and seventeenth centuries.

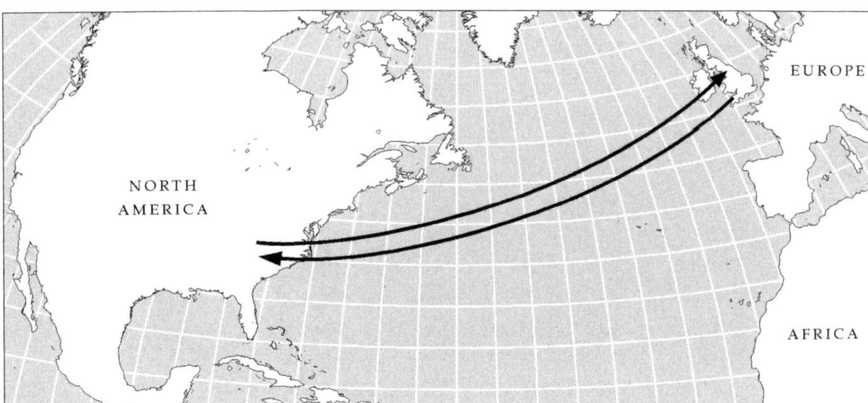

Map 4. Transatlantic flux ca. the eighteenth century.

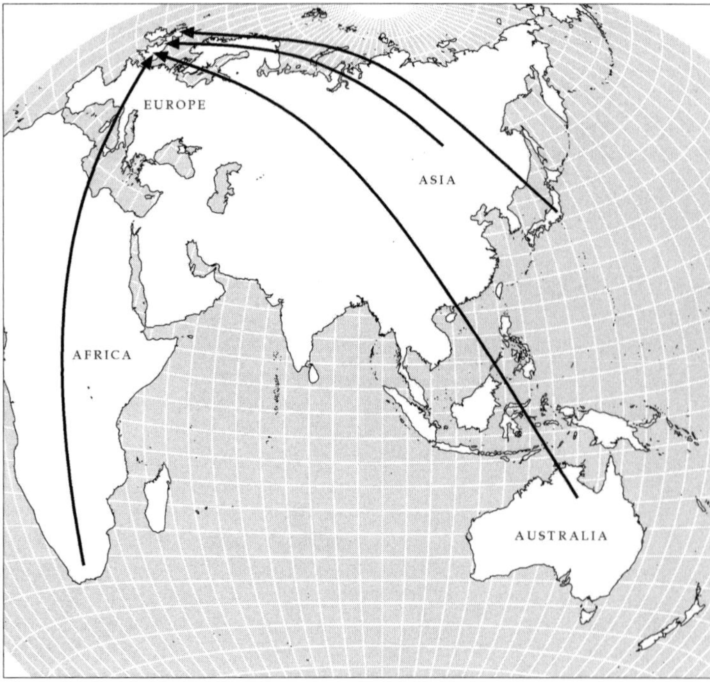

Map 5. Early introduction of Asian and Southern Hemisphere plants, ca. 1760 to 1800.

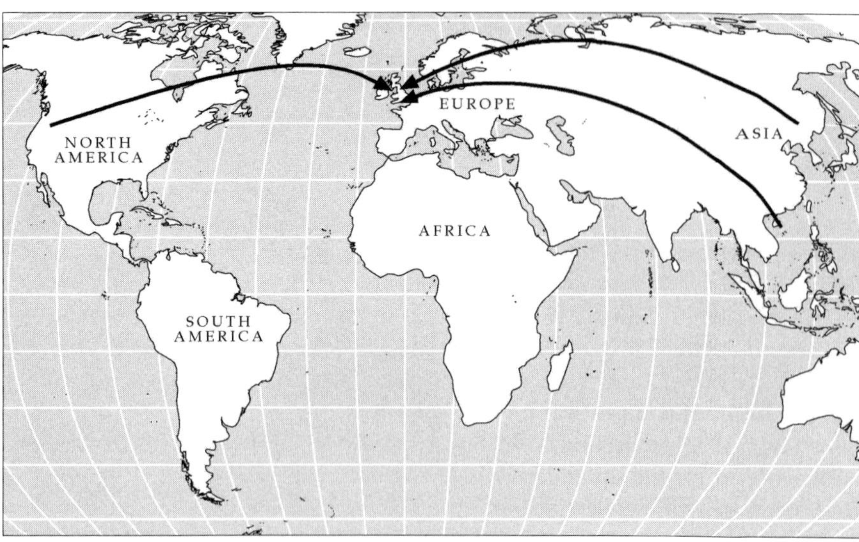

Map 6. Crossover of introductions from the western United States and the East in Britain, mid-nineteenth century.

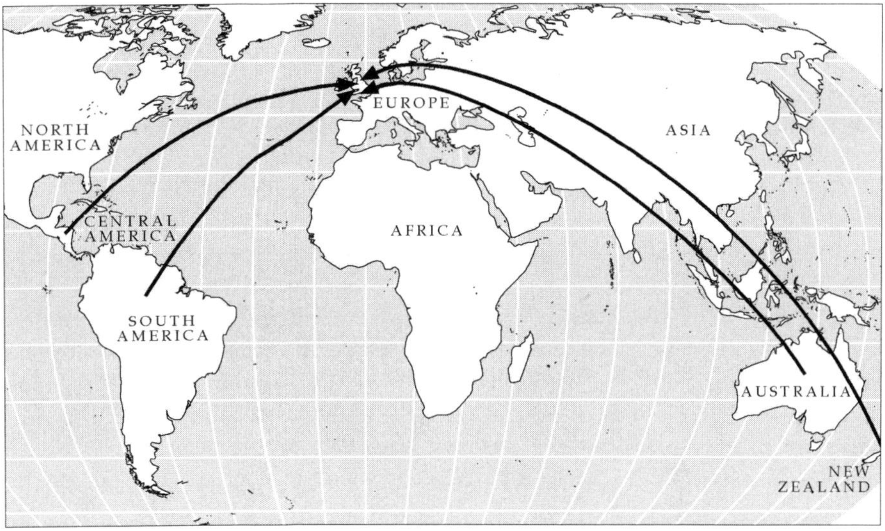

Map 7. Increasing importance of South Africa, Australia, and New Zealand, as well as Central and South America, in the second half of the nineteenth century.

Toward the end of the eighteenth century, the great flood of American imports slowly waned. Plants from the Far East and the Southern Hemisphere, including South Africa, started making their appearance during this same epoch, trickling in slowly. Small numbers of plants from China and Japan eventually arrived in England and Europe after a roundabout journey (Map 5). The initial phase of the fourth transformation, to the Asia–Southern Hemisphere period, was beginning.

The last significant introductions from North America were from the western states to Britain in the mid-1830s. Collecting still continued, but on a reduced scale, to the 1870s. It was succeeded by the flow of new plants from Asia, which picked up very fast. From the mid-1840s, Asia was the source of an enormous number of new species for about 90 years. This was the fourth transformation (Map 6).

Exploration of South Africa, Australia, and New Zealand accelerated as the nineteenth century progressed. Central and South American species also began arriving with increasing frequency (Map 7; see Appendix 3).

World War II and postwar political upheavals put an end to this phase. There have been modest collections since but nothing on the previous scale. In the recent past, concern for the squandering of their national heritage has led countries to ban foreigners from collecting and exporting their plants except under very controlled circumstances. Stringent international treaties such as CITES (Convention on International Trade in Endangered Species of Wild Fauna and Flora) govern these matters.

Until about 1900, almost all the discoveries were sent to Europe, and only later found their way to the United States. There were, however, some exceptions to this. David Fairchild collected for the U.S. Department of Agriculture, Bureau of Plant Industry, and Ernest Wilson and Dr. Charles Sprague Sargent collected for the Arnold Arboretum in Massachusetts. Dr. George Rogers Hall sent new plants directly from Japan to nurseries on Long Island in the 1860s.

There were also a few other cases of direct transmission to the United States. Oakes Ames, aristocratic professor of botany at Harvard and George Plimpton's grandfather, collected orchids in the Philippines in the early 1900s. Ames' influence was immense, stimulating a genius like Richard Evans Schultes (see Chapter 12) to devote his life to collecting Amazonian plants and elucidating their uses.

The profound replacement of native plants in North American gardens by discoveries from Asia and the Southern Hemisphere started slowly at first and then gathered steam. The balance began tipping by about 1800. It continued to shift slowly for another 40 years and was really complete by about 1870.

The vast extent of the North American continent, the great range of climatic variation, and the absence of centralized horticultural institutions probably meant that these new plants were distributed more slowly than in England or on the continent of Europe, but visitors to Alaska or Louisiana from other parts of the country nowadays will see familiar flowers in the local gardens and nurseries. The process is almost complete.

CHAPTER 2

THE ENGLISH GARDEN STARTS TO GROW

IN 1771 BENJAMIN FRANKLIN wrote, "Britain, it is said, produced originally nothing but sloes (p. 176)."[11] Franklin could be outrageous at times, but he was essentially correct. The source of the sloe, the blackthorn, or *Prunus spinosa*, was indeed a native species. Sloes produce an excellent gin.

Successive waves of conquerors started the process of taking new plant species to the British Isles. The alien plants thrived and seemed to have been there always. No one knew they had come from somewhere else after the invaders were forgotten. British weather is unattractive, cool, damp, and unpredictable, but the climate is ideal for gardens. There are seldom great extremes of temperature, and in most years there is enough rain. Only plants from much farther south in Europe were not particularly successful.

As discussed in Chapter 1, the story of how England's flora was enhanced has several recognizable phases. It is traditional to start with the Romans. Transfers of plants may well have taken place previously across the Channel, North Sea, and Irish Sea, but nothing is known about them.

ROMAN PERIOD

For the Roman legions, serving in Britain was a hardship. To make life a little more comfortable, they took some favorite foods with them.[12] Plum, sweet cherry, sweet chestnut, fig, and medlar trees flourished in the cool northern climate. Fig trees produced fruit if planted close to a warm wall. Grapevines also did fairly well, but the olive tree could not survive.

The Romans also imported cabbage, onions, leeks, asparagus, lettuce, opium poppy, hellebore, rue, sage, and summer savory among others. What could be more English than cabbage?

They did not import many flowers. Roman gardens featured strong architectural structure and the sort of plants that stood up to this treatment, rather than flowers. Box hedges, *Buxus sempervirens*, are a good example. Box is native to much of Europe, including England. *Acanthus mollis* and *A. spinosus* came at this time. Both of these are sculptural in nature rather than merely pretty. *Helleborus niger*, the Christmas rose, can also be traced to Roman times (Figure 2).

Smaller flowers grew in pots and containers. Even a plant as beautiful as the rose was cultivated mainly for its perfume.

Patrician Romans adored roses. They made wreaths of them for their hair, stuffed the petals into pillows, used them in scents and cosmetics, and put them into many dishes (Figure 3). Roses were so important that three towns in northern Italy devoted their land and energies solely to growing them. If the supply ran short, roses were imported from Egypt and Carthage. No one now can say how the sailors kept the blossoms fresh during the long sea voyages.[13]

Pliny the Younger, Seneca, and other writers incorporated many hints and instructions about growing roses in their treatises. *Rosa gallica* and a form of *R. alba* are mentioned in Pliny's texts (Figure 4). Virgil (Publius Vergilius) talked about the "twice-bearing roses of Paestum" in his *Georgics*. Experts believe that this was a reference to *R. damascena.*

DARK AGES

Little is known about gardens or planting in the islands after the Romans left Britain in 410 C.E. The country was roughly divided into three main segments: Britons in the west, mixed Angle and Briton in the north, and primarily Angle in the southeast. The terms "Angle" and "Saxon" are now almost interchangeable, but they were different tribes in the beginning. Without the Roman conquerors' firm hand, these groups began to fight with each other very soon. Gardens were far from anyone's mind in those chaotic conditions. People were lucky if they could grow enough food to eat.

"Britons" were the ancestors of modern Scots, Welsh, and Irish people. "Celtic" is another term for these groups, but modern historians are questioning the accuracy of that designation.

The Romans had reached very far into the land, going north as far as the Scottish border. The emperor Hadrian started to build his wall in 122 C.E. to keep the Picts out of northern England. The ancient Britons used woad to tattoo or paint their faces blue, hence the name "Pict." ("Pict" is a corruption of the past participle of the Latin verb *pingere*, "to paint.") They must have been a fearsome sight to unsuspecting invaders. A few hundred years later the Picts were no longer identifiable, having been overtaken by the Scots.

MONASTERIES

Christianity reached Britain early. Missionaries from Gaul crossed to Britain early in the fourth century and established a Christian hierarchy. Monasteries were built.

In the sixth century, the rebellious Irish monk Columba was excommunicated and left Ireland for Scotland in his coracle, landing on the island of Iona in 563

C.E. His missionary efforts there led to the growth of literacy and religion in the western parts of the country.

The papal envoy Augustine was sent to England from Rome in 597 C.E. to seize control of the Church and straighten out this rather remote and independent people. He made the British bishops swear adherence to the rules of St. Peter. Augustine chose Canterbury to be his headquarters because, for political reasons, London was not available to him.

The church's emphasis on literacy made it the only institution able to maintain some remnants of a higher civilization. They maintained the use of Latin and preserved fragments from the Roman Empire at its peak. Monks and other religious men and women grew herbs, or "simples," in enclosed gardens separately from their crops and vegetables. Flowers were cultivated if some part of the plant had a medicinal function. Modern domestic gardens are still at the back of houses or partially enclosed by the building, reflecting the ancient monastic tradition.

SAXONS AND VIKINGS

The Dark Ages were probably not as benighted as we have been led to believe, but there was a long hiatus in the transfer of plants between the continent of Europe and the British Isles. Germanic and Scandinavian tribes all crossed to England in this period, leading to the change from the predominantly dark-haired native Britons to the fairer Anglo-Saxons.

The Saxons' increasing domination of the British Isles took at least two centuries and was achieved in increasing steps, not in a cataclysmic conquest. Possibly the new settlers took fruit and vegetable seeds with them, but nothing is actually known.

Saxons and their British slaves gathered wildflowers and herbs for fragrance and medicinal use. The herbs and rushes were strewn over the floors to smother the usual unpleasant mess. This is known from the bardic accounts of how the great halls of the Saxon lords were furnished.

The Saxon word for a garden was "garth." The imported leek played such a significant role in the British diet that a garden was later known as a "leac garth."[12] The Welsh adopted the leek as a national symbol. Shakespeare had fun with this in *Henry V.*

In the ninth century, massive Viking raids began. The Vikings crossed the North Sea and attacked the northern and eastern parts of Britain. They eventually tired of frequent "commuting" between Britain and Scandinavia and created permanent settlements in the north of England, soon after related groups had taken over northern France. Settlements like this often meant that new plants were introduced.

Abbot Aelfric of Eynsham made a list of known plants in the England of his day, in 995. It was a sophisticated piece of work in Latin, with about 200 entries. He did not distinguish between local native plants and those that had come from the Continent. Few knew what had gone on 500 years before.[12]

NORMAN INVASION

Familiar herbs like lavender, indelibly associated with English gardens such as Hidcote Manor and Sissinghurst Castle, are not native to the British Isles but come from regions around the Mediterranean. Lavender has been in England for such a long time that few people know it is of French origin and was brought by the Normans. The word is derived from the Latin *lavare*, "to wash." Wet clothes were laid across the sweet-smelling lavender bushes to dry.

Most of the plants coming in the wake of the Norman invasion were varieties that had been in cultivation for a long time. It is known, however, that at least two varieties of wildflower did cross the Channel: carnations, *Dianthus caryophyllus*, and wallflowers, *Erysimum cheiri*. The seeds were probably embedded in the mud attached to the stone blocks imported by the Normans, but it was not long before people began to grow these flowers purposely. Their warm colors and rich scent have always made them favorites.[12]

The Normans were descendants of the Vikings who had formed communities in their conquered territory somewhat earlier. The word "Norman" is a corruption of "Norseman." The Normans developed a rich culture after they settled in the northern part of France, Normandy. They made use of many plants in the region and also may have brought back others from the Baltic region and parts of Russia where they had extensive trade.

Britain seemed very backward to them. The Saxons and remaining native Britons grew rye, a little wheat, and a few oats. They had peas, beans, cabbage, onions, leeks, and carrots. Their carrots were the old dark purple kind, transferred by the Romans from the Middle East. The familiar large orange carrot came several centuries later from the Netherlands.

Cowslips, daffodils, bluebells, and violets grew in the woods. Poppies came up in the fields and were a pest. The peasants used flowering hawthorn to make hedges.

CASTLE GARDENS

The Normans protected themselves with impregnable castles. Norman castles were a series of concentric circles protecting an inner core. Once the kingdom was secured, the invaders settled down to life in their new country, unaware that they were gradually becoming Englishmen. The language we now know and use as English arose even while the Normans believed that the only civilized language was French.

They imported lilies, pears, and apples. The apple is so quintessentially English that it is difficult to think of England without it.

There was not much space within the castle enclosures. Castle gardens were pocket handkerchief–sized spaces much like the monasteries' enclosed gardens, *hortus conclusus*. Pictures of some of these tiny gardens show the "flowery meads" that were the style of the time. Briar roses, carnations (gillyflowers), wild daffodils, and poppies can be seen. Another important flower of the period was the Madonna lily, *Lilium candidum.*

The lily was known in antiquity and found across the whole of Europe. It had been used in pagan festivals, but the Venerable Bede (ca. 672–735), the seventh-century abbot of Jarrow in Northumberland, managed the logical and theological sleight of hand needed to make it into a Christian symbol of the Virgin Mary. Bede was famous for writing the first history of England.[14]

Ladies sat on a grassy mound in the *hortus conclusus* to chat with the other women or their knightly lovers. These quiet enclosures also contained fruit trees. Orchards in the modern sense had not yet been developed. Troubadours sang their haunting songs under the trees. These idyllic scenes were restricted to a few months each year when it was warm enough to be outdoors.[15]

There are many pictures of early roses, red and white, with single petals, the red ones perhaps being a gallica. Roses were used as a medicament in the summer, and *Rosa gallica* is known as the "apothecary rose." In one particular text it was said to be good for "inflamed brains."[13] By this the author may have meant simply a headache. Before the days of aspirin, a headache was a nasty experience.

Larger, more open gardens emerged several centuries later, after the countryside had been pacified and one no longer had to live in a fortified dwelling. Life was still extraordinarily dangerous, but it became possible to walk safely in the open air among trees and shrubs.

CRUSADES

The First Crusade was in 1096. The Pope declared that anyone going to wrest the Holy Land back from the infidels would be forgiven all his sins. This was a powerful incentive. Thousands marched. In all there would be nine crusades, finally ending in 1272.

Survivors returning from each episode of this continuing war coincided with the arrival of new plants, but there is little or no evidence to suggest that the crusaders themselves brought them. Still, it is easy to see how the myth of their bringing flowers back from the East would arise.

One exception to this was the introduction of the windflower, *Anemone nemorosa.* The stories are somewhat apocryphal, but one or more pilgrims probably collected the seeds. The Knights Templar grew them at their abbey in Yorkshire, Ripston.

These flowers were favorites of the Byzantines and later the Turks. The stimulus to collect them may have been their prominent display in the gardens of Constantinople. Anemones had been modified and cultivated in Turkey for a long time. Campbell-Culver points out that this anemone was the first flower known to be grown specifically for its beauty.[12] It had no known real or imaginary therapeutic qualities.

Houseleeks, *Sempervivum tectorum*; the black mulberry, *Morus nigra*; marigolds, *Calendula officinalis*; wall germander, *Teucrium chamaedrys*; and hyssop, *Hyssopus officinalis* all reached England in the twelfth century.

MEDIEVAL PERIOD

The long trajectory from Roman times through the Dark Ages and the Norman invasion was culminating as England took a recognizable modern form. According to some historians, the clashing and eventual melding of residual Roman, Germanic, and Christian cultures in Europe from about 300 C.E. during this phase may be called the "early or proto-Middle Ages." A great deal more was going on than we usually imagine.[16]

What is commonly known as the "Medieval" period began in the twelfth century and lasted until the middle or end of the thirteenth century. "Late Middle Ages" is the term applied to the fourteenth and fifteenth centuries. These distinctions are arbitrary but useful as a form of shorthand.

Life expanded a bit in the thirteenth century. It was the century of the Magna Carta and the Plantagenets. The wars with France slowed for a time, and the interminable civil wars also ended. It was possible to follow civil pursuits. The development of banking meant that trade became easier, and the prosperity resulting from successful trade led to improved social conditions.

Producing food was still overwhelmingly important, but more flowers appeared to be grown simply because they were beautiful. Fruit trees had both qualities, being useful for food but also attractive when they were in blossom.

Quince, *Cydonia oblonga*, was a case in point. A marmalade (*mermelada* in Portuguese) is a preserve made from quince. Only later did it come to mean a preserve made from oranges or other citrus fruit. Quince probably came to England with Eleanor of Castile in 1254 when she married Edward I, Edward "Longshanks." Eleanor (1241–1290) liked marmalade. *Cydonia* is not the same genus as the ornamental quinces, *Chaenomeles*. The perfume of the ripe *Cydonia* fruit is exceptional.

The Castilian Eleanor was from the same family as Eleanor of Aquitaine, a redoubtable queen in her own right. Eleanor of Castile resembled the first Eleanor in her competence and strength of character. Keeping clean by bathing was one of her innovations.

Because she lacked any private means and may not have even had a dowry, she set out to accumulate property for herself in a most determined and business-like way. This was not unheard of for a noblewoman, but it was somewhat unusual. Sweet rocket, *Hesperis matronalis*, came from Castile during her reign. Hollyhock, *Alcea rosea*, that quintessentially English stalwart of cottage gardens, was also probably imported into England in the thirteenth century.

The red and white roses of Lancaster and York in the sixteenth-century English Wars of the Roses were from the Continent and not native. Edmund, Earl of Lancaster, brought the red *Rosa gallica*, often referred to as the variety *officinalis* in the horticultural literature, back from Provence in 1277. Edward IV named the white *R. alba* as the emblem of York in the 1460s. Probably a natural hybrid of *R. canina* and *R. damascena*, *R. alba* originated in the Mediterranean.[13]

The fourteenth century started out as prosperous as the period that preceded it but was devastated by the Black Death in the 1340s and 1350s. *Yersinia pestis*, the bacterium that causes plague, traveled to Europe in the blood of the fleas infesting black rats. Merchant ships coming from the East via Venice brought the rats, which first spread throughout the Continent and then reached the British Isles.

In the period of recovery from the plague's devastation, very few new plants came into England. Yet records show that many important plants arrived in England close to the time the plague broke out. They could well have been transported in the same ships that carried the rats. The Oriental plane, *Platanus orientalis*; almond, *Prunus dulcis*; pomegranate, *Punica granatum*; and two kinds of crocus, *Crocus sativus* (saffron) and *C. vernus* (Dutch crocus) are recorded in about 1350, and Italian cypress, *Cupressus sempervirens*, was evident by 1375.

THE RENAISSANCE

For most of the fifteenth century, the important scholarly action was around the Mediterranean. When the Turks sacked Constantinople in 1453, Byzantine scholars fled to Venice. They took their precious Greek manuscripts with them and ushered in a new epoch known as the Renaissance.

Since the fall of Rome in the fifth century, European scholars had been using whatever books had been available before then. Many were written in Latin, but a substantial number were translated from the Greek. Ignorant scribes created numerous mistakes in the texts as they copied from one corrupt manuscript to another. Only a few priests and monks were left who knew Greek, and that number dwindled as the years went by. Now they could compare their tired old translations with the originals and revel in lots of "new" books. Famous scholars like Erasmus led the movement. This all coincided with the development of printing with moveable type. The new learning stimulated objective scholarship and enquiry.

This in turn stimulated exploration. Centuries of trade along the Silk Road had whetted Western merchants' appetite for more spices, gold, silver, and jewels. Shortening the route and making it more convenient by going directly to China and India would lead to greater profit, and the kings of Spain and Portugal each intended to seize the expected benefits for themselves.

Geography and cartography made huge strides in this new atmosphere of enquiry, but the old notion that the world was flat and had finite edges died hard. It led to the idea that there was a way to get around the north by a route known as the "Northwest Passage" and go straight through from Europe to Asia. Epochal voyages started out because the sponsors were determined to find this passage first.

Kings and princes encouraged their admirals to seek new routes to the Orient. Some headed north to the Arctic; some went farther east and south than ever before. Then finally one truly adventurous genius went west across the Atlantic Ocean. Close to the end of the century, this intrepid explorer found a New World.

CHAPTER 3

BROADENING HORIZONS

Results of the Colonization of North America

THE EXISTENCE OF THE NEW WORLD, an entire hemisphere previously unknown to the Europeans, turned people's thinking upside down. If something of this magnitude had been unknown to the ancients, what else might be in store? The amazing discoveries challenged both Christian dogmas and the intellectual heritage of antiquity. The profusion of plants and animals, different from anything ever seen before, made a mockery of the established belief about the number of species left after the biblical flood. Finding so many plants and animals that had not been seen by the ancients led people to question the classifications of Theophrastus and of Aristotle, whose work had survived in Latin translation and been available to scholars even before the Renaissance.

These immense discoveries caused the existing intellectual framework to fragment. It became acceptable to question authority and to find out for oneself what was actually happening in the natural world. Everything was on the table now. Many long-held concepts were swept away without regard to their position in the scheme of things.

The initial observers in the New World were quite alarmed by their findings. One way they protected themselves from accusations of heresy was to use the native name for new and unknown plants and creatures, as if to say, "Don't blame me—I am only reporting what I see." *Potato* was one such word, as were *moose* and *tobacco*.

Of course, Christopher Columbus (1451–1506) was looking for much more valuable things than mere plants, but he collected a few, principally the edible kind. His lieutenants and the sailors all helped to disseminate the foods they had discovered after they returned to Spain. The plants that he and other Spanish and Portuguese explorers carried back to Europe would become essential foods throughout the world. Corn (maize), beans, and squash had been cultivated by native people on the mainland of the New World for millennia. Corn seeds, easy to transport and grow, were soon distributed around the known world.

The warmer European countries adopted corn first. Within about 50 years it had penetrated the Philippines and reached China, and in Africa it became a staple for centuries.* The potato was discovered during the conquest of Peru by Francisco Pizarro (1475–1541). Sir Walter Raleigh (1552–1618) introduced it to England in Queen Elizabeth's time, late in the sixteenth century.

Columbus was also the first European to encounter the pineapple, an edible species of bromeliad that he found in the Bahamas. Pineapples would later be taken to the other parts of the New World and eventually to Hawaii.

In those original voyages, no ornamental plants of any note were taken back to Spain. Spain did have very sophisticated gardens based on Moorish tradition, with both fruit trees and ornamental plants growing in profusion and jasmine and roses grown for their beauty and perfume. Someone like Columbus, with access to royal palaces, would have seen such gardens and appreciated their beauty, but he and his crew had not sailed west in search of flowers. They really cared only for gold, silver, precious stones, and the lucrative spices of the Orient. To them, the flowers of this new land were simply weeds.

It did not occur to them to waste precious cargo space on such things. Their ships were tiny and could not take on any cargo that was not essential. Even packing the new food seeds might have meant that something else could not go.

THINGS WOULD NEVER BE THE SAME AGAIN

It has become commonplace to refer to Columbus' discoveries and their sequelae as a collision between cultures. The arrogant hubris of the Europeans led to the destruction and almost complete erasure of once-flourishing societies, whose peoples were usually characterized as "savages." In spite of that, some foods and medicines gleaned from these despised peoples' ancient knowledge still sustain the rest of the world to this day. Truly, one might say that things would never be the same again.

In the sixteenth century the English, French, Dutch, and Spanish explored the eastern sections of North America, and long-term colonization began in the following century. Although North America was eventually dominated by the English, at the outset this outcome was by no means certain or predictable, and all parties vigorously pursued the acquisition and protection of their territories and the resources they found there.

It did not take long for conflict to develop over who owned the exciting New World. In 1493 Pope Alexander VI, tiring of the squabbles between the kings of Spain and Portugal, took it upon himself to divide the New World into two segments by simply drawing an arbitrary line down the Atlantic, giving Spain the unclaimed territories west of the line and Portugal those to the east. In 1494, the

*Gavin Menzies in *1421* questions this received view about the spread of corn. In his opinion, based on his research, the great Chinese voyages of 1421 had brought corn back and it had already been distributed by the time Columbus found the New World. Very few people agree with Menzies' interpretations.

staffs of both the Spanish and the Portuguese courts met in Tordesillas to renegotiate the line and establish the actual boundaries.

As the pope's first line really had shortchanged the Portuguese, one of their first adjustments was to move the line 300 leagues west, so that it was no longer in the middle of the ocean. The boundaries of the old Treaty of Tordesillas in 1494 are still not clear, but the pope essentially assigned the western part of South America to Spain and the eastern part to Portugal. Brazil remained a Portuguese territory.

Spain and Portugal were united in one matter: keeping the English and the French out of their exciting new territories. The English did not sit on the sidelines for long, watching other countries enrich themselves so easily. They eyed the northern sector of the New World warily, as nothing much had been said about that region. By staying to the north, both England and France hoped to avoid conflicts with the Iberian claimants.

John Cabot (ca. 1450–1498), born in Genoa as Giovanni Caboto, settled in England in 1490, then offered the Tudor king Henry VII his services as an explorer. In 1497 Cabot set out across the Atlantic Ocean from Bristol and claimed North America for Henry. At first it seemed that the new land would only be an adjunct to wider cod-fishing grounds off the coast of Newfoundland (a territory further explored by Cabot's son Sebastian), but that would soon change.

European economic activity was stimulated by the new discoveries. The vast amount of new *specie* had major effects on Spanish and Portuguese currency and trade balances. Imports from the Americas and exports to the colonists meant that ports with easier access to the Atlantic Ocean increased in significance. Mediterranean ports like Genoa lost favor while ports that faced west, such as Lisbon and Antwerp, prospered.

THE THIRD TRANSFORMATION

The colonization of North America launched an even more extensive horticultural transformation of the British Isles and Europe. Over the course of two centuries, new and exotic plants would flow eastward across the ocean in an almost unending stream. This movement occurred against a backdrop of continued exploration and discovery in the farther reaches of Europe and the Near East.

Many new plants entered the British Isles and became an integral part of the flora. By the time emigration to North America quickened, the familiar plants in the emigrants' possessions consisted almost entirely of these recent newcomers (Table 1).

Preserving living plants during the weeks and months it took to cross the Atlantic Ocean was a major problem. This problem would not be fully solved until the mid-nineteenth century. To circumvent the difficulties, the colonists took seed with them and sent the seed of new plants back. Even seed was not free of problems. Rats and mice ate it voraciously, and it was vulnerable to mold and rot, but relatively speaking there was a greater chance of success with seed than with living plants.

Table 1. Cultivated Plants in Early North American Gardens[1]

Plant	Common Name	Introduced
Acorus calamus	Sweet flag	y
Agrimonia eupatoria	Agrimony	y
Ajuga reptans	Bugle	y
Allium sativum	Garlic	y
Anchusa officinalis	Bugloss	y
Angelica archangelica	Angelica	n
Anthemis nobilis	Chamomile	y
Anthriscus cerefolium	Chervil	n
Arctostaphylos uva-ursi	Bearberry	n
Atriplex hortensis	Arach/orach	n
Berberis	Barberry	y
Beta	Beetroot	n
Borago officinalis	Borage	n
Brassica oleracea	Cabbage	y
Carum carvi	Caraway	y
Chelidonium majus	Celandine	y
Cichorium intybus	Chicory	y
C. endivia	Endive	y
Cnicus benedictus	Blessed thistle	y
Cynara scolymus	Artichoke	n
Daucus carota subsp. *sativa*	Carrot	y
Echium vulgare	Vipers bugloss	y
Fragaria	Strawberry	y
Galium verum	Lady's bedstraw	y
Geum urbanum	Herb bennet	y
G. rivale	Chocolate root	y
Glechoma hederacea	Ground ivy	y
Helianthus tuberosus	Jerusalem artichoke	n
Malus	Apple	y
Melissa officinalis	Balm/balsam	y
Myrica pensylvanica	Bayberry	n
Nepeta cataria	Catmint	y
Ocimum basilicum	Basil	y
Ophioglossum vulgatum	Adders tongue	y
Phaseolus	Beans	y
Pimpinella anisum	Anise	?
Polygonum hydropiper	Arsemart/waterpepper	?
Potentilla reptans	Cinquefoil	y
Prunus cerasus	Cherry	y
Ribes rubrum	Currant	y

(continued)

Table 1. Cultivated Plants in Early North American Gardens *(continued)*

Plant	Common Name	Introduced
R. uva-crispa	Gooseberry	y
Rubus chamaemorus	Cloudberry	n
R. fruticosus	Blackberry	y
R. idaeus	Raspberry	y
Rumex sanguineus	Bloodwort	n
Sagittaria latifolia	Archer	y
Salvia sclarea	Clary	n
Sambucus canadensis	Elderberries	y
Sanguinaria canadensis	Bloodroot	n
Sanguisorba officinalis	Burnet	n
Saponaria officinalis	Bouncing bet/soapwort	y
Satureja calamintha	Calamint	y
Smyrnium olusatrum	Alexanders	n
Stachys officinalis	Betony	n
Vaccinium angustifolium	Blueberry	n
V. myrtillus	Bilberry	n
V. oxycoccus	Cranberry	n

n, native; *y*, yes.

1. Adapted from Ann Leighton, 1970, *Early American Gardens for Meate and Medicine*, Boston: Houghton Mifflin. Whenever possible, the nomenclature represents the current, accepted names for plant taxa.

Queen Elizabeth I was kept busy signing a number of charters authorizing the formation of private companies of political and strategic importance. The ones that concern us here are the East India Company, chartered in 1600, and the Virginia Company of 1606. A major step in English colonization was the 1620 arrival of the Pilgrims in New England. A larger number of Puritans followed 10 years later. The Jamestown colony of Virginia did not survive, but more colonists reached Virginia and the Carolinas after 1660.

The North American colonial garden evolved as a mixture of plants. Some were local, and some came from the Old World. Strict utility governed the choices at first. The Puritans were an austere people, despising frivolity in any form (see Table 1).

Hobhouse believes that some of them might have read John Frampton's English rendition of Nicholas Monardes' work.[17] In 1569 the Spanish physician Nicholas Monardes had published a very exciting book, *Joyfull Newes out of the Newe Found World* (Frampton's English title) about the "Divers Herbs, Trees, Plantes, Oyles & Stone" from the new land. The English version appeared in 1574.[18] Monardes never actually went to America, but he cultivated the plants brought back by Spanish explorers in his Seville garden.

One of the new spices Monardes mentioned in his book was sassafras (*Sassafras albidum*). He had received it from a Frenchman in 1566 and used the French name to describe it. The plant was discovered in Florida by a French party that visited the southeast Atlantic coast from 1562 to 1566, following the "Rivière de May" (later named the St. Johns River). Monardes may also have cultivated discoveries of René de Laudonnière and Jacques Le Moyne.

When John Gerard issued his *Herball* in 1597, he had already co-opted some of Monardes' "newes."[19] Gerard was Lord Burghley's head gardener. He listed the American sunflower, for example. A few travelers took Gerard's *Herball* with them to the new country. Even if they knew neither of these works, there was considerable general discussion about the discoveries of the Spaniards and the French in the New World. There was a good deal of ferment.

The emigrants took as much food and seed as they thought they would need to start farming in the New World. Flowers would not have been important except as remedies. Many useful plants came across the Atlantic from England and were naturalized in the Americas. Though these plants left Plymouth, England, in the Pilgrim ships, quite a few of them had originated in other parts of Europe.

Hollyhocks went, as well as stocks, tansy, "gillyflowers," columbines, wallflowers, lavender, rosemary, and clary (Figure 5). Their provenance is complex. Most of these plants had been in England for 200 to 300 years and were no longer considered to be imports.

One of the very earliest plants taken from England to North America was the medicinal southernwood (*Artemisia abrotanum*), grown in England since the Roman occupation. Its vernacular name indicates that the populace knew it was from somewhere else. Angelica (*Angelica archangelica*), which had been taken to England from Russia fairly recently, was another.

The edible apple (*Malus*) went west very early. America's modern sentimental veneration of apple pie has its basis in fact: apple pies helped keep the settlers from starvation when it was hard to find other food. The herbs catmint (*Nepeta*) and borage (*Borago*) were attractive as well as useful and occupied a sort of halfway niche between medicine and food.

As the settlers struggled to survive in eastern North America, for many years nature was the enemy. The Pilgrims found New England bitingly cold in the winter and fearfully hot in the summer. Very few men or women had energy to spare for such whims as collecting new shrubs or flowers, and they fiercely rejected the godless notion of decorating the churches. For them wildflowers were primarily weeds. The trees were obstacles to be cleared, though their wood was useful as fuel and for building.

Their new landscape was heavily wooded. Forest surrounded any settlement more than a few miles from the sea. It was perhaps fortunate that the principal trees were either the same as the ones they had left behind or very similar. Unlike the Spanish in the southern parts of the continent, the English settlers did not have the severe cognitive shock of everything being totally strange.

They had arrived in what would later be termed the Eastern Deciduous Forest

THE FLORISTIC PROVINCES

By classifying the flora of each continent, and regions within the continents, into floristic provinces, modern plant geography permits scholars to understand how climate and terrain affect the growth and development of genera and species. We intuitively recognize the obvious differences between the provinces when we associate palm trees with the tropics and edelweiss with the Alps. Scientists use the terms as a simplified way of referring to complex ecological concepts.

The artificially defined provinces have fluid boundaries. One of the conditions of plant segregation is that the edges are in constant flux. Slight shifts in climate alone, such as a succession of droughts or change in annual temperature ranges, are sufficient to make distinct changes. Light, shade, and the availability of water all influence the growth of various species close to one another.

Spontaneous fires play an important role in the cycle of destruction and restoration in forests. Buried seeds need light and moisture to develop. If the neighboring flora has moved too close to their ground and cut off these elements, the forest trees will recede to the point at which their seeds can germinate and survive. Phytoarchaeology can tell us where this sort of retrenchment took place hundreds of years before and can reestablish the old boundaries.

Today the effects of human habitation and destruction clearly dominate. Quite probably, today's frontiers between the provinces are very different from what they were almost 400 years ago.

Floristic Province (see sidebar), which extended as far south as Georgia, north and east to the future Maine, and west to Pennsylvania.[20] The chief tree species of this province, beech, oak, and hickory, were familiar to them. Sugar maple and basswood resembled the sycamore and lime trees they knew in England, so they gave them these same familiar, comforting names.

The narrow stretch along the Atlantic and Gulf seaboards (which would be designated the Coastal Plain Province) appeared more exotic. Settlers here marveled at oak trees that did not shed their leaves, the bald cypress (*Taxodium distichum*) of coastal swamps that despite its conifer appearance did lose its leaves, and the striking feathery plumes of the longleaf pine, *Pinus palustris*. Even stranger to the newcomers were the palms, mangroves, and sassafras trees of the future state of Florida (first explored by the Spanish, later by the English and French), typical of the West India Province.

English settlers who ventured north and east into Maine and Canada found species more consonant with their notion of a forest: the black spruce, balsam fir, tamarack, and associated understory plants of the region that would be called the Northern Conifer Province.

In those first days, colonists viewed the native plants with one basic consideration in mind: did the plant threaten their survival, or could it be used as food, shelter, or fodder for livestock? Among the new and exotic plants there were some that the newcomers had to learn to avoid. Poison ivy and poison sumac, so beautiful to look at, induced a severe rash. Settlers who forged ties with the native people, such as the Dutchman van der Donck, benefited from the natives' knowledge of the plants, climate, and seasons.

Even though the newcomers had little time or inclination to collect or nurture new plants for their ornamental value, there are surviving references to the pleasure they took in some of the native wild roses. Colonists planted the wild damask rose around their houses for its sweet perfume. They also cultivated the native pasture rose, *Rosa carolina*; the smooth or meadow rose, *R. blanda*; *R. palustris* (known as swamp rose, shining rose, or northeastern rose); and the climbing prairie rose, *R. setigera*. These were largely abandoned once new roses from Europe began coming across the Atlantic.

In spite of the spartan conditions in the new settlements and the immense difficulties they faced in establishing themselves in an alien land, the settlers sent enough specimens of the very rich flora back to England to stimulate considerable demand. Once the colonists had consolidated their hold on the new land and the plants and seeds they brought with them had become part of their new heritage, the dominant flow of plants across the Atlantic shifted from westward to eastward, to an eager audience of European scientists, gardeners, and the wealthy and curious. These shifts were taking place in several parts of the New World, not only in New England but farther north in the French territories.

In a future century, as Americans' tastes became more sophisticated, this eastward flow would once again be balanced by a renewed westward export of traditional European trees and shrubs. This became a two-way flux.

Taking stock at this juncture, it is instructive to examine a table prepared by the twentieth-century American botanical historian Joseph Ewan (1909–1999), showing that before 1600 there were 30 species of plants from America in English gardens (Table 2).[21] These included the tulip trees, columbine, nicotiana, sassafras, yucca, and sunflowers.

The "French" and "African" marigolds (*Tagetes*) recorded in England in the 1530s were actually from Mexico and Central America, brought through Spain, North Africa, and France in the train of the returning conquistadors. Parkinson records that the nasturtium, *Tropaeolum minus*, had been growing in England since the 1530s.[7] Someone had taken it there from Spain, but its true origin, Peru, was apparently not recognized. Most of the plants Ewan listed were introduced in 1588 by Thomas Hariot (1560–1621). Hariot went to Virginia with Sir Walter Raleigh on the Roanoke expedition and wrote *A Briefe and True Report of the New Found Land of Virginia*, the first book by an Englishman devoted to the flora and fauna of North America.[22]

Yucca gloriosa appears on this list. Gerard grew it in his garden in Holborn. It is known he received it from Thomas Edwards, an apothecary in Exeter, Devonshire,

Table 2. North American Plants in Europe before 1600[1]

Plant Name	Known As[2]	Associated With	Date
Agave virginica	Silk grass	Hariot, T.	1588
Anaphalis margaritacea	Cudweede of America	Gerard, J.	1599
Apios americana	Openauk	Hariot, T.	1588
Aquilegia canadensis		L'Obel, M.	1581
Asclepias syriaca	Apocynum syriacum	Clusius, C.	1576
A. variegata	Wausauke	White, John	1585–1587
Canna indica	Indian reede	Gerard, J.	1599
Castanea dentata	Chestnut	Hariot, T.	1588
Commelina virginica	Branched spiderwort	Gerard, J.	1599
Cucurbita pepo	Cucurbita indica rotunda	Dalechamps, J.	1587
Diospyros virginiana	Medlar	Hariot, T.	1588
Helianthus annuus	Planta solis	Hariot, T.	1588
H. tuberosus	Flower of the sunne	Gerard, J.	1599
Juglans nigra	Black walnut	Hariot, T.	1588
Juniperus virginiana	Cedar	Hariot, T.	1588
Liriodendron tulipifera	Rakiock	Hariot, T.	1588
Morus rubra	Mulberrie	Hariot, T.	1588
Nicotiana rustica	Uppowoc	Hariot, T.	1588
Opuntia humifusa	Metaquesunnauk	Hariot, T.	1588
Rhus typhina	Shoemake	Hariot, T.	1588
Sabatia stellaris	Gentian	White, John	1585–1587
Sarracenia flava	Limonio congeneris altera	Dalechamps, J.	1587
S. purpurea	Limonio congener	Clusius, C.	1576
Sassafras albidum	Sassafras	Monardes, N.	1569
Smilax bona-nox vel aff.	Tsinaw	Hariot, T.	1588
Taxodium distichum	Cypres	Hariot, T.	1588
Thuja occidentalis	Arborvitae	Belon, P.	1588
Tradescantia virginiana	Spiderwort without branches	Gerard, J.	1599
Yucca gloriosa	Iucca	Gerard, J.	1599
Zea mays	Pagatowr	Hariot, T.	1588

1. Adapted from Joseph Ewan, 1969, *A Short History of Botany in the United States,* New York: Hafner Publications. Whenever possible, the nomenclature represents the current, accepted names for plant taxa.
2. As in Ewan.

but no one knows how Edwards obtained it. The most likely possibility is that travelers from Roanoke Island brought specimens back with them. They probably landed at Bristol, not too far from Devonshire.

Yucca species have a wide range. Mexico is their principal home, but some species can tolerate cold better and may be found in the United States as far north as New England. Nowadays there is argument as to what constitutes *Y. gloriosa.*

Specialists believe it is a natural hybrid between *Y. filamentosa* and *Y. aloifolia.*[23]

While the Pilgrims struggled so mightily in North America, back in England it was becoming possible to devote oneself to gardening. Beautiful gardens need a lot of attention and time to develop, and one cannot concentrate on this if every year or two one must go off to do battle. After centuries of conflict, long periods of peace were maintained by Elizabeth I and, later, her cousin James I, formerly James VI of Scotland and son of Mary, Queen of Scots.

In 1605, James I ordered the formation of the Worshipful Company of Gardeners. From then on, gardening became a formal profession with a royal imprimatur and its own guild.

Francis Bacon, Lord Verulam (1561–1626), wrote about the value of a garden in his *Essays* (1597): " . . . it is the Purest of Humane pleasures. It is the Greatest Refreshment to the Spirits of Man; Without which, Buildings and Palaces are but Grosse Handy-works (p. 266)."[24] Bacon was very influential, even though he was often in political hot water at court. He was Lord Burghley's cousin and a courtier who often infuriated Queen Elizabeth but who also charmed her with his brilliance. His *Novum Organum* was the theoretical underpinning of modern science, finally casting away the old dependence on dogma and syllogism. The Royal Society was founded to study nature empirically because of his insights.

Bacon was writing for the wealthy few who could afford not only greenhouses but also the 30 acres he recommended as the minimum garden size. He was an intensely practical man. After his brief opening sentences, he jumps right in with " . . .there ought to be gardens, for all the moneths [*sic*] of the year: In which, severally, Things of Beauty, may then be in Season (pp. 266–267)."[24] Bacon then listed plants he assumed everyone would know. He recommended the use of evergreens in winter: holly, ivy, bay trees, cypress, yew, "pineapple trees," fir, orange trees, lemon trees, and various herbs.

We do not know what he meant by "pineapple trees." It could have been the fruit found by Columbus, but it also could have been one of the newly imported northern European conifers with cones resembling apples.

For other seasons, Bacon recommended (among others) the cornelian cherry, tulips, double peony, and oriental hyacinth, without indicating that these were in any way unusual. Clearly, many imported plants were already commonplace in England. He did not specifically say that the shelter of an orangery was essential, but the pineapples, oranges, and lemons could not have withstood the English winter outdoors.

The ancient rivalry between England and France continued to be played out in the New World. The French claimed the northernmost section of the Atlantic region for themselves as "New France" (now Canada), and French cartographers and naturalists energetically explored the New World.

France dispatched professional naturalists to the new countries in the name of the king. During this period the kings of France were relatively nondescript in comparison to the great administrator who followed them, Louis XIV, whose reign began in 1643.

In the 1530s, during the reign of Francis I, Jacques Cartier recorded his findings in the northern part of North America. The French cartographer André Thévet described Newfoundland in 1557. Laudonnière and Le Moyne described the West Indies as well as Central America and eastern Canada during their many voyages in the 1560s. French colonists had the same struggle in the New World as the Anglo-Saxons to the south, though the French settlers' lot was much more difficult. Their territory had an even less hospitable climate.

Although France was not yet constituted in the modern form that we know, it had a highly centralized structure, with power concentrated in the royal court. Outlying provinces stubbornly resisted complying with royal power and went their own way. Paris' writ did not run in the Loire.

Initially the French were ahead of the English in botany and horticulture. Louis XIV, "Le Roi Soleil," reigned from 1643 to 1715. He and some of his wealthier nobles built palaces and landscaped their grounds on a monumental scale, with *grande allées*, canals, fountains, *parterres*, and shrubberies. Nothing like them had been seen since ancient times. New plants were an essential ingredient of their wizardry.

The king's Controleur General des bâtiments du Roi, André Le Nôtre (1613–1700), was much more than just a landscape architect. He was also skilled in horticulture, writing three books on the subject. Serving under him were specialist gardeners such as Jean-Baptiste de La Quintinie, keeper of the royal fruit and vegetable grounds at Versailles. Le Roi Soleil had to have fresh melon and peaches every day of the year. De La Quintinie worked out how to accomplish this.

In contrast, the English were not building in the lavish manner of the French nobility at that time. Travel to the colonies was driven by commercial interests or those of the church rather than sponsorship by the monarchy. Private individuals went if they could afford it.

Imports from America altered the French flora as radically as that of England. Jean Robin (1550–1629) and his son Vespasien (1579–1662) were the gardeners to several French kings, initially Henri IV and then Louis XIII. They quickly propagated the new arrivals. This bounty and their skill made them dominant in European horticulture. As Richard Gorer put it, when England's royal gardeners, the Tradescants, wanted to go on a "shopping trip" for their masters, they went to Paris.[25]

A beautiful and fragrant American tree, the black locust, was later named *Robinia pseudoacacia* in honor of the Robins. Native to the Southeast, it became naturalized throughout New England and Canada and also reached France eventually. The wood is extremely valuable in shipbuilding, as *Robinia* pegs swell when wet, making a better seal than iron rivets.

King Henri IV sent one of his sea captains, Samuel de Champlain (1567–1635), to explore New France in 1603 and report back to him. Champlain established a garden at Port Royal (now Annapolis Royal) in Nova Scotia. He grew American and European plants and probably sent seed to Jean Robin in Paris. Champlain remained in North America until 1629. He published accounts of his voyages as early as 1613 and again in 1632.[26] Lake Champlain, shared by Vermont and New York, was named in his honor.

As an example of how fast things were moving in an age when travel was still very clumsy and slow, the French naturalist-physician Pierre Belon (1517–1564) wrote about the American eastern arborvitae, *Thuja occidentalis*, in 1558. He had seen this tree on one of his journeys. Less than 20 years later, Carolus Clusius (a great contributor to modern botany who will be discussed at length in Chapter 4) saw specimens of *Thuja* growing in Henri IV's gardens at Fontainebleau, just outside Paris. The seed had probably come from New France. John Tradescant, who not only served the crown but became one of the best-known professional English gardeners of the seventeenth century, took the tree to England from France around 1630 (see Chapter 5).

The deluge of new foreign plants impelled scientists and interested amateurs to find names and patterns among the imports, thus spurring the development of botany into a major science. (Other individual sciences, such as chemistry and physics, took much longer to be separated out from the mass of "natural philosophy.") The Swiss-French botanist Gaspard Bauhin (1560–1624) made the best attempt to gather all the known information in one volume, *Pinax Theatri Botanici*, published in Basel in 1623. Bauhin covered more than 6,000 plants, giving each one a unique name.[27] He was frustrated by the numerous polynomials and synonyms, but the book survived and was used even as late as the mid-eighteenth century, when the great botanist Carolus Linnaeus consulted it.

Although the Netherlands played an active role in exploration and settlement, little is recorded of their participation in the flow of New World plants to Europe; their botanical genius would be displayed in later centuries with collections from Asia. One record is that of Adriaen van der Donck (ca. 1618–1655), a young lawyer educated at the University of Leiden, who noted the blooming imports colonists had planted. Dutch colonists, backed by the Dutch West India Company, established New Amsterdam in what is now Manhattan. Henrik Hudson landed there in 1609; the colonists followed in 1624, and in 1642 van der Donck described five or six kinds of flowers growing in the gardens. Van der Donck noted crown imperials, fritillaries, roses, carnations, and tulips.[28] None of these was originally native to the Netherlands, yet the colonists had taken them to the New World to remind them of home.

His 1655 *A Description of the New Netherlands* was the first book published by a resident of what would become New York State.[21] Van der Donck opposed the heavy-handed rule of Peter Stuyvesant. As a result of his work, a more tolerant and liberal attitude was fostered in New Amsterdam and inherited by New York. The island of Manhattan was ceded to England in 1664.

The effects of the new finds were widely distributed throughout Europe, even in countries that had not sent explorers themselves. There was immense interest in many German-speaking provinces. Vienna was a leader with the gardens at Schönbrunn. Clusius spent time there and created a physic garden for the emperor. The cities of Berlin and, much later, St. Petersburg also developed major botanical institutions (see Chapter 15).

While new plants from the Americas engaged scientists and gardeners, the migration of plants within the Eastern Hemisphere continued. As detailed in the next chapter, one import led to a craze still well-known even to those who take no interest in gardening: the "tulipomania" boom and crash of the early 1600s. One man who played a major role in that drama left his mark on botany in general.

CHAPTER 4

EXCITING CHANGE IN THE OLD WORLD

Clusius and Tulipomania

ONE PROBLEM THAT THE DISCOVERY of the New World did not solve was how to get to the Far East more easily or rapidly. The dream of an easy trip through the Northwest Passage fizzled. Merchants still had to rely on caravans of pack animals slowly grinding their way through central Asia to reach the eastern Mediterranean countries, with all the old risks, delays, and expenses that entailed.

Because the merchants still yearned for a better route, European explorers continued to seek direct sea routes to India and China, but, in the end, even the routes they found did not make the journey that much faster.

A few managed to reach India overland, but Portuguese sailors penetrated the Indian Ocean and the South China Sea by sailing farther to the south and east. In 1498, not long after Columbus went west, Vasco da Gama (1469–1524) was the first Portuguese to reach India by sea. He did it by co-opting a native fisherman and his boat.

In 1516, the Portuguese captain Raphael Pestrello took over the island of Malacca, then sailed to China on a junk. By 1537 three small Portuguese settlements were established on Chinese islands. The Portuguese already controlled Goa in southern India, and they were determined to breach China's defenses by any means they could.

The well-educated Portuguese physician Garcia da Orta (ca. 1499—1568) wrote the first western book about Indian and Chinese plants in 1563, *Colóquios dos simples e drogas he cousas medicinais da India.*[29] Garcia was a crypto-Jew. He moved to Goa in 1534 and practiced medicine very successfully there. Despite this admirable record, eight years after his death officers of the Inquisition had his bones dug up and re-interred with those of the other heretics.

His book was translated into Spanish and illustrated by Christoval Acosta in 1578. Acosta used a new title, *Tractate de las drogas y medicinas de las Indias*, but the content was da Orta's. Somewhat later Clusius translated it into Latin and had new pictures made. These different versions helped to disseminate the important new information.

The only plant known to have been sent home by the Portuguese was the Chinese sweet orange, in the 1540s. That orange tree was sent from Goa and survived in Lisbon for more than a century. Subsequently, the Portuguese took orange trees to Brazil, where they became a major crop.

PLANTS FROM THE CONTINENT OF EUROPE AND THE NEAR EAST

The exploration of routes to the Far East was still quite tentative. The events in Europe itself were much more relevant and absorbing at the time. Europe was undergoing the usual religious strife throughout the sixteenth century, Protestants against Catholics, and fighting for territory was widespread. The worst threat to the West came from the Ottoman Empire, but in addition Spain and France were still trying to get England to yield its wealth to them. All these upheavals were very unpleasant on a day-to-day basis, but the constant movement and travel had valuable side effects.

The English flora benefited from the fluid conditions and started to expand again in the sixteenth century. Soldiers and envoys brought booty back from central Europe, Russia, and the eastern Mediterranean. That often included plants that attracted them.

Plant movements among European kingdoms also increased. The natural habitat of snowdrops, *Galanthus nivalis*, ranged across Europe to Ukraine; these flowers arrived in England in 1500. The Turkscap lily, *Lilium chalcedonicum*, came from Greece in about 1530. Yellow crocus, *Crocus flavus*, those indispensable little corms, appeared in about 1579, imported from southern Europe. The quintessentially British bluebells, *Hyacinthoides italica*, reached England from the Mediterranean region in the early 1600s.

The rock rose, *Cistus salviifolius*; goat's rue, *Galega officinalis*; winter cherry, *Physalis alkekengi*; and other now very familiar plants were also in this wave. The Norway maple came at about this time, probably not from Norway itself. James Sutherland, Regius Keeper or "Intendant," recorded it in 1683 for the first time at the Edinburgh Botanical Garden in his book *Hortus Medicus Edinburgensis*.[30]

Horse chestnut trees appeared in Vienna soon after Mattioli described them in Constantinople in 1557. Clusius was growing them by 1576. Their decorative potential was discovered very early. Their role is purely ornamental. The timber is not useful, and the seeds are not fit for human consumption. By 1633 John Tradescant had a tree in Lambeth, and by 1699 Henry Wise was using them extensively in Windsor Great Park.

Wise (1653–1738) was among the great master gardeners who worked for English royalty. He laid out the early grounds at Blenheim Palace after Queen Anne gave the estate to John Churchill, Duke of Marlborough.

All these plants were from the nearer reaches of Europe. Many of them reached England directly and indirectly. More than one scholar received new species from an adventurous Dutch gardener, William (Guillaume) Boel, who collected plants in Spain. John Parkinson (1567–1650) of England, author of the 1640 *Theatricum Botanicum*, received Boel's specimens of the lovely blue lily *Scilla peruviana*; the spring snowflake, *Leucojum vernum*; the Spanish love-in-a-mist, *Nigella hispanica*; *Armeria latifolia*; and the dwarf morning glory, *Convolvulus tricolor*.[31]

SUBSEQUENT ARRIVAL OF OTHER COMMONLY GROWN GARDEN FLOWERS

Iris, laburnum, clematis, grape hyacinth, different varieties of lavender, edible aromatic herbs, sweet William, delphinium, and *Primula auricula*, all standard in any modern English garden, arrived from various parts of Europe and western Asia in the seventeenth century. The other tree that affected future landscape design, the Cedar of Lebanon, *Cedrus libani*, also came into England at the end of the seventeenth century. Its origin was in western Asia. Lilac, *Syringa vulgaris*, arrived in western Europe and England from Turkey at about this time.

Turkish gardening was already very advanced. All the varieties that went to the West had been cultivated for many years. These were not the wild species found by plant explorers.

Islamic gardens had been developed in western Asia over centuries. The Persian "paradise" garden was a model for gardens all over the Islamic world. The concept was taken to Spain after the Muslim conquest, giving rise to such great gardens as those at the Alhambra.

In Turkey the Ottomans had overcome the Byzantines and introduced many of their own customs. However, John Harvey, an English historian of architecture and horticulture, considered that the Turkish skill in cultivating and manipulating plants was derived from the Islamic tradition and not grafted onto a Byzantine tradition.[32]

Byzantine culture was directly descended from Greek and Roman antecedents. One reason Harvey held his views was that the tulip did not appear in Byzantine imagery, whereas it was known to the Persians. He also referred to the Turkic dynasties in central Asia and their lists of favorite plants. No doubt there is room for discussion on this point, as the Byzantines had long experience with plants themselves. The information is scanty, but Byzantine gardens were constructed in the Greco-Roman style.

The first rhododendrons to be grown in England were present before 1656. Tradescant had *Rhododendron hirsutum* in his Lambeth garden.[25] Its common name was the "alpine rose." (*Rhodos* is Greek for rose, and *dendron* is Greek for tree.)

Gerard knew about it but had not seen the actual plant. The other European rhododendron, *Rhododendron ferrugineum*, was not noted in England before some American ones began to arrive.

A very small number of plants began to come from South Africa and Asia at about the same time. The first pelargonium and agapanthus both arrived in England from the Cape of Good Hope in the mid-1630s (Figure 6). Dutch colonists at the Cape of Good Hope sent these and other plants to their masters in the Netherlands. From there the new species traveled to England. The first plants from China and Japan appeared in 1700.

Henry Wise, part owner of the Brompton Nursery, was active in this sphere, too. He continued to run the busy nursery that he and his partner George London had taken over after the four founders sold out or died. The Brompton Nursery was extremely large and dominated English garden making for 50 years. Gorer reports that Wise grew a Japanese *Narcissus* at his nursery and the "Guernsey lily" later in the year. The latter was probably a Japanese *Lycoris*. This information is supported by the list of Wise's plants that has survived. This overlap of plants gradually coming from all over the globe would continue.

RECOGNIZING GEOGRAPHICAL VARIATION IN FLORAS

It was the age of the great herbalists: Gaspard Bauhin, Hieronymus Bock, Otto Brunfels, Rembert Dodoens, Leonhart Fuchs, Konrad Gesner, Matthias de L'Obel, and Pietro Andrea Mattioli on the continent of Europe and John Gerard and William Turner in England. (Gerard was not as scholarly as the others, but his *Herball* has persisted into modern times as a quaint anachronism, and this justifies keeping his name in front of the public.) Their efforts on the Continent and in Britain preceded the beginning of critical science. The catalogs and lists supplied a basis for review and judgment as more plants arrived to confound received notions.

In the sixteenth century, medical training involved immersion in botany. The texts were copied from the ancients. Medicine as a profession was considered to be distinct from surgery. Modern surgery owes everything to Andreas Vesalius, who had challenged antique notions of anatomy and surgery by examining wounded soldiers and human corpses. Medicine by contrast remained hidebound and mired in ancient ways much longer. Critical thinking as we know it did not yet exist. One started with the ancient text and went from there.

No matter how incredible and superstitious an entry seemed, it stayed put for centuries. For years medicine subscribed to the Doctrine of Signatures, in which the fancied resemblance of the plant to some human attribute was considered to be its modus operandi. Perhaps the most egregious example was the mandrake, *Mandragora officinarum*, a forked root that suggested the shape of the human body and was believed to be a powerful aphrodisiac. Rembert Dodoens was the earliest scholar to question this notion.

The herbalists embodied the principal core of medical knowledge, until a more scientific attitude separated the medical from the botanical at the end of the seventeenth century. They described European plants and the foreign ones as they appeared. Producing new woodcuts was laborious and expensive. It was tempting to use ones that already existed, but Mattioli and some of the others commissioned artists to make woodcuts from life.

Now, for the first time, doctors and students were able to express how puzzling it was that the plants seen in the books by masters such as Dioscorides, author of the five-volume *De Materia Medica*, and those that they used to treat patients in Germany or France did not match.[33] Previously any questions about this were treated as impious or heretical. How could some young whippersnapper presume to know more than the ancient masters?

Pedanius Dioscorides, probably a Roman military physician, lived in the first century C.E. Both the text and the images from his manuscripts, *De Materia Medica*, remained in circulation throughout the Dark Ages. When new textbooks of *materia medica* (nowadays known as pharmacology) and botany were written later, the authors simply used his pictures.

Newly inquisitive physicians realized that Dioscorides had worked in Greece and Asia Minor and had illustrated plants that grew there. Pierre Belon, a French physician who went to Crete and Turkey in the 1540s, and Leonhard Rauwolf, a German apothecary from Augsburg who traveled to the Holy Land in the 1570s, both hoped to find those plants and perhaps develop new and better medicines. (The large *Rauvolfia* genus of tropical trees and shrubs is named for Rauwolf. An extract of the root of *R. serpentina* was one of the earliest modern treatments for high blood pressure.)

Rauwolf brought back many useful plants and seeds from the Holy Land, including sesame seed used to decorate loaves of bread. His attitude was revolutionary in its day.

CLUSIUS

The preeminent Flemish botanist Charles de l'Écluse (1526–1609), better known by his Latin name, Carolus Clusius, reintroduced ranunculus and anemone into northern European gardens, also the fritillary crown imperial, after traveling in the Balkans. According to Dr. Christopher Trew (1695–1765), a botanically minded physician in Nuremburg who employed the great artist Georg Ehret to illustrate his *Hortus*,[34] ranunculus had probably been introduced into Europe during the Crusades but then been neglected and forgotten. The anemone had reached England soon after the Crusades (see Chapter 2). Ranunculus and anemone were very popular in the Ottoman countries and had been carefully bred and propagated there for many years.[12,32]

Clusius spent some time in England after leaving Vienna. He met Sir Francis Drake, and they became friends. This may be why he distributed the potato across Europe, taking it to France, Flanders, the Netherlands, and Germany. His taxonomic work was so thorough that it still stands today. With Dodoens, he could be called the "father of modern botany."

In Clusius' time there were no formal postal services, but he would persuade itinerant booksellers to take seeds and bulbs to his correspondents, together with instructions on how to raise the plants. This was a very good idea. The booksellers were literate and could be relied upon to carry out his requests carefully.

Clusius began his career by collecting plants in the Iberian Peninsula for two years in the early 1550s. Under his given name, de l'Écluse had studied medicine at Montpellier and taught himself seven languages in addition to Latin. He came from a wealthy Flemish family and traveled with two of the Fugger brothers on an extended tour of Europe. The Fuggers were the leading bankers of the fifteenth and sixteenth centuries, successors to the Lombard dynasties.

De l'Écluse's father was a nobleman on equal terms with the Fuggers, but politics and religion led to his losing all his money. The de l'Écluses were Protestants. Unlike many children of wealthy families, Charles de l'Écluse was not too proud to earn his own living, and even took a humble position if nothing else were available.

While in Spain he collected over 200 species of plants. This was a major accomplishment. The Spanish authorities had little or no interest in botany. It was hard to get permission to travel about the countryside. Brigands were always a threat, and even if one had official permission to move about, local magistrates were not always ready to honor this. We are apt to forget that our idea of a cohesive, centralized country is very modern. Laws made in Madrid or Paris had very little sway in the provinces. There really was no "France," "Spain," or "Germany" as we now know them.

In other journeys, he took the fritillary, daffodils, and the "florentine iris" (*Iris germanica* var. *florentina*) back to Flanders from Turkey. While in England he saw *I. xiphioides* (now *I. latifolia*) in Bristol. It was known as the "English iris," but Clusius listed it as the "Spanish iris" in his *History of Spanish Plants.*[35] English merchants had taken the plant back to Bristol after buying wine in Spain, and it had flourished in the new environment. This plant needs a cool, moist soil, and England could easily supply that requirement.

He even introduced the tuberose, though this came to him circuitously via India from its source in Mexico. Did Wordsworth know that daffodils came from Turkey?

TULIPS

Any history of plant importation must highlight the extraordinary story of the tulip's impact on Europe. European travelers in the Levant discovered the tulip and sent back many specimens. Reactions to this plant gave rise to some of the most extraordinary repercussions in history. Tulips became the rage in western Europe, starting with the Turkish bulbs carried to Vienna and thence distributed all over Europe. They were very popular in the Netherlands and cultivated there on a huge scale, in part because the flat, open land was ideal for this.[28] The Dutch ascendancy in tulips dates from Clusius. He was invited to become the director of the Leiden Botanical Garden in 1592, and part of the charge was to take his collection of foreign plants with him.

Clusius was particularly fond of bulbs. He corresponded with his colleagues all over Europe and distributed many specimens of bulbs and seed that he had collected in the Mediterranean lands and regions such as Thrace.

Clusius intended to control the distribution of the precious bulbs, but thieves broke into the garden and decimated his tulips. It was partly their dastardly activity that opened up the market. Suddenly tulips were everywhere. Extensive commercial growing grounds sprang up in the area around Haarlem. Wealthy men bought thousands of bulbs a year from these nurseries.

"Tulipomania" was a frenzy out of proportion to the actual value of the plants. Rich and poor alike gambled more than they could afford on finding the perfect tulip bulb (Figure 7). The variegations in the petals' color fascinated everyone, but they did not know how this occurred and so could not duplicate the patterns at will. The financial bubble, which had been growing since the 1620s, peaked and burst in 1637. A great many people were bankrupted.

The tulip was probably first taken to England from Flanders by Protestant Huguenot weavers. The Spanish empire included much of the Low Countries, Flanders, and the Netherlands. In the 1570s, seven Dutch provinces broke away to create an independent Dutch Republic, but it did not survive for long.

The Huguenots were fleeing persecution by His Most Catholic Majesty, Philip II of Spain. One of these refugees was the herbalist Matthias de l'Obel, for whom the lobelia is named. L'Obel found work as head gardener to Baron Zouche, Lord Burghley's protégé.

Flemish gardens were legendary, but almost all of them were destroyed by bitter civil wars. The tulip bulb was a godsend for refugees, being highly portable and very valuable. It was easily turned into cash, a necessity during unsettled times. The majority of the tulips in French gardens came from Flanders.

The plant, which is native to western and central Asia, reached Europe before 1559, the year Konrad Gesner described it for the first time in Johannes Heinrich Herwart's garden in Augsburg. Herwart was a senior official at the Bavarian court.

The Flemish diplomat Ogier Ghislain (Ghiselin) de Busbecq is said to have taken the tulip back to Vienna from Turkey, but it is possible that this is an error. He left for Constantinople in 1554, yet only referred to tulips 20 years later.

Anna Pavord believes that Pierre Belon brought the bulbs back from a journey to the Levant, which started in 1546.[28] Even though the claims for de Busbecq are not well substantiated, Belon is also an unlikely candidate.

Belon did spend a fair amount of time in Crete. The Romans knew Crete was a rich source of plants, illustrating the fact that collecting plants is an old activity. Wild tulips can still be found there, but it is likely that the first ones to reach Europe were cultivated.

Belon made exquisite observations, drawing flowers and seeds, but he did not take anything home to France. Richard Hakluyt, the English maritime historian, referred to tulips in 1578. No one knows how Herr Herwart in Bavaria got his tulips.

Two separate waves of French Huguenots fled from France to England. The first occurred in 1572, with the St. Bartholomew's Day Massacre. About 100 years later, in 1685, Louis XIV prompted a new exodus of Huguenots when he revoked the Edict of Nantes. They, too, took tulips.*

---

*Winston Churchill's mother, Jenny Jerome, was descended from a Huguenot family that later immigrated to America.

CHAPTER 5

ENGLISH BOTANY AND HORTICULTURE

BISHOP COMPTON AND SEVENTEENTH-CENTURY ENGLISH BOTANY

By the end of the seventeenth century, England had been through a crucible but was fairly stable again. The Civil War divided the country, but Cromwell and the Puritans had come and gone. The republic faded away. The House of Stuart was restored. London was recovering from the Great Plague of 1665 and from the Great Fire in 1666.

Systematic botany advanced in the seventeenth century. Campbell-Culver considered the seventeenth century to be the crucial period for botany and horticulture in England.[12] The immense numbers of new plants challenged men to fit them into a cohesive theory of nature. Creating patterns from large numbers of isolated facts required an operating hypothesis and a vocabulary with which to express the ideas. This took a number of iterations, but the pressure exerted by the piling up of so much new material made some kind of systemization important.

The work of the early English herbalists, "proto-botanists" one might call them, William Turner (ca. 1508–1568), Thomas Tusser (ca. 1520–1580), and John Gerard (1545–1612), was augmented by that of the next generation.* Agnes Arber considered Turner to be the "father of English botany."[36] These men's studies established botany as a distinct science in England. The Bishop of London, Henry Compton (1632–1713), played an important role in this. He used the garden at Fulham Palace as a repository for new and exotic plants, and he used his religious authority to send explorers to the New World. The bishop often had many plants to spare and distributed them through the Brompton Nursery of George London and Henry Wise. They sold them, and thus the plants entered general circulation. George London (ca. 1640–1714) had been the bishop's head gardener.

Bishop Compton was the son of the Earl of Northampton, a favorite of Charles II. Once a soldier, he had had a religious epiphany in middle life and entered the church. Gardening was so important to him that he relished a period when the

*William Thackeray chose the name of Thomas Tusser for a parson in his novel, *The History of Henry Esmond.*

king relieved him of his episcopal duties, leaving him a lot of free time, and was annoyed when the ban was rescinded.

Compton dispatched the Reverend John Banister (1650–1692) as a missionary to the new colonies in 1678, where he helped to found the College of William and Mary in Williamsburg with colonial magnate William Byrd. He had studied botany at Oxford with Robert Morison, who recommended him to the bishop. Then, as now, networking paid off.

Banister took the first specimens of *Magnolia virginiana* and *Rhododendron viscosum* back to England. By the time Bishop Compton died in 1713, more than 400 species of American plants were growing in England.

RISE OF SCIENTIFIC BOTANY

I believe that scientific botany developed when it did because of the influx of imported plants. There were chairs of botany at British universities before there were chairs of chemistry or physics. Robert Morison (1620–1683), an arrogant and quarrelsome man according to Howard Reed, held the first chair in botany ever endowed at an English university.[37] He took up his duties at Magdalen College, Oxford, in 1669.

The University Botanical Garden was created across the road from Magdalen and is still there. Morison wrote several books on the classification of plants, basing his scheme on the structure of the fruit. Even if the particular ideas were later discredited, work like Morison's showed that herbalism had given way to a more modern form of botanical scholarship.

John Parkinson (1567–1650) wrote *Paradiso in Sole*, "Park in Sun" (a pun on his name), in 1629.[38] A little more than 10 years later, in 1640, he brought out *Theatricum Botanicum*.[31] This listed 3,800 different plants, many from Virginia, Canada, and New England. One of these Virginia plants was clearly an orchid, illustrated 60 years before the tropical genera were found (Figure 8).

Parkinson made a brave stab at naming it, calling it *Calceolus Mariae Americanus*. According to Reinikka, an authority on the history of the orchid, this plant was a *Cypripedium*. A similar species, *C. calceolus*, was common in England, and this may be why Parkinson chose the name. *Calceolus* comes from the Latin word for a slipper.

The Reverend John Ray (1627–1705) was an extremely significant scholar. He laid the foundation for systematic classification and inspired succeeding generations of botanists. Apart from his error in believing that trees and herbaceous plants were different in kind, his other insights survived very well.

He divided the herbs into monocotyledons and dicotyledons, an idea promoted by the prominent botanist Nehemiah Grew. This division is still in use. Ray used the terms "petal" and "pollen" for the first time, copying Joachim Jung of Hamburg, who died before his work could be published.

Ray wrote numerous books, one of them an account of his travels through the Low Countries in which he cataloged the native plants he collected and saw. For about 10 years he traveled with a wealthy student of his, Mr. Willughby. Ray had resigned his post at Catharine Hall in Cambridge over a religious conflict of conscience. Quite suddenly he was without any resources, even cut off from using the library. Willughby came to his aid.

In 1686 he published the first part of a three-volume text, *Historia Generalis Plantarum*, which influenced many students and scholars. The book came out over a period of years, the last volume in 1704. It is considered to be the first textbook of modern systematic botany.[39]

Ray emphasized the importance of all the plant's structure to reach a classification. Returning to his work and ideas led to the eventual overthrow of Linnaeus' artificial system more than a century later (see Chapter 15). Morison's successor, Jacob Bobart (the younger), sent Ray much of Banister's catalog with its pictures. This work listed 18,000 plants.

Leonard Plukenet (1641–1706) was another major figure of the period. Like so many contemporary botanists, he was a physician. Queen Mary appointed him Royal Botanist and put him in charge of her garden at Hampton Court. He used many of Banister's plants to illustrate his book *Phytographia*. It referred to almost 1,000 American plants.

Nehemiah Grew (1641–1712), James Petiver (1658–1718), and Sir Hans Sloane (1660–1753) all made further contributions to the new science. Grew startled the scientific world by correctly ascribing plants' male sexual function to pollen in *The Anatomy of Plants*.[40] Petiver was Apothecary to the Charterhouse and was actively involved in the Chelsea Physic Garden.

William Sherard (1659–1728), a wealthy and very keen amateur gardener in Surrey who endowed the chair of botany at Oxford, decided to reissue *Pinax Theatri Botanici* by Gaspard Bauhin (1560–1624), a Swiss scholar, and incorporate the wealth of new plants that had been collected since its first appearance in 1623.[27] The task of editing and reissuing Bauhin's work was so enormous that Sherard and his assistants could not finish it.

It is quite probable that Sherard's interest in plants and knowledge of botany were stimulated by being the British consul at Smyrna in Turkey for 13 years, between 1703 and 1716. The Turkish gardens were very impressive. He may have sent Turkish plants to his brother James in England and like-minded friends.

Some of the blame for Sherard and his assistants' failure to edit and reissue Bauhin's work lay with Sir Hans Sloane (Figure 9). Sloane's wealth and scholarly bent were crucial to the development of botany. He supported the London Society of Apothecaries' Chelsea Physic Garden financially. His only request was that in return they send him 50 plant analyses per annum.

As a young man Sloane had spent two years in Jamaica. He discovered epiphytic orchids while there but, because he had not seen such plants before, included them in plant families with which he was familiar. For example, Reinikka notes that Sloane classified *Stenorrhynchus speciosus* as a cardamom, *Erythroides plantaginea* as *Orchis elatior*, and *Vanilla claviculata* as a form of cactus.[41]

Sloane became a fashionable physician and married a very rich woman. This enabled him to buy many of his contemporaries' collections, such as that of James Petiver. These were scattered all over London and other places. In the end he had so much material that the British Museum was founded to house it all. For various reasons, often purely personal, he did not always give permission for people to examine the herbaria.

Sherard, Ray, and another Oxford botanist, Johann Jacob Dillenius, waited in vain but never did get to see them. Sherard and his colleagues believed that some of the plants they had not seen might have helped them identify the ones they already had.

Identifying and classifying the new discoveries presented a challenge. The old method of adding more and more adjectives to the name was clumsy and unwieldy. Scholars attempted to absorb the ever-growing supply of new material but did not have the tools required.

They had to wait almost half a century for a definitive solution. Carolus Linnaeus (1707–1778) published the first edition of *Genera Plantarum* in 1737 (see Chapter 15).[42] This book laid out the significance of the genus as the stabilizing basis of classification. *Species Plantarum* (1753) would give every plant a unique binomial label. For the first time, there would be a way out of the difficulty. Here was a system that could accommodate all the known plants but that also stretched as new plants were constantly added.

THE TRADESCANTS

The twin movements of plants from Europe to the British Isles and from America to the British Isles affected the development of gardening and nurseries in addition to science. The careers of the Tradescants are a case in point. They availed themselves of the opportunities that opened up across the span of two generations. Their activities epitomize the transitions in English horticulture and thus warrant a little special attention.

The Tradescants, father and son, were probably the two best-known professional English gardeners of the seventeenth century. At least two monographs have been devoted to them.[43,44] John Tradescant (the elder) was probably born in 1570 and died in 1638 (Figure 10). His son, John (the younger), was born in 1608 and died in 1662 (Figure 11).

These men were responsible for much of the rapid change in English gardens because they brought new plants into England themselves. John Tradescant the Elder created the gardens at Hatfield House, family seat of the Cecils, and eventually was taken on by King Charles I to remake the royal gardens.

Tradescant wanted to make his master's gardens the best in the land and anxiously looked for novelties all the time. At that period, the place to find new plants was France. The relative paucity of fine trees and shrubs in England led him to correspond with the Robins, his counterparts in France. Jean Robin (1550–1629) and his son Vespasien (1579–1662) were Henri IV's gardeners. Tradescent the Elder visited France more than once.

He also traveled in the Low Countries and Russia and brought many plants back. During one trip, he bought more than 800 tulip bulbs in the Netherlands for Hatfield. In 1618, Tradescant sailed for Russia with the English ambassador, Sir Edward Wootton. He found a rose, *Rosa acicularis*, and angelica, *Angelica archangelica*. Neither was unique to the north. Closely allied varieties could be found in England, but it was his first taste of collecting plants in strange places.

North Africa was even more exotic, and Tradescant thrived on these adventures. He joined a punitive naval expedition to North Africa in 1621. The British navy was trying to quell the Barbary pirates. He went to see Algiers and perhaps look for plants. Although he indeed saw a good many wonderful flowers, the plants commonly attributed to his travels probably did not return with him. Two plants that he is known to have taken back to London from North Africa were the apricot and Persian lilac.

He also found the rock rose, *Cistus* spp., and smilax, *Smilax aspera*. The Chian turpentine tree, *Pistacia terebinthus*, and *Trifolium stellatum*, a clover, both came from this period. Virginia creeper, *Parthenocissus quinquefolia*, and *Rudbeckia laciniata*, the cutleaf coneflower, from the eastern part of North America were introduced about that time. The cutleaf coneflower's formal name is anachronistic; Rudbeck, Linnaeus' teacher, lived many years after the plant was discovered.

The American discoveries fitted Tradescant's needs very well. The American sycamore, *Platanus occidentalis*, was an imposing tree well suited to a great estate. He published his catalog in 1657. It included 30 species from Virginia.[45]

Tradescant created a private garden at his property in south Lambeth, just south of the river Thames. There he could indulge his overwhelming passion for all sorts of new and untried plants. He accumulated so many plants that he had the foundation of a botanical garden, but he lacked the scientific skill to organize his material. The earliest systematic botanical gardens in Europe had been started at Pisa and Padua, major universities teaching medicine. Tradescant's garden was simply stuffed with everything he could find.*

Tradescant's son, John, joined the new Gardeners' Company, started by King James I, quite early. He went to America himself three times, in 1637, 1642, and 1654. The last date coincided with Oliver Cromwell's new regime, the Commonwealth. In light of the Tradescants' connection with the late king, it may have made a lot of sense to disappear for a while. Many illustrious folk somehow managed to pretend they had never had anything to do with Charles I, and, amazingly, everyone managed to believe it. Some of this sleight of hand is disclosed in Samuel Pepys' *Diary*.

Young Tradescant sent some plants directly to England from Virginia. Red maple, *Acer rubrum*, and tulip trees, *Liriodendron tulipifera*, arrived in London between 1637 and 1654.

---

*He also collected "curiosities" like a magpie. After the death of John Tradescant (the younger), Elias Ashmole bought all these Tradescant relics and took them to Oxford where they became the nucleus of the Ashmolean Museum.

THE MOVEMENT OF TREES

Other trees from the eastern seaboard of North America, black locust, bald cypress, black walnut, maple, sycamore, and plane, had been arriving in England since the end of the sixteenth century.*

The American trees were used as timber, as windbreaks, as a source of food, and, finally, for purely ornamental purposes. Some plants survived the journey and took hold; others failed, but the survivors contributed to a permanent change in England's appearance, and gradually in the appearance of the rest of the British Isles.

All these new discoveries were a blessing for landowners. Commercially useful timber was a crucial source of wealth for the aristocracy. Large portions of their estates consisted of woods or forest. For centuries their income came from agriculture and timber. (In the technical sense, "timber," as distinct from wood, means tall, relatively straight trees.) Even those estates devoted to timber still had a very handsome appearance. The timberlands created a rich setting for the pleasure grounds so brilliantly exploited by Lancelot "Capability" Brown, an eighteenth-century landscape architect who has been called "England's greatest gardener."

Although beautiful plants were being imported, most of the landowners were more interested in the plants of immense economic value that were also arriving. These included the ones from Europe as well as North America. The tree that a landscape architect might use as an accent in the lawn or as a focal point for a vista also provided a new source of income for nurseries.

In ancient parlance, a forest was a place for hunting. The "legal forest," which was in fact a game preserve, was not necessarily the same as the physical forest. Many ancient forests have specific names, such as the Forest of Arden or Sherwood Forest. The trees provided a habitat for the game. William the Conqueror founded the New Forest in Hampshire in 1079 as a hunting preserve. Today trees constitute the meaning of "forest."

From the earliest days of the Normans, forest management was more sophisticated than has been thought.[9] It was not all left up to nature. Foresters knew that certain trees would respond to controlled cutting by regenerating rapidly. This was called "coppicing," from the French *couper*, "to cut."

"Pollarding," trimming the terminal branches in the autumn to encourage new shoots in the spring, is associated with the native willow trees. The familiar weeping willow, *Salix babylonica*, is not native but was imported from western Asia at the end of the seventeenth century. Its habitat extends from China westward. Weeping willows are not pollarded. They are purely ornamental.

Many types of tree were logged for building and fuel. Because cutting the trees down was done laboriously by hand, enough wood was spared to allow regenera-

*The common walnut, *Juglans regia*, has a very extensive range from southern Europe through Asia Minor to China. It had been growing in Britain for many centuries and naturalized quite freely.

tion, but greed led the owners to reach a very low point. Not only did the owners want to raise cash, but in the highlands of Scotland they wanted to enclose as much land as possible for highly profitable sheep. The terrible toll this took on the quality of the land was only very slowly recognized.

When the kings kept insisting on more and more timber for ships, even the best forests dwindled, although they never reached bottom as has been erroneously believed. There is no primary growth left and has not been for centuries. The tallest and most useful trees were removed very early. Whatever was left was mostly secondary growth. The old oak trees and tall pines were cut down largely for ships.

Charles I had to raise money again and again for ships. The king was ultimately executed in 1649 after the Civil War, which had its roots in resentment against the "ship money."

Scotland was equally denuded of trees, for the same reasons as England. Drainage of agricultural land became abysmal, and many trees could not survive. Small farmers and poor laborers had been driven off, in "highland clearance," allowing the big landowners to enclose their land and promote sheep farming.

Things were a little different on the Continent. The nature of their soil and the threat of inundation by the sea led the farmers in the Low Countries to be more innovative. As a result, agriculture was much more advanced in the Netherlands than in the British Isles at the end of the seventeenth century. Gradually, their practices were adopted by forward-thinking British landowners. The use of turnips as winter fodder for cattle led to a great improvement in animal husbandry. Another valuable Dutch practice was planting willow trees near water to stabilize streams and riverbanks and act as windbreaks.

Eventually reforestation followed, but now on a planned basis. The climate of opinion very clearly indicated that reforestation was a good thing. Landowners planted new trees intentionally, if intermittently.

The arrival of a novel tree, the Norway spruce, *Picea abies*, was the sentinel event that started the transformation of the landscape in the fourteenth century, together with white poplar, *Populus alba*, and the Oriental plane tree, *Platanus orientalis*. The spruce came from northern Europe and Scandinavia, the poplar from a broad range of countries around southern Europe, and the plane from southeastern Europe and western Asia, principally Turkey.

The European larch, *Larix decidua*, was one of the most successful transplants. Soon after its arrival, probably around 1550, it was a rarity, but within a century its stout qualities were fully appreciated. It thrived particularly well in Scotland and is still a mainstay of the British timber industry. The Duke of Atholl planted a million pine and larch trees in the early nineteenth century.[46] If saplings were sixpence each, this would have cost the duke £25,000.

John Gould Veitch (see Chapter 10) took the Japanese larch, *Larix leptolepis*, back to England in 1861 as an ornamental tree. This larch was also very valuable as a source of timber. In the late 1880s, the Duke of Atholl's foresters planted a small group of the European and Japanese larches close together. A spontaneous hybrid between the two, the Dunkeld variety, turned out to be the best timber tree of all, almost completely trouble free. The crossing occurred in the village of Dunkeld on the duke's estate, hence the name.

The discovery of the amazingly tall trees in the Pacific Northwest led to an additional and triumphant phase in Britain's reforestation, especially in Scotland.

Three hundred years after the first colonies in America were established, a child growing up in London had no idea that the common plane trees in the streets were not native. *Platanus acerifolia* (= *P. hispanica*) is known as the London plane tree. The best guess about the origin of this natural hybrid is that it is half American and half European. The question is where the cross occurred. *Platanus orientalis* (from southeastern Europe and Asia Minor) and *P. occidentalis* (from North America) were both growing in the Tradescant garden, but *P. occidentalis* did not flower in England. It is assumed that cross-fertilization between these two trees to create *P. hispanica* took place somewhere else, perhaps Spain, in view of the plant's classification. A specimen of *P. hispanica* was growing in the Oxford University Botanical Garden in 1666, received three years earlier.

NURSERIES IN SCOTLAND

Scottish nurseries began in response to the need for trees. At first one man supplied everything that was needed, in the sole operation of its kind. For more than 30 years, from the 1730s to the 1760s, Robert Dickson sold the Duke of Atholl and other Scottish noblemen their saplings. His nursery was at Hassendeanburn in Roxburghshire and remained a significant enterprise for over 100 years.

Many of the trees were native to Scotland, such as Scots pine, oak, beech, and ash, but Dickson's son Archibald also offered laburnum, Italian black poplar (Lombardy poplar), and Austrian lime trees.[47] The Italian black poplar owed its rapid spread to Dickson's activities. Dickson sold fruit trees and farm necessities as a sideline.

Dickson is more than just a footnote to this narrative. One of his apprentices, John Veitch, became perhaps the most successful English nurseryman ever known, starting a dynasty that ended only in 1917. Veitch grew up in the Scottish border country where Dickson had acres of young trees in cultivation. The vista of endless trees captivated John's imagination when he was a boy and determined his profession. He would be a gardener and a nurseryman.

More will be said about Veitch and his family as the story unfolds (see Chapter 13). They and many other nineteenth-century nurserymen played an enormous role in the transformations to come.

A FEW OTHER SEVENTEENTH- AND EARLY EIGHTEENTH-CENTURY TRAVELS

Late in the seventeenth century a young English gentleman, off on his grand tour, decided to go to Greece with a companion, Dr. James Spon. Sir George Wheler was well educated, and he wrote a book about his journey after he returned.[48] He traveled from 1675 to 1676, and the book, *Journey into Greece*, appeared in 1682, when he was just 31.

Wheler and Spon left for Greece from Venice. Wheler sent almost 200 specimens back to England, though most of them were already known. *Journey into Greece* was important in its day. Continental botanists knew it and used it when they went east.

The French botanist, Joseph Pitton de Tournefort (1656–1708), explored the Levant and region around the Black Sea for Louis XIV in 1700. His charge was also to find the plants of Dioscorides. He collected *Rhododendron ponticum* during that journey, but this plant did not reach England until 1763. Once there, *R. ponticum* did very well in England. It naturalized rapidly and became hard to eradicate.

Tournefort was born in Provence but rose to be the director of the Jardin des Plantes in Paris. He had been a prodigious plant collector in the Levant and had given a lot of thought to classification. In his opinion the genus was the most important subdivision. His system was very artificial but lasted until Linnaeus came up with a better one.

Tournefort's rhododendron was the forerunner of a series of glorious plants that generated great passion for the next two centuries and continues to do so. The rhododendron is second only to the rose and orchid in the intensity of its adherents' devotion (see Chapter 8).

DECORATING THE GARDEN

Other flowering shrubs such as azalea, ground cover such as ivy, flowering trees such as magnolias, and great ornamental trees like the maple, red oak, and sycamore came to England from the eastern part of the United States in the first half of the eighteenth century.

These arrivals coincided with changes in taste. By about 1750, Lancelot "Capability" Brown was in his heyday. The landscape was to become "natural," asymmetrical sweeping expanses of grass dotted with specimen trees, preferably including a lake. Detailed flower beds, redolent of the previous century, were condemned as old-fashioned and fussy.

The interest in tulips died down both because of this change and because, in the English mind, the flower had come from France, not Turkey or the Netherlands. It suffered the fate of all things French.

At the end of the eighteenth century and into the nineteenth century, handsome new conifers, smaller ornamental shrubs, and the many herbaceous plants now totally absorbed into the British garden scene were sent from the Pacific Northwest. "Exotic" came to mean a plant from the New World, usually America and Canada and, to a lesser extent, Central and South America.

During this latter period, David Douglas contributed many new annuals, mainly from California (see Chapter 7). Poppies, baby blue eyes, forget-me-nots, and lupins were a revelation. Gardeners looking for novelties seized upon them eagerly. They could be propagated en masse and planted out in great swaths of color, quite unlike the standard iris, phlox, lilies, and primroses of earlier years. The horrors of "carpet bedding" were to spring from this innocent innovation.

THE ROYAL GARDENS AT KEW

SIR JOSEPH BANKS

Joseph Banks (1743–1820) was the dominant presence in British science, particularly botany, in the second half of the eighteenth century (Figure 12).[49] The son of a wealthy manufacturer, Banks knew what he wanted to do very early after university and had the temperament and money to support his ideas. He wanted to be in charge.

Banks was not a great thinker or originator, but he was an administrator of genius. He employed Daniel C. Solander, a Swedish botanist who had studied with Linnaeus, as his assistant and attached himself to Captain Cook's expedition to Tahiti and the Southern Hemisphere in 1768 in the *Endeavour* at great inconvenience to many others on the ship, blithely disregarding their legitimate grievances. It was the right thing to do, taking artists and botanists to the unknown places and getting firsthand information.

Very little stood in his way for long. It was fortunate that he had a powerful vision for the public good. As the world opened up further, he wanted to find plants and animals that would benefit England. It was his idea to send collectors in the royal name proactively and not just wait for the new finds to come to him.

Banks recognized the potential in the royal garden at Kew and developed the Royal Botanic Gardens into a leading scientific establishment. The Enlightenment was in full swing. Discovering the rest of the world and filling in the gaps was very important.

There was a strong link with India already, and America was well under control from his point of view. That left him with Asia and the Southern Hemisphere.

His brave exploits in Australia gave him enormous credibility when he went to the king and his ministers for grants to send collectors to faraway places. As will be seen, he also paid some of them out of his own pocket when he deemed it necessary. After Banks' death, Kew was not the same for 20 years.

WILLIAM AND JOSEPH HOOKER

The Horticultural Society watched the deterioration at Kew with some dismay. There was no sense of competition, but society members were concerned about the loss of an important botanical resource. In 1841, when it appeared that the government was going to destroy the garden completely, the officers decided to act. One of them suggested the appointment of the rising star in Glasgow, Sir William Jackson Hooker (1785–1865). Hooker came from the University of Glasgow where he taught medicine and *materia medica* (Figure 13).

The Hookers' influence on Kew and the course of English botany was incalculable. Sir William founded a small dynasty of botanists. His son, Joseph Dalton Hooker (1817–1911), took over his position at Kew after his death (Figure 14). Joseph's daughter then married William Thiselton-Dyer, the next incumbent in that post. The marriage was a classic instance of how young men can succeed professionally regardless of their merit. It was fortunate that Thiselton-Dyer was very good at his job though rigid and unyielding to his staff. He had a passion for uniforms and other emblems of rank and servitude. When the first women gardeners were employed at Kew, he forced them to wear very inappropriate and uncomfortable garments.[50]

William Hooker had two sons and three daughters. He had married the banker Dawson Turner's daughter. Turner was very interested in botany. The Hooker's elder son died young after a marriage that did not fit his parents' expectations. Their next son, Joseph Dalton, was everything that one could desire. He, too, studied medicine at Glasgow.*

In spite of his medical training, the only thing Joseph wanted to do was to explore for plants, rather like Frank Kingdon Ward 50 years later. Both young men had grown up awed by the glamour of plant exploration.

While still very young, he spent four years in the Antarctic, from 1839 to 1843, as the botanist on Captain James Ross' expedition seeking the magnetic South Pole. They wintered in New Zealand and Tasmania. After his return, he published *Flora Antarctica* (1847), *Flora Novae-Zelandiae* (1852–1855), and *Flora Tasmaniae* (1855–1859), each in two volumes.[51-53] This expedition had been privately financed by Felix Booth, the gin distiller.

Then, in 1848, Joseph went to India to collect plants (see Chapter 8). Some years after he returned to work at Kew, he wrote the *Flora of British India.* He was physically a very strong man, tough and agile even until his 80s. From time to time he went on other plant-hunting trips, including one to Africa. The "dark continent" attracted him because of its mystery and dreadful allure. His interest was piqued when Dr. Welwitsch sent him a specimen of *Welwitschia* from South Africa, but he also liked the informality of camping and "wanted to taste the delights of savagery again (p. 155)."[7]

---

*The daughters did all that could be expected of them in the 1840s. They married well.

CHAPTER 6

THE BEGINNING OF AMERICAN BOTANY AND HORTICULTURE

AMERICAN BOTANY STARTED WITH Englishmen coming to explore for themselves. It grew into a formidable indigenous discipline. It is possible only to examine a few highlights and look at a few representative places and epochs. Once again, the press of new material stimulated the growth of a science. Physicians also led the way in the new country because of the age-old connection between medicine and botany.

In 1700 the full size of the continent remained to be learned. Shallow penetration of both the east and west coasts could not offer any idea of the interior. The country had many climatic zones, so there was not one but many floras.

The American imports generated immense interest in England, chiefly London. Curiosity drove a few Englishmen to visit North America specifically to learn as much as they could about its phenomenal vegetation instead of sitting back passively, waiting for someone else to collect the plants and send them home. Bishop Compton had done his work well.

Other countries were also enthusiastic about the new finds. The French needed to replenish their timber trees. They sent someone to America. The Swedes were curious, too. This was at a time when Sweden was quite dominant in Europe and not the much less influential nation it is now. The star botanist of the age, Carolus Linnaeus, was Swedish. He never went to America from Sweden but sent at least one of his students, Pehr Kalm.

American botany developed out of these efforts. It was still a little too early for the colonists themselves to recognize what they had because there was insufficient leisure. The first systematic exploring was done by men from the Continent.

The cost of the journey was covered in various ways. Selling an illustrated book could help to recoup the cost, though preparing these volumes was a slow and frustrating process. Selling plants could also help. Serious amateurs like Peter Collinson paid good money for the plants when they reached him in London. Collinson (1694–1768) was a wholesale importer of linen (i.e., "in trade," hence not a "gentleman"), but he was very significant in the transatlantic flower trade. His interest and financial support were invaluable. He had men at the wharf every day, loading and unloading his bales of cloth. They could meet the ships and expedite the transfer of plants and seeds. As soon as new plants and seeds arrived, his men were able to take them off the ships quickly and deliver them to him in good order.

Collinson shared the new imports with interested acquaintances and the men who subscribed to syndicates to cover the costs of the collecting. One of the most influential was Robert James, eighth Baron Petre of Thorndon Hall. Although still quite young, Lord Petre devoted himself to botany. He stocked his very large garden with plants from America, Europe, and Asia. The estate had more than 30,000 trees and shrubs to be tended. He was fortunate in having a very capable gardener, James Gordon. Gordon later opened a successful nursery of his own.

The participation of a wealthy nobleman in the various syndicates to cover the costs of importing plants was very beneficial. Lord Petre's glamour as one of the "quality" helped Collinson to bring in more merchants and middle-class enthusiasts to help defray the cost.

Documenting the findings was essential. Books incorporating the new information were used to teach both students and amateurs. Colonial scholars began to prepare their own texts almost as soon as the Englishmen got started. There was considerable overlap.

This was a very important development. The books from that age are not only exceptionally handsome, but they are also most informative. In some cases they depict plants that no longer exist in the wild.

The colonial educated classes had more leisure to observe and study natural history in the wilderness than farmers. Clergymen frequently provided careful descriptions of plants and animals, a tribute not only to their piety but also to a more thorough education. Medical students were obliged to study botany to compound the drugs of the period. With the initial impetus from England, as in the work of Mark Catesby and, later, John Banister, the colonial residents soon became active.

Increasing English and other northern European settlement in eastern North America and wider expansion of the colonies went hand in hand with the attempts to reproduce European life in the wilderness. Isolated outposts and small villages developed into towns and cities as the population slowly moved westward.

INDIVIDUALS IMPORTANT TO THE DEVELOPMENT OF AMERICAN BOTANY

MARK CATESBY

Queen Anne, daughter of James II and the last of the Stuart monarchs, assumed the throne in 1702 after the deaths of William and Mary. The Duke of Marlborough, Winston Churchill's formidable ancestor, began beating the French all over

Europe soon afterwards. The relative peace led to some stability and an increased possibility of exploration. Exchanges between Britain and the North American colonies increased.

An outstanding English naturalist, Mark Catesby (1683–1749), whose work was seminal, spent several years in the southeastern states collecting both animals and plants. His books showed the full richness and variety of the American wilderness for the first time. The drawings can be compared with those of Audubon for beauty and accuracy. Many people enjoy the paintings for their artistic merit without knowing their full context.

As a boy, Catesby was introduced to John Ray's work by his uncle Nicholas Jekyll. He later collected for Ray's assistant, Samuel Dale. The nurseryman Thomas Fairchild (1667–1729) employed him, too, offering the choicest new American discoveries at the well-known Fairchild nursery in Hoxton.* The Royal Society also backed Catesby's research by giving him a handsome annual retainer.

He visited America twice, staying more than seven years the first time. One of his sisters lived in Williamsburg. Catesby published the first section of his book about the Carolinas in 1730, *The Natural History of Carolina, Florida and the Bahama Islands.*[54] It had pictures of the ruffed grouse, pigeon of passage, and "mock bird," as well as many trees and shrubs. Catesby introduced the catalpa tree, *Catalpa bignonioides*, using the native Cherokee name "catalpa." Catesby's *Natural History* was considered so important that several new editions were issued, including a few that were unauthorized. Linnaeus used it as a reference. After his death, *Hortus Europae-Americanus*, which is based in part on his collection of plants in the West Indies after his second visit to America, appeared.[55]

There were other sources of specimens from Florida, Georgia, and South Carolina. Colonial settlers like Edmund Bohun, Robert Ellis, and William Halsteed all sent plants and seed to James Petiver in London. Hannah Williams included butterflies, reptiles, and shells in her consignments.

When Joseph Lord, who was born in the colonies and had graduated from Harvard College, moved to South Carolina and met Mrs. Williams, she suggested he, too, collect for Petiver, and he did so faithfully for many years. Because he was a literate and well-educated man, he added detailed notes about where he found the plants and how they were used by the Native Americans.

Halsteed, in his turn, encouraged John Lawson to collect as well. He was a surveyor and thus traveled very widely. Lawson was murdered by Tuscarora Indians.

PHILIP MILLER

Philip Miller (1691–1771) must be mentioned here. He did not go to America in person, but his work was crucial to the success of the American discoveries. The

---

*Fairchild remains significant as he was the first Englishman to cross two separate species of plant intentionally as early as 1720, *Dianthus caryophyllus* × *D. barbatus* (the carnation and the sweet William).

Chelsea Physic Garden was extremely important during this epoch, fostering the new and exotic plants (see Chapter 15). Sloane's investment had been successful. He had bought the estate at Chelsea and presented it to the Apothecaries' Company in 1673.

Miller took charge of the garden in 1722 with the title of "Gardener" rather than "Horti Praefectus" (Figure 15). That was the year Catesby left for his second stay in the Carolinas. The appointment of Miller was a very good choice, but in the end the ambiguity of the two leading positions at the garden led to his downfall. In modern terms, Gardener was perhaps the chief operating officer compared with Horti Praefectus being the chief executive officer.

Miller published one of the very first useful botanical encyclopedias, *The Gardeners and Florists Dictionary*, in 1724.[56] It went into numerous editions and became the workhorse of reference manuals. He was well read in botany and chose the acclaimed classification system of de Tournefort (see "France and America," this chapter) to organize his plants in the first seven editions. After 1750, he turned to Linnaean principles. By then Tournefort's insistence on the flower's corolla as the basis of classification had been discredited.

Approximately halfway through the eighteenth century, Linnaeus began to publish his work (see Chapter 15). In 1753, he published *Species Plantarum*, giving each organism a dual label.[57] Every known plant was assigned to a genus, and then distinguished by its own species name. Plants in other genera might receive the same specific name if warranted, such as *scandens* or *pulchella*, but the generic name served to keep it unique.

Miller's father had owned a market garden in London, and Miller himself started out as a florist. Like so many other successful gardeners in London, the Millers were originally from Scotland. Young Scottish hopefuls who came to London looking for work were made welcome and given every assistance. The flow of Scottish garden workers was so vigorous that the English journeymen complained of being squeezed out.

Miller sent useful seeds and plants to the colonial officials. The quintessentially American cotton industry in the southern states was said to be started by Miller. The cotton seed came from the Duchess of Beaufort's estate. She in turn had received it from her estates in Jamaica.

In addition to the American plants, Miller grew 37 Chinese plants at Chelsea. Even in the first segment of the eighteenth century, the Asian influence was creeping in, illustrating the gradual overlap taking place (see Chapter 9).

The Chelsea Physic Garden declined badly after Miller died. The loss of Sloane's funds and Miller's skill were both serious.

JOHN CLAYTON

John Clayton (1694–1773) was born in England but spent the largest part of his adult life in Virginia. It is unfortunate that many details about his life and career are missing, but the main outlines were pieced together in a monograph published in 1963.[58] He collected more than 500 new specimens and sent them to England, where they are still to be found in herbaria. The stimulus probably came from Catesby. Clayton rode with Catesby when the latter was exploring for plants in the Blue Ridge.

Clayton was important because he wrote the first flora of North America, *Flora Virginica.* He sent the manuscript to the Netherlands to be reviewed by Johan Frederik Gronovius, a Dutch physician and scholar. Egged on by Linnaeus, Gronovius published Clayton's work under his own name in 1739 (part i) and 1743 (part ii).[59] Copyright did not exist in those days, and integrity was fairly thin on the ground. Clayton's manuscripts have been lost, as have his notebooks and personal papers. Agitators burned his house down in 1787.

THE BARTRAMS AND THEIR CIRCLE

The leading colonial plant collectors were the Bartrams, father and son, from Kingsessing in Pennsylvania. The Bartrams, John (1699–1777) and his fifth son, William, were some of the earliest Americans who searched new territory to find plants suitable for an English garden.

The Bartrams were Friends (Quakers). John Bartram supplied Peter Collinson in London with new and amazing plants (Figure 16). This business relationship continued for almost 30 years, until Collinson's death in 1768. Collinson, also a Friend, had a passion for gardening. His garden in Mill Hill, then a village northwest of London, now a London borough, was crammed with exotic growth. Collinson's house and grounds are now the Mill Hill School, a well-known private boarding school for boys.

Collinson kept up with many other American plantsmen, but Bartram was his principal source. For instance, Collinson received seeds of the umbrella tree, *Magnolia tripetala*; paw-paw, *Asimina triloba*; Washington thorn, *Crataegus phaenopyrum*; and sourwood tree, *Oxydendrum arboreum*, from John Bartram.

The Bartram botanical garden near Philadelphia was a repository of the plants John and his son William brought back from New York, Ohio, and Florida. It was the first of its kind in North America, not laid out as a pleasure ground but as a working property. One could compare it to a merchant's warehouse.

The Bartram garden contained about 2,000 plants at its peak. There appeared to be no organization to the plantings, but later scholars have shown that Bartram planted them in the same groupings in which he had found them in nature. This was actually a very sophisticated method of arranging the plants.

John Bartram had been a farmer and was self-taught. John's wife had frequently railed against his neglect of the farm by wasting time collecting "weeds." He laboriously taught himself Latin to learn botany and correspond with educated men in England. George III rewarded his efforts by naming him "Botanist to the King" in 1765.

Naturally, so much worldly success led to competition and rivalry, both at home and from the mother country. This never seems to have worried John Bartram. Most annoyances simply slid off him, including the nagging of his wife. William Young and the Scottish John Fraser infringed most noticeably on his activities. They attracted the favorable attention of the London patrons. Another Scot, John Lyon, was working farther south but in much the same territory of the East Coast as the others.

Fraser (1750–1811) had a nursery in Chelsea where he worked with his son. He was strongly influenced by Miller's successor at the Chelsea Physic Garden, William Forsyth, in choosing to explore for plants. The Frasers started in Newfoundland, avoiding the new republic. John Fraser's plants arrived in England in very good condition, long before the Wardian case was invented (see Chapter 7). This success is attributed to his using sphagnum moss from the swamps of Newfoundland.

Fraser's reputation extended as far as Russia. Catherine the Great collected living treasures as well as works of art. Fraser took a shipload of American plants to Russia for her, and her son named him Royal Botanical Collector.

William Darlington, a physician in Philadelphia who had studied with Professor Benjamin Barton of Philadelphia, published the Bartram correspondence, *Memorials of John Bartram and Humphrey Marshall*, in 1849.[60] Marshall was one of John Bartram's cousins who often assisted him.

Letters between Bartram and Collinson have also been preserved. Collinson wanted to be sure that Bartram received the respect he deserved and gave him quaint advice about dressing properly. He was to wear suitable clothes when visiting wealthy or aristocratic people.

Unlike his father, William Bartram (1739–1823) could be called a professional botanist and plant collector. He continued to collect plants in the South and eventually wrote *Travels through North and South Carolina*.[61] William had made very desultory attempts to find a career, and his father was quite worried about him. Once he started collecting plants, all this ceased. He became tenacious in his zeal.

Father and son also went farther afield and traveled in the Carolinas, Georgia, and Tennessee. Their most famous find was the Franklinia tree, *Franklinia alatamaha*, named for the patriot Benjamin Franklin. They saw the tree in 1765 near Fort Barrington on the banks of the Altamaha River, merely noting it in passing. William went back eight years later to collect specimens and fruit. John Lyon collected specimens after Bartram's last visit. It has never again been seen in the wild.

William sent the *Franklinia* to his patron in London, Dr. John Fothergill. The latter presented a *Franklinia* tree to Kew in 1774.[62]

While the Bartrams were still sending glorious specimens to London, finding almost a dozen new genera, English scientists were beginning to be much more excited about Banks' and his assistant Daniel C. Solander's discoveries in the Southern Hemisphere. Banks and Solander did ultimately settle down and analyze the Bartrams' collection after returning from their Australian adventures. *Franklinia* was only the most exciting of the new finds, but genera later known as *Balduina, Bejaria, Chapmannia, Chaptalia, Elliottia, Glottidium, Macranthera, Mayaca, Pinckneya*, and *Polypteris* were also in the packages.

Bartram's sons suffered during the War of Independence. John Jr., who inherited the garden, was a pacifist, as befitted a Friend, and opposed to the war, as did his brothers. The brothers earned a living by turning the garden into a nursery. The business continued until 1850, even after the brothers had died. In 1891 the City of Philadelphia acquired the property and saved the garden from development.

BENJAMIN SMITH BARTON

The Bartrams, Clayton, Catesby, and Benjamin Smith Barton were all acquainted with each other and corresponded frequently. Though not a creative man, Barton (1766–1815), a professor of medicine at the University of Pennsylvania since 1789 who taught *materia medica* (pharmacology) and natural history, was important because as a scholar he fostered the study of North American plants in a systematic manner.

In 1803 he published the first American textbook of botany, *Elements of Botany*. It was a rather mediocre book, but still important as a milestone. He was not in good enough health to undertake the rigorous journeys needed to collect new plants, so when *Flora Virginica* came out he was very happy.[21] Barton wanted to create a Flora of North America. It was he who supported Frederick Pursh and Thomas Nuttall, pushing them to explore in Canada and the Great Lakes region.

Barton accepted the Linnaean sexual system of classification, fully aware that it had shortcomings. He knew about Tournefort's system using the structure of the corolla and the natural system of de Jussieu, which was dependent on seed morphology. His pupils and disciples spread the Linnaean teaching widely, though in the case of Almira Lincoln, the use of blunt terms for the plants' sexual organs had to be delicately shrouded in euphemism. Mrs. Lincoln prepared a textbook for use in ladies' seminaries, which went through many editions after it was issued in 1831.* [63]

Although a great deal more development took place in American botany for the rest of the nineteenth century, in a way this was a watershed period.

EUROPEAN EXPLORERS AND COLLECTORS IN NORTH AMERICA

The American Revolution (War of Independence) in the 1770s reduced the export of plants to Britain. The psychological and political aspects of the war affected the movement of trees and flowers indirectly. The blockades preventing shipping from going into or out of ports also prevented plants from being exported. In spite of these impediments, lovers of plants still managed to keep in touch and get a few things through. Thomas Jefferson's *Garden Book* indicates he managed to improve his property during those hard times.[64]

Continuing hostilities eventually cooled even these warm connections with England. Foreign visitors were preferred for a time. The Americans turned to France.

*Her sister was Emma Willard, a leading figure in early women's education.

FRANCE AND AMERICA

The New World was very important to France. It offered territory and riches. The French did not cease exploring the known world but also expanded their exploration farther and farther westward.

Systematic botany developed rapidly in France. Since the discovery of the New World, the French kings appointed a chief botanist, usually the best in the land but not always. At the end of the seventeenth century Louis XIV's leading botanist, Joseph Pitton de Tournefort, was among those most outraged by Grew's insights about pollen and its sexual function. Tournefort developed a not unreasonable classification system based on what he had observed himself in the flowers and fruits of plants. Although it was a slight improvement over the views of the period, he did not spot the sexual nature of plants. Tournefort's expedition to the eastern Mediterranean left in 1700 and lasted for two years. An artist, Claude Aubriet (ca. 1665–1742), was attached to the party. The *Aubrietia* was found during that expedition and named in his honor.

Nepotism in the selection of botanists was never far away, but in one case it worked very well, the de Jussieus. The family of de Jussieu was exceptional in that several generations each produced noteworthy botanists. Antoine de Jussieu (1686–1758) was the first one to go to Paris, in the reign of Louis XIV. He taught his brother Bernard (1699–1777), who became a masterful teacher revered by all his students (Figure 17). Bernard in turn taught his nephew, Antoine-Laurent. André Michaux also studied with Bernard de Jussieu.

French botanists followed their own ideas about organizing plants into genera and families. Linnaeus had thought in terms of static pigeonholes. The French transformed this into the more dynamic modern discipline and created the word "biology." Michel Adanson (1727–1806); Bernard de Jussieu; Georges-Louis Leclerc, Comte de Buffon (1707–1788); and Jean-Baptiste Lamarck (1744–1829) were all founders of this new science. They worked during the Enlightenment, when finding a rational explanation for phenomena was the basis of all intellectual endeavors.

Louis XV and XVI both encouraged botanical exploration and collection in the New World to replenish the French forests and sent explorers specifically for this purpose. France faced one of the same problems as England, a shortage of useful wood for ships and building. Naturalists had previously sent trees and shrubs from Québec back to France at approximately the same time as the English had sent plants back from other parts of North America.

THE MICHAUX FAMILY

André Michaux (1746–1803) and his son, 15-year-old François André, with an equally juvenile assistant, Pierre Paul Saunier, were sent to America in 1785 by

Louis XVI. They returned to France in 1796 to find their world destroyed by the French Revolution.

Michaux began his career by collecting under the auspices of Marie Antoinette and Louis XVI's brother, "Monsieur." In 1782, he went east to Persia and Mesopotamia.

Michaux spent 11 years in America. A student of Bernard de Jussieu, distinguished royal botanist, Michaux was the first professional French plantsman to go the New World. The French needed new and reliable trees to replace the millions of acres of forest decimated by war. They also needed new sources of food. Ornamental plants were very low on the list.

The Michauxs finally sent more than 60,000 living specimens and countless seeds to France. Michaux established two nurseries as a basis for his operations, one in South Carolina and one in New Jersey. Seventeen-year-old Saunier ran the New Jersey nursery on his own very effectively. Much of that land is now part of the Hoboken Cemetery. The branch in South Carolina was managed by Louis-Augustin Bosc d'Antic. Bosc succeeded very well in South Carolina, ending up as the French consul. When he finally returned to France, he was a noted horticulturist.

Michaux made numerous collecting trips, going south and east as far as Cape Canaveral in Florida. Many now familiar trees and shrubs were first collected by Michaux. The yellowwood, or virgilia, *Cladrastis kentukea*, came from the banks of the Cumberland River in Tennessee. He and his son documented the rare Oconee bells (*Shortia galacifolia*). They saw the flame azalea, *Rhododendron calendulaceum*, in North Carolina and the mountain stewartia (*Stewartia ovata*) nearby.

Unfortunately, many of his discoveries were lost in the turmoil of revolution and the carelessness of those who should have been faithful custodians at the Jardin des Plantes. His return journey to France was a disaster. A massive shipwreck almost killed him. His notebooks were lost, and many of his carefully dried samples swept away. The revolutionary authorities expressed some mild regret but never paid him what he was owed.

He must have been a truly remarkable man, because he picked himself up, dried out what was left of his specimens, and, a couple of years later, volunteered to go to Africa to explore some more. He died of an infection in Madagascar. Louis Claude Richard, a teacher of botany, compiled *Flora Boreali-Americana* based on Michaux's collections and published it in 1803, the year Michaux died.

Michaux was not alone in receiving such shabby treatment. Père Armand David (1826–1900), a missionary who was distinguished in both botany and zoology, sent thousands of specimens from China 60 years later, which moldered in the basement of the Jardin des Plantes for a generation before anyone grudgingly began to take a look at them. David found the unusual animal in China now known as the Père David deer.

Other French naturalists were also welcomed in America and left records of their work. These included Palisot de Beauvois, Alire Raffeneau Delile, Jacques Gerard Milbert, and Claude Robin.

ROSES AND DAHLIAS AT MALMAISON

Twenty years after the French revolution, the Empress Josephine Bonaparte, née Marie Josèphe Rose Tascher de la Pagerie Beauharnais, established herself at Malmaison following her divorce from Napoleon in 1809. As consolation she devoted herself to growing roses. Josephine was careless about money and lavished huge sums on her garden. In 1811, she spent over £700 for plants, a very large sum indeed for the day. Napoleon rolled his eyes but paid.

In 1810, while the Napoleonic wars were raging, the London nurseryman John Kennedy, of Lee and Kennedy's Vineyard Nursery in Hammersmith, took a specimen of the new China rose, 'Hume's Blush Tea-scented China,' to Malmaison for her. The English government gave him a special passport for the occasion.

In return, James Lee brought back a few dahlia tubers.[65] This Mexican plant was sent to Europe in the belief that the tubers had nutritional value and could be substituted for the potato. Josephine was entranced by the flowers and devoted a great deal of space to develop them. She was extremely possessive about them. When one of her entourage bribed a gardener to give him some, she tore out the whole dahlia bed out and never grew them again.

Josephine employed expert gardeners and commissioned Pierre-Joseph Redouté to paint the flowers in her Malmaison garden. He continued to do this even after her death in 1814, recording all her roses. The first volume of *Les Roses* by Claude-Antoine Thory incorporating Redouté's paintings was published in 1817. The others continued to appear until 1827.[66] The garden at Malmaison also had rare and exceptional lilies, which Redouté painted, *Les Liliacées*.[67]

FRENCH REVOLUTION AND WARS

About 15 years after the American Revolution, the French populace rose against their masters in 1789. Gardening went on even in spite of the revolution and the subsequent war with England. English and French rose fanciers still exchanged seeds and cuttings in 1800, only now the United States began to be part of the process.

OTHER EUROPEAN EXPLORERS

Pehr Kalm (1716–1779), from Åbo, Finland, was a pupil of Linnaeus. He spent about three years in North America, from 1748 to 1751, and took back a relatively small collection of 325 new species for Linnaeus to classify. Linnaeus named the mountain laurel *Kalmia latifolia* in his honor. Kalm ended up impoverished as result of his travels. When he first went to America, he was able to stay with Swedish colonists who had settled near Philadelphia, but eventually he had to use his own money to remain abroad.

Another distinguished European botanist in America was Constantine Samuel Rafinesque (1783–1840), born in Constantinople. He became a professor at Transylvania University in Lexington, Kentucky, far from centers of intellectual stimulation, but died destitute and forgotten in Philadelphia. Rafinesque created 6,700 binomials and proposed 2,700 generic names for plants.

Rafinesque was a lone eccentric, a foreigner among hearty Anglo-Saxons and thus easy to ignore. Born in a suburb of Constantinople to French parents, he finally immigrated to the United States in 1815. Some of his work was valid, based on his collection of 40,000 plants in his Kentucky garden and massive herbaria, but his logic unraveled as he grew older and his classifications became quite unreliable. Rafinesque was held in high esteem on the continent of Europe and awarded prizes, but in the cold light of modern science his contributions are somewhat meager.

COLONIAL AND POSTCOLONIAL PLANTATIONS AND ESTATES

THE SOUTH

Plantations in the South combined the ugly realities of cotton, sugarcane, or tobacco cropping with pleasant residential grounds for the family of the owners. A semitropical climate allowed landowners to grow crops that flourished in the heat. The use of slaves spread from the Caribbean islands. Native and imported plants were used in the gardens. They were purchased from nurseries, either local or far away. Flowers were often the province of the women in the family, but the shopping was done by the owner himself on rare trips into town. The actual work of making things grow was done by slaves or servants.

Plantations developed close to navigable rivers. The cheapest way to transport the bulky cotton or tobacco bales was by water. In many cases there was no other route to the property. Meadows and gardens sloped down to the rivers very attractively, and water was never a problem. Most of it had to be carried by hand, but some irrigation systems were developed.

Southern cities were laid out elegantly. Trees and shrubs were integrated into their design. Savannah, Georgia, and Charleston, South Carolina, remain as shining examples. Forward-looking governors and other officials seized the opportunity to create something new.

Georgia

Colonel James Oglethorpe, the founder of Georgia, worked hard to find useful plants and promote the economy of his province. He experimented with tea, coconut, and cotton, planting trial plats of tea, coconut, and cotton in Savannah in 1733. Philip Miller, Gardener of the Chelsea Physic Garden, sent the cotton seeds from London in 1732. The Duchess of Beaufort had given them to Miller from her estate at Badminton. She in turn had received them from the West Indies.[68] Tobacco was native to the South and together with cotton was at the root of the plantation system.

Modern names for long-staple cotton include the terms "Egyptian" and "Sea Island." These labels are inaccurate, but reflect the worldwide distribution of the fiber and the success of cotton in Georgia. The cotton plant, *Gossypium hirsutum*, is not native to Egypt. It is found in several parts of the world including Florida, but not in Egypt. Muslin, the cotton cloth made in Mosul (now in Iraq), was already very well known by the eighteenth century.

Colonel Oglethorpe allowed religious refugees from Austria to establish a silk industry, but it failed as so many others did at different times. The reason for these failures varied but usually came down to the wrong diet. Silkworms in China eat the leaves of *Morus alba*, the white mulberry. American farmers used the red mulberry, *Morus rubra*, to feed the silkworms. This tree is native to Georgia, but silkworms cannot thrive on its leaves.

Oglethorpe hoped to import tropical plants through the West Indies. He appointed William Houstoun to be the physician and naturalist for the colony. Houstoun was a capable botanist and went to Jamaica in preparation for this task but died before too many plants could be sent.

One has to admire Colonel Oglethorpe. Georgia was the last resort of all sorts of riffraff that no one else wanted in England or the other colonies, yet he constantly had excellent ideas and was extremely conscientious.

Virginia

At the end of the seventeenth century, Virginia was still new and unknown. It took courage to move there from the comfort of England and more courage to go into the wilderness that menaced it on all sides. The great English novelist William Thackeray was fascinated by the period and the new colonies. He set one of his best novels, *The History of Henry Esmond*, partly in Virginia.

Some of the best-known colonial remnants are at Williamsburg in Virginia. There are fragments of both public and private gardens. The university and the Crown, in the person of the governor, possessed especially fine properties. A prosperous attorney by the name of Wythe owned a handsome house and garden right near the Governor's Palace.

When Williamsburg was restored in the 1930s, the gardens were carefully recreated as well. The box hedges, originally from England, had grown over 12 feet tall in the 250 years since they were planted.

As Americans became wealthier, those who had lived abroad for long periods of time tried to re-create the splendor and refinement they had seen in European capitals and country estates on their return. Inevitably, they wanted the same trees and shrubs for their own properties. The classic example is Thomas Jefferson (1743-1826). Jefferson led the way in garden design and management at his plantation Monticello in Virginia.

The U.S. Congress sent Jefferson to France as an ambassador. John Adams and Benjamin Franklin had been ambassadors in France before him, but neither of them had an estate or plantation and thus was not stimulated to import new plants. Abigail Adams could barely keep the family farm in Massachusetts going.

Jefferson imported numerous useful and ornamental plants from Europe, many of them after his service as the American ambassador in France. Living in France allowed him to travel throughout the rest of western Europe. He often worked in concert with his Italian neighbor, Phillip Maffei. The olive tree defeated them both. It could not flourish in Virginia's climate.

All the results of Jefferson's work were recorded meticulously in his *Garden Book*.[64] These books have allowed Monticello to be accurately restored to its former condition.

Southern Roses

Cherokee Rose

Subtropical and even tropical plants thrived in the warm climate. The story of the Cherokee rose, a white climbing rose originally from China, epitomizes the way in which imports affected the local flora.

This rose, *Rosa laevigata*, reached England from China in the 1780s (Figure 18). It then appeared in North America in about 1792. It is thought that the seeds were sent to Thomas Jefferson at Monticello. The Cherokee rose found conditions in the southern states ideal.

There is a persistent myth that it escaped from cultivation and naturalized, rapidly spreading to Tennessee and other states. Twenty years later it had become a wild plant, with an American name falsely attributing it to the native Cherokee. More recent research indicates that the plant's habitat was extended by human agency.[69] Farmers and others found it to be reliable as well as attractive, and so they planted it widely. As pests go, this is a pretty good one to have.

Early Example of American Hybridizing

Using two imported roses, *Rosa moschata* and *R. chinensis*, John Champneys of Charleston, South Carolina, created the first variety known to have been hybridized in America, 'Champneys' Pink Cluster,' in 1810.[70] A florist in Charleston, Philippe Noisette, sent a fine blush variety of this rose to France as 'Blush Noisette.' This started the class of roses that continue to be known as Noisettes.

THE NORTH

The northern states depended on agriculture but on a much smaller scale than the South. Without slave labor, a farmer had to depend on himself, his family, and a few hired hands or indentured servants to work the land. Only very wealthy families could cultivate extensive holdings.

Some of those wealthy families laid out handsome estates in Massachusetts, New York, and Pennsylvania. The wealth usually came from the mother country, and the models for the estates were the memories of English designs. Colonial gardens were slightly coarser copies of the English originals. They were also somewhat out of date. Fashion took a long time to catch up in remote colonies. The owners of these estates imported trees and shrubs from England but did not bring landscape architects.

William Hamilton, a rich landowner who lived on his estate "The Woodlands," a very significant property in Pennsylvania on the Schuylkill River, imported trees and shrubs from "home," probably after 1780 and into the early nineteenth century. He also bought American plants from Bartram's nursery, which was located near his estate.

To the hard-working plum and apple trees, he added stately Lombardy poplars (*Populus nigra* 'Italica') and the sturdy Norway maple (*Acer platanoides*), which grew very well. These trees, or their descendants, survive in some cemeteries.

The Lombardy poplar is hard to reproduce but will grow from cuttings. Hamilton was very generous. He gave cuttings to friends all up and down the eastern seaboard, thus dispersing the tree very widely. The trees can grow over 100 feet tall. Unfortunately, the tree thrived too well, and it is now considered a nuisance.

Hamilton employed a young Scottish gardener, John Lyon (1765–1814), who had emigrated from Scotland to America looking for a better future. When Hamilton wanted to find more plants, he sent Lyon into the Carolinas. With this start, Lyon became a freelance collector and went into business for himself.

He did this in a very large way. The disappearance of *Franklinia* from the wild most probably reflects excessive collecting and encroachment on its habitat by development. Lyons was the last person to see it in its native habitat. In 1806, the sale of Lyon's plants in London was the biggest anyone had ever seen, with 550 lots. Lyon grew rich but died alone in the American wilderness, without really having been able to enjoy his wealth.

He sent back many plants that were already known but were in high demand, but the fetterbush, *Pieris floribunda*, was his original discovery. He exported large numbers of oakleaf hydrangea, *Hydrangea quercifolia*, from Florida. The shrub had been discovered by William Bartram and was sent back to England by both Bartram and Hamilton. Lyon also exported the yellowwood, *Cladastris kentukea*, originally found by Michaux.

Christopher Gore (1758–1827) imported the European sycamore maple, *Acer pseudoplatanus*, for his elegant property Gore Place, near the Charles River in Massachusetts. Gore had fought in the War of Independence and spent eight years in England as Jefferson's chargé d'affaires.

BOTANY AND HORTICULTURE IN COLONIAL AND POSTCOLONIAL AMERICA

In the newly independent United States, remote external events such as the French turmoil were very important but had little day-to-day effect. Postcolonial America was getting into its stride. All the necessary adjuncts of a sovereign, preindustrial society were emerging. The banking system was one prerequisite, developed by Washington's protégé Alexander Hamilton. Scientific pursuits went hand in hand with increased development and downright exploitation of the new territories.

Satellite businesses such as nurseries began in response to the needs of both wealthy plantation owners and estate owners. There was increasing interest in foreign plants whether from France, England, or, with the passage of time, other parts of the world. This was driven by public demand. Commercial needs and the dictates of fashion moved business very quickly. Science had to catch up later.

American scientists were trying to comprehend the magnitude of the American continental floras and name the new plants. A lot of time passed between the finding of the new specimens, their distribution to taxonomists, and the publication of results. These time lags distort the smooth trends one seeks in a narrative of this sort.

The Lewis and Clark expedition overloaded Washington physically. There was literally nowhere to store the bundles of specimens. When the results were finally published formally, there must have been a sense of anticlimax. This was old news.

David Douglas (1799–1834) (see Chapter 7), who explored in the Pacific Northwest independently and sent everything back to London, probably had his specimens named and the results published more quickly than explorers in their own country. He could not match the resources of a major expedition and so sent back smaller parcels, but his contributions were still large and important.

THE SOUTH

There were nurseries in the Carolinas such as Thomas Arnott's and John Watson's by the end of the eighteenth century. Nurserymen were active in seeking new and novel plants for their customers. While the English naturalists were busy collecting American plants to supply their home market, the colonials were already lusting after even more exotic imports.

One nursery in Richmond, Virginia, Minton Collins, which was active in the last quarter of the eighteenth century, offered "Garden and Grass seeds with a choice collection of flower roots and seeds just imported." What were these imports? They included Persian iris, "Red, White, and Blue Hyacinths," "Large

Nestertions" from Peru, and "Gerenium" from South Africa, "the best of all sorts," secured through a "very particular friend in London."[70]

Denise Wiles Adams' painstaking research has shown that seven other nurseries in the region also sold hyacinths.[70]

Dr. Alexander Garden (1730–1791), a Scottish physician who had settled in Charleston, originally known as Charlestown, collected seeds and shrubs in his adopted country and sent many specimens to Peter Collinson in London. He also corresponded with the Bartrams. Linnaeus, in another of his muddled attributions, used the name *Gardenia* in the doctor's honor. The only problem was that gardenias are not native to the Americas. Dr. Garden was the first to send seeds of the native witch alder to London. This was named *Fothergilla gardenii* for another English patron of horticulture, Dr. John Fothergill.

THE NORTH

One of the first recorded nurseries in the northeast was Robert Prince's at Flushing on Long Island in New York. He opened in 1737 and was a pioneer in importing plants from other colonies and other countries.

Prince's sons split their inheritance. William Prince continued his part under the grandiose title of "Linnaean Botanic Gardens" until it ended in 1865. Contemporary remarks indicated that this was hyperbolic. The nursery was a rather untidy and unprepossessing place by then.[71]

The Princes dealt in Asian plants and tropical flowers from Africa and Australia, which reached the northern United States via London, but they also exported many American plants to England.

Although the Princes were said to be the first professional nurserymen in the Atlantic states, E. Davies, a Boston firm, advertised seeds and plants in the *Boston Gazette* for the week between February 29 and March 7, 1719.[70] (Leap year was not restricted to the even-numbered years before the Gregorian calendar was adopted in 1751.) This discovery puts Davies in business 18 years before Prince, but there is no other reference to Davies whereas Prince's nurseries continued for many generations.

Philadelphia was the center of scientific botanical activity in America for many years. The American Philosophical Society, originally proposed by Benjamin Franklin, and a series of remarkable men were responsible for this. The Bartrams thrived because of this atmosphere but also contributed to its richness. There were many "firsts" in Philadelphia, such as Humphrey Marshall's book *Arbustum Americanum*.[72] It appeared in 1788, and was said to be the first consistently American flora. Marshall was John Bartram's cousin.

Bernard M'Mahon (1775–1816) was the leading commercial nurseryman in the city. One of the earliest citations for his nursery was in 1804.[62] He was a literate and scholarly man from Ireland, more than just a gardener. M'Mahon ran a busy shop where the city's botanists and horticulturists congregated to gossip and find out what was happening. It took on some of the characteristics of a club. He devoted himself to learning about the new plants from the rest of the country, and his catalog listed about a thousand species.

M'Mahon sent American seeds and plants to England. M'Mahon is remembered today for the *American Gardener's Calendar*, published in 1806.[73] This is probably the first significant horticultural text in America for the general public. Two years after M'Mahon's death, Thomas Nuttall named the Oregon grapeholly, *Mahonia aquifolium*, in his honor. It is also known as *Berberis aquifolium*.

David Landreth was another early nurseryman in Philadelphia. His nursery opened in 1784. Landreth was born in Northumberland but immigrated to America as a young man. His son, David Jr., continued the business.

New York

Botany in New York was moving forward at the same time as the developments on the rest of the continent. New York City was essentially Manhattan at that period. Flushing was a separate village on the edge of Long Island just as Brooklyn was an independent city. Thomas Hogg and G. Thorburn and Son had nurseries in New York City in the early 1800s. It is possible they or Prince supplied David Hosack with his plants.

Dr. Hosack (1769–1835), an American physician trained in Scotland, created the Elgin Botanical Garden on the east side of the city in 1801 (Figure 19). It was the first public botanical garden in the United States. He needed it to teach the students at King's College, predecessor of Columbia University.

Before returning to the United States, Hosack had spent some time in London working with William Curtis, of *Curtis's Botanical Magazine*, and Sir James Smith, president of the Linnean Society. This conscientious attention to botany put him in the forefront of young physicians in his day, and the students benefited from his concern. He provided them with modern textbooks and access to both dried and living plant material.

Hosack concentrated on medically useful plants, but he also grew *Gleditsia sinensis* and *Aucuba japonica* just to please the visitors. Hosack had been a contemporary of Benjamin Barton at Edinburgh University.

Although he was said to be an easygoing man and very convivial, he took his teaching seriously. Through his instruction of Amos Eaton (1776–1842) and Eaton's teaching of John Torrey (1796–1873), American botany was significantly influenced for a generation.*

Amos Eaton and John Torrey were two of the leading scientists in their time. Eaton published the first American *Manual of Botany* in 1817.[74] The book went through eight editions and was a bible to students all over the country. Eaton clung to the outmoded sexual system of classifying plants in spite of the natural system coming into vogue. Asa Gray (1810–1888) built on this work when he pro-

*In addition to these achievements, Hosack is remembered for having attended Aaron Burr and Alexander Hamilton at their famous duel.

duced his own *Manual of Botany*, another perennial favorite that went through many editions.[75]

Minor legal problems of a civil nature, probably unpaid debt, landed Eaton in jail for a time. The prison fiscal officer had a son, John Torrey, who eagerly learned everything Eaton had to teach him. He would go on to a remarkable career in geology and systematic botany and published the monumental *Flora of North America*.[76] His student in turn was Asa Gray, who had studied medicine in upstate New York.

Torrey enlisted Gray as coauthor of *Flora of North America* very early, after trying to get Nuttall to help him. Although the latter was out of work, he would not join the team. Nuttall was only willing to assist Torrey and Gray by describing a few hundred of the plants he had found. This lack of cooperation may have been one of the reasons that Gray did not like Nuttall.

Torrey and Gray worked together on the *Flora* for 30 years but published only two volumes, both in the 1840s. Gray took several positions teaching botany but really wanted the chair at the University of Michigan. Once there he spent several months buying botanical books in Europe, slowing his work on Torrey's *Flora* still further. Torrey and Gray also did botanical studies of the proposed transcontinental railroad route. Moreover, both men were deluged with material from the Wilkes United States Exploring Expedition in the South Seas and other surveys of the continent. The government approached Gray to join the exploring team, but he chose to go to Harvard instead. By then the second volume of the *Flora* was ready.

Horticulture developed very rapidly and effectively in upstate New York over the next few years. By the 1850s, Rochester was known as the "flower city." Ellwanger and Barry had the leading establishment, sending plants all over the country on the new railroads. Vick's Nursery became an institution. James Vick started more than one horticultural magazine.[77]

Rochester's nurseries specialized in seeds. This was in part due to the proximity of the Shakers. The sect pioneered many valuable horticultural practices. One of them was selling seeds in packets of uniform weight.

Massachusetts

New England was also a center of science and learning. Harvard College in Cambridge had rapidly lifted itself out of the religious groove and begun embracing broader disciplines.

When Asa Gray moved to Harvard, he began to collect rare plants to help Torrey. By this time, Torrey had the most comprehensive herbarium in North America. When Gray moved to Harvard, he contributed his personal botanical library to the college, thus founding one of the great scientific libraries in America.

Gray's career spanned a great part of the nineteenth century (Figure 20). His work consolidated the reputation of the botany department at Harvard. Gray was Fisher Professor of Natural History at Harvard College and director of the Botanic Garden at Cambridge.

Plants continued to come from the Pacific coast, needing names. It was the age of the heroic round-the-world surveying voyages. The Belcher voyage with Richard Brinsley Hinds and George Barclay was one of these. It lasted from 1835 to 1842. Edward Belcher took over the command of the *Sulphur* when Captain Beechey became indisposed. Belcher was born in Halifax but entered the Royal Navy and rose through its ranks. Each time one of these epic journeys ended, hundreds of new plants were deposited with some hapless taxonomist, who then had to describe them and name them. It was exhilarating but exhausting.

Among Gray's many contributions was the insight that similar plants were to be found in widely separated places such as the eastern United States and Japan because of ancient climatic, geological, and evolutionary changes.

Gray pored over *Flora Japonica*, written by Philipp Franz von Siebold (1796–1866), a German surgeon who had lived intermittently in Japan and amassed a remarkable collection of Japanese plants (see Chapter 10), and Joseph Gerhard Zuccarini (1797–1848) after Siebold returned to Europe permanently.[78] Some of the Japanese species were not only similar to American ones but actually identical. This was a great mystery. There was no contiguous landmass along which plants could migrate.

Maples, hydrangeas, and azaleas are among the most prominent species that appear simultaneously on the two continents. Gray suggested that many millennia before, the plants had all been growing in proximity on an undifferentiated landmass, but with the shifts of the tectonic plates and rising of oceans, the continents had separated, taking a portion of the plants with them. Once separated and isolated, the plants developed new species independently.

In order to establish the validity of this theory, geological movements need to be correlated with the development of the flowering plants. Alfred Wegener's theory of tectonic plates had not been published yet and even when it did appear, most scientists poured scorn on it. Gray was speculating constructively.

A current geological time line indicates that early flowering plants were beginning to develop about 100,000,000 years ago, just as the North American and Eurasian continents were separating.[79] South America, Antarctica, and Australia are said to have split away from each other about 50,000,000 years ago. These findings lend some credence to Gray's ideas.

Gray and Charles Darwin (1809–1882) were in frequent communication. It was a fortunate coincidence that Gray received a collection of Japanese plants to classify in the late 1850s while Darwin was preparing *The Origin of Species*. The idea of parallel evolution based on a possible common ancestor agreed with what Gray observed in the plant specimens. In May 1856 Darwin wrote to Gray, "America might be related to Eastern Asia (always excluding Arctic forms) by a genus having the *same* species confined to these two regions; or it might be related by the genus having different species, the genus itself not being found elsewhere."[80]

Some of Gray's work assisted Darwin in writing *The Origin of Species*. In his turn, Gray wrote a book about Darwin and his ideas, *Darwinia*.

Charles Wright, a member of the North Pacific Exploring and Surveying Expedition in the early 1850s, sent Gray about 10,000 specimens to examine and analyze. Some of them were familiar to Gray. Siebold's work had made him think of a connection between the floras, but seeing the Japanese plants again brought it into clearer focus. He spent a lot of time he really could not spare to develop a consistent theory that has never been refuted. It was published in 1859, the same year as Darwin's work.[81] In the end, however, Gray and Darwin disagreed about evolution, after all their mutual collaboration.

Living Japanese plants did very well in the northeastern United States. The direct horticultural exchange between Japan and the United States benefited from Gray's work. Commerce began during the Civil War in 1861, but once the war ended it increased rapidly.

Later in the nineteenth century, another important Bostonian, Charles Sprague Sargent (1841–1927), visited Japan several times, bringing back cherry trees and other favorites.[62] He was the director of the Arnold Arboretum of Harvard University (see Chapter 15).

Sargent had not shown early signs of academic or scholarly ability but grew into his position as director of the new arboretum. The fact that he had inherited great wealth had made him a very attractive candidate. There was no money for a salary when the arboretum was started.

He was able to attract excellent scholars to work with him and initiated valuable publications. Asa Gray helped him in many ways, but the vision and inspiration came from within him.

In 1893, he returned to the United States from Japan with 200 species of plants and more than 1,000 dried specimens. A crabapple, *Malus sargentii*, was previously quite unknown, as was the cherry, *Prunus sargentii*. He brought *Rhododendron kaempferi*, the torch azalea, at the same time. It is hard to imagine American gardens without these stalwarts.

CHAPTER 7

ATLANTIC CROSSINGS

ALMOST AS MANY plants were now going from east to west as from west to east. During the period that Continental plants were being absorbed into English gardens, mid- to late seventeenth century, the English and other northern Europeans continued to immigrate to the New World. As stories about the difficulties of life in the American colonies began to trickle back to England, new colonists knew that they had to take as much as they could to help them survive in the strange environment. The seeds of edible plants were essential, but a few flowers made them feel a bit more at home. Both the fruit and the flowers had been in England long enough to seem as though they were native. People forgot very quickly that much of what they carried to the New World had only recently come into England from parts of the Old World, such as Russia and the Near East.

Plants imported into America were to have a tiered history, later ones superimposed on these earlier ones. Even many weeds in North America were displaced by early European imports. Some of them were garden plants that escaped cultivation but managed to succeed in the wild environment.

Other weeds came in with imported grain and were inadvertently planted together with the wheat or oats. In California the oats themselves escaped cultivation and became weeds in their own right, helping to submerge the native grasses and clovers.

THE MOVEMENT WEST

The earliest explorers had found Florida, the Carolinas, Tennessee, and other southern sections of the eastern seaboard. The Bartrams had traveled in the Southeast, as had the Michaux family. Spanish sailors had been cast ashore on the Gulf of Mexico, and Spanish missionaries had established parishes in New Mexico and Texas.

With the first essentials under their belts, the settlers set out to explore their extraordinary new land. They came from an island in which distances were modest and comprehensible. They were not accustomed to horizons stretching so far away. Adapting to this seemingly endless country needed a new outlook and perspective. Farming was the principle activity of the newcomers, but there was always someone who preferred to be less constrained and travel around. Being free from the backbreaking toil and overwhelming tasks made up for the uncertainties of exploration and the privation it entailed.

European settlement pushed out into what is now the Midwest. Ohio was carved out of the Western Reserve of Connecticut. George Washington had surveyed the territory in the 1760s. Even usually unobservant farmers noticed new types of vegetation.

Botanists had cautiously moved north from the plains, up into Michigan and Minnesota, where they encountered great conifer forests, extending as far east as Maine. For farmers, this only meant they had to cut down a different sort of tree.

Naturalists also moved south and west, collecting shrubs and trees all the time. They even reached the Far West before migrating settlers managed to get there. Susan McKelvey's magnum opus, *Botanical Exploration of the Trans-Mississippi West, 1790–1850*, lays out a panorama of this period.[82]

LEWIS AND CLARK

The most important North American exploration of the early nineteenth century was undoubtedly that of Lewis and Clark from 1804–1806, organized by Thomas Jefferson. In South America Alexander von Humboldt and Aimé Bonpland were also making history at this time (see Chapter 12).

Neither Meriwether Lewis (1774–1809) nor William Clark (1770–1838) was a botanist, but Jefferson asked Benjamin Barton to give Lewis a quick course in natural history and Native American customs, with enough information to make his plant collecting more effective. Barton was not very cooperative. His health was said to be poor, and he may have been uncomfortable much of the time, but some penetrating criticisms of his self-aggrandizing behavior have come down to the modern era.

When the material from Lewis and Clark's expedition was received, Bernard M'Mahon was given some of the seeds to germinate and propagate. He was also charged with distributing them to his colleagues. Jefferson was one of his customers, and they were very friendly. Another large portion went to William Hamilton.

Most of the professional nurseries in the country at that time did not have the scope and facilities of a large private estate. The new American government had clearly not thought about the results of its great expedition. It had organized the journeys but had not yet established the institutions necessary to do the lengthy follow-up and testing. The Smithsonian Institution would be created as a result of this recognition.

MENZIES AND DOUGLAS

At the beginning of the nineteenth century two Scottish naturalists, both exceptional collectors, followed fairly closely upon one another. It fell to David Douglas (1799–1834) actually to introduce many of the handsome plants that Archibald Menzies (1754–1842) had noted in the Pacific Northwest 20 years earlier.

Neither of them spent very much time in the more settled Northeast. They sent all their discoveries to England. Supplying American gardeners was not part of their instructions.

Menzies (Figure 21) explored the northeastern section of North America only briefly.[83] His acquaintance with America was principally on the Pacific coast. He visited Alaska and the Pacific Northwest between 1791 and 1795, first with Captain James Colnett and then with Captain George Vancouver.

Menzies came from a family of gardeners in Perthshire, Scotland. His father was in charge of the grounds at Plean Castle (Castle Menzies). Menzies' father taught him gardening and later sent him to Edinburgh when he was 14. The young Menzies studied medicine under the direction of Dr. John Hope, essentially as an apprentice, though he is said to have attended lectures at the university.

Hope (1725–1786) was a well-trained physician and botanist who had spent time in Paris with Bernard de Jussieu. He was appointed superintendent of the Royal Botanic Gardens in 1761. At the same time, he was Regius Professor of Botany at the university. Menzies was one of his most prominent students. (William Roxburgh, who distinguished himself in India, was also trained at Edinburgh [see Chapter 8].) He played a very important role in the expansion and improvement of the Gardens.

Menzies' elder brother was a gardener at the Royal Botanic Gardens. Menzies' training combined botany with medicine, the usual system at the time.

After his apprenticeship, he joined the Royal Navy as an assistant surgeon. The navy sent him to North America in the 1780s, and he was posted to the naval station at Halifax. He began to collect new plants while there and sent them to Sir Joseph Banks in London to be identified. These efforts impressed Banks.

Banks was always on the lookout for enthusiastic and energetic young men. The Earl of Sandwich, one of Banks' friends, was First Lord of the Admiralty at the time. Banks was able to have Menzies attached to the *Prince of Wales*, under the command of Captain James Colnett. The ship was going to go around the world, seeking profitable furs. They left in 1786 and returned in 1789. One of Menzies' principal duties was to find a particular spruce plant. It was used to make a tea that protected sailors from scurvy.

The voyage took them around Africa to Australia, on to New Zealand, north to Hawaii, and finally up the west coast of North America. The only part of the coast of Alaska Menzies saw during that voyage was Nootka.

In 1792, Menzies set off with Captain Vancouver on HMS *Discovery* and explored the North Pacific. He collected in brief interludes. Vancouver did not allow him to stay on land for more than a few hours. They also fell out over the maintenance of the specimens. At one point Vancouver accused Menzies of mutiny. That charge was dropped once they went back to London. The final indignity was that Vancouver kept Menzies' journal and would not allow him to have it after they returned to England.

Menzies brought the first seeds of the monkey puzzle tree, *Araucaria araucana*, back from this voyage.[83] About 10 years earlier a Spanish captain, Don Francisco Dendariarena, had found the tree in Chile while looking for timber to build ships.

In addition to the monkey puzzle tree, Menzies found another of the three striking trees that are associated with this voyage: the giant redwood in California. Controversy over the name of the amazing ancient redwood tree lasted almost a century. After a great deal of taxonomic tinkering, this redwood was finally named *Sequoiadendron giganteum* in 1939. The monkey puzzle tree was an *Araucaria, A. araucana* (= *A. imbricata*). The author of a monograph on Menzies, Clive Justice, notes that the trees were not immediately used at Kew nor in English gardens as ornamental specimens.[83] That had to wait until many years later. Then the monkey puzzle tree made up for lost time. It became ubiquitous in suburban gardens, a tiresome cliché.

Menzies collected the third tree in 1795 on the Vancouver voyage, the southern beech from New Zealand, *Nothofagus menziesii.* Other collectors 40 and 50 years later, such as William Lobb, sent by the firm of Veitch, would round out knowledge of the *Nothofagus* by finding more species in Chile. Chile and its offshore islands yielded many more beeches.

Menzies is said to have taken some unusually large seeds off a table at a dinner party in Valparaiso, sown them on board ship, and given the seedlings to Kew once back in London. They turned out to be the first *Araucaria* in England. Some of this may be true. There is more about the tree in Chapter 12.

An island in the straits off what is now British Columbia was named for Banks, possibly at the suggestion of Menzies who owed so much to his patron. *Menziesia ferruginea*, a "false azalea," came from the island and is one of the plants bearing Menzies' name.

Later he may have regretted his loyalty. Banks preempted all his notes and journals, not even mentioning that Menzies had collected the remarkable specimens. The collection was for the greater glory of Kew, but Banks did bask undeservedly in some of the reflected glow.

James Edward Smith, who afterwards formed the Linnean Society of London, worked on the early Menzies material from his first voyage and issued papers intermittently from 1802 to 1825. Menzies' collection ended up in fragments, each of various botanists getting a piece of it. His collection was finally put back together in 1990 and can now be studied in its entirety.

Menzies was distressed by losing many of his specimens on the voyages back to England. On one voyage, he lost everything. Banks had commissioned structures he called "garden hutches" for the decks of several navy ships, hoping they would protect and safeguard the specimens, but there were too many design flaws for the hutches to work.

The principal flaw was too little light, but it does show that Banks was always thinking and at least trying to solve problems. The ships' captains were livid about having to put up with these things because they took up too much space. Vancouver reassigned the sailor who was supposed to monitor the plants. With no one paying attention, the plants became waterlogged in a storm, and they all died.

The Wardian case, described for the first time in 1842, would change everything.[7] Nathaniel Ward (1791–1868) was a London physician and amateur botanist who wanted to protect his delicate fern plants from the dirty air in the city.

Ward found that ferns enclosed in a watertight wood and glass cabinet survived the dreadful pollution of London's air as early as 1829. Moisture enclosed at the outset continued to condense and recirculate without evaporating. It was not long before plant collectors took up the idea and sent their precious cargo back in similar cases.

Ships' captains liked them because the cases relieved them of any responsibility for the care of new plants on board. Without the Wardian case, many of the significant imports of the nineteenth century, such as tropical orchids, would never have made it back to England in good condition.

Once use of the case was perfected, the spoilage and rate of loss dropped almost to nil. The glass sides allowed sunlight to enter the case. The problems encountered by earlier plant collectors were almost completely overcome.

It is a testament to Menzies' skill that so many of his specimens did survive. He found 24 species and one genus, though at least eight other species were lost. Several subspecies were also attributed to him. Eighteen plants from Hawaii bear his name, though not all of them are in cultivation.

The Oregon grapeholly, *Mahonia aquifolium*, and flowering currant, *Ribes sanguineum*, are perhaps the best known of his shrubs. Pacific dogwood, *Cornus nuttallii*, was named for Thomas Nuttall, who traveled in the same territory 40 years later. Menzies found it first, long before Nuttall.

Menzies particularly prized ferns. The common sword fern used so often by florists, *Polystichum munitum*, recalls his work. Britain's legacy of evergreen narrow-leafed conifers, known to him as "pines," would have been enough to corroborate his reputation and establish the validity of his work.

Menzies returned to London and set up practice as a physician and surgeon, surviving more than 50 years after his exciting travels. Sir William Hooker, professor of botany at the University of Glasgow, began to study Menzies' herbarium sheets in 1830. He incorporated Menzies' plants in *Flora Boreali-Americana*, published from 1829–1840.[84] Menzies died in 1846 not knowing that his botanical work would be properly recognized and that he would ultimately receive his due.

Menzies was never really involved with the American wars. David Douglas (Figure 22) also missed the worst of it. Seed for some of the most important trees was collected by Douglas just after this tense period. Both he and Menzies introduced very important species of trees into the British Isles, once again changing the landscape. Like Menzies, Douglas concentrated on the Pacific Northwest. His very first assignment was to look for orchard trees in the Northeast, but he did not return to that part of the country. He later introduced a number of the plants that Menzies had seen and named.

Douglas had a much less auspicious beginning than Menzies. Menzies' family were estate gardeners, and while no one at that level was well paid or considered to be other than a working man, at least his parents could manage to support him in a medical apprenticeship. Douglas' father was a jobbing stonemason, a little lower in the social scale than Menzies' father.

Douglas was born in the village of Scone, Perthshire. All contemporary accounts indicate that, as a child, he was obstinate and recalcitrant. He hated school and did nothing he was told. His mother had come from a slightly superior family and tried to instill the idea of improvement and respectability in her children. David's behavior led only to her cold disapproval. Tantrums do not get you very far in a devout Presbyterian household.

Finally, in desperation, his parents apprenticed him to William Beattie, head gardener at Scone Palace. This turned out to be a good move. He loved nature and paid very close attention to his master.

He began to settle down and behave better. As he grew older, he remained shy and touchy, and he ultimately managed to antagonize almost everyone whose approval he sought and whom he previously admired. For the present the study of plants absorbed him totally.

From Scone he went on to Sir Robert Preston's estate at Valleyfield. The Prestons had a greenhouse with tropical palms, orchids, and ferns. These exotic plants were a revelation. It is said he bombarded the head gardener Alexander Stewart with endless questions. Stewart obtained permission for young Douglas to use Sir Robert's library. When Glasgow University announced there was an opening for a gardener in its botanical garden, Stewart recommended Douglas.

A year after Douglas went to Glasgow, William Jackson Hooker (1785–1865) became Regius Professor of Botany at the university (see Chapter 5). This coincidence was most important for Douglas' future career. William Morwood, one of Douglas' biographers, considers that Hooker became a surrogate father in Douglas' eyes.[85]

Douglas did everything Hooker wanted to perfection and never complained about any discomfort during their often grueling joint expeditions. He idolized William Hooker. No letters to his own father in Scone have been found, but there are numerous ones to Hooker and his family.

Hooker sent Douglas to London when the Horticultural Society needed a plant collector for the Pacific Northwest. That journey was postponed, but after a period of training at the society's grounds in Chiswick, Douglas was dispatched to New York to purchase fruit trees. This bizarre assignment stemmed from flagrant conflict of interest in the society's president, an owner of large orchards.

As soon as Douglas was allowed to explore the western section of North America, he got into his stride and sent back more specimens than any other collector, although many of the plants he found in the Pacific Northwest had been discovered earlier by Meriwether Lewis. Hooker had trained him how to preserve the specimens. He did this so carefully and conscientiously that his herbarium sheets are still bright and vivid. The Douglas fir, *Pseudotsuga menziesii*, may be the best known tree he introduced, but it is just a tiny fraction of his achievements.

The formal name indicates the fact that Archibald Menzies had seen the tree first, more than 20 years before Douglas. The suffix *-ii* or *-iae* shows that Menzies discovered the plant. The suffix *-iana* is an honorific, to indicate respect and admiration for Menzies (or another collector). These minor refinements are not always observed. Douglas introduced many plants that had been seen by Menzies first, but no one knew that until a long time later.

Scottish estates were enriched by the North American conifers. The Scottish landscape was changed by Douglas firs and California coast redwood trees.

Douglas introduced more than 670 species from California alone. He sent back seven of the 17 pine species native to California. *Pinus radiata*, *P. coulteri*, and *P. sabiniana* are just three of the better-known ones. He found *Ribes speciosum* (flowering currant), California bay, California poppies, lupins, penstemons, evening primroses, forget-me-nots, and the sole peony in the Western Hemisphere; these are now growing all over the temperate world.

Conditions in the remote Pacific Northwest were harsh. Douglas left some notes in his diary about one harrowing situation he found.[86] His gun needed repairs. He found the gunsmith Jacques Finlay at the old Establishment between the Columbia and Spokane rivers. Finlay's sons accompanied him.

". . . Mr Finlay received me most kindly, regretting at the same time that he had not a morsel of food to offer me, he and his family having been subsisting for several, at least six, weeks on the roots of Phalangium squamash [*Scilla esculenta*, Bot. Mag. 28, t. 1574. 1813], on those of *Lewisia redeviva* [*Lewisia rediviva*, Fl. Amer. Sept. (Pursh) 368] and on a black lichen [*Lichen jubatus*, now recognized as *Bryoria fremontii*] that grows on the Pines (p. 230)."[86] The native name for the phalangium was "camass."

The Finlays were from France and knew almost no English. In his turn, Douglas knew almost no French, but hunger needs no translation. Douglas had shot some game just before he reached them and gave the Finlays half of it. Not only was this plain human charity, but he commented that there was no one else for 800 miles around who could repair his gun.

Douglas almost always collected on his own, very seldom taking an assistant with him. On the few occasions when someone was attached to him over his objections, things went wrong. Nothing went more disastrously wrong than his last months in Hawaii.

He spent six months there, finding more than 2,000 specimens, but wanted to make one last expedition before returning to England. Because of bad weather as his ship made for Hilo, he asked to be put ashore early. The captain insisted he take a black man ashore whose master, a missionary, had somehow forgotten to take the servant with him when he left the ship previously. Douglas really did not want the responsibility but had very little choice. The missionaries were to be his hosts in Hawaii.

All these events are reconstructed in hindsight. No one knows anything beyond a few bald facts. Douglas was slowed by having to cope with the frightened black man. He decided to leave the servant John at a lodge he found on the slopes of Mauna Kea. This additional leg of the journey made him change his timetable. He then took off by himself and reached Edward Gurney's cattle factory much higher up the mountains.

Gurney, an ex-convict, trapped cattle and skinned them for their hides. He showed Douglas the traps, deep pits so skillfully disguised that even the native people were fearful of them.

Douglas was later found in a pit, gored to death by a bull that had fallen in before him. Many questions were asked about this tragedy, especially because his money was missing, but in the end Douglas was buried in Honolulu in 1834, and no one was accused of murdering him.

His little dog Billy, which went everywhere with him, survived and went back to England. A clerk in the Foreign Office took him home.

William Hooker worked diligently on the North American specimens he received principally from David Douglas and Thomas Drummond (d. 1835), another Scotsman who was roughly the same age as Douglas, although his date of birth has not been recorded (see Chapter 13). Drummond explored in Texas and the Southwest. Hooker published *Flora Boreali-Americana* in two volumes and 12 parts between 1829 and 1840. This predated his appointment to run the Royal Botanic Gardens at Kew.

Douglas' death in 1834 at the age of 35 was essentially the close of an era. Systematic exploration of the United States flora continued, but the great flush of initial discoveries tapered off.

THOMAS NUTTALL

Thomas Nuttall (1786–1859) was an Englishman who covered an immense amount of territory in the western United States (Figure 23).[87] Starting out as a journeyman printer in Liverpool apprenticed to his uncle, he emigrated from Liverpool to Philadelphia in 1808. The uncle was quite annoyed by this quixotic move and ceased contact with him for a long time. Thomas had always been interested in plants, and once he arrived in the United States he made this his work.

There is perhaps an apocryphal story of his meeting Professor Barton in Philadelphia because of an unknown flower he had found, but whatever the ostensible reason, Barton felt he could make use of this enthusiastic young man and turned him into a botanist and plant collector.

Barton contracted with Nuttall to explore unknown territories in northwestern New York, toward Lake Ontario, and later to examine the Great Lakes area. Later, when he went to the western regions, Nuttall was supposed to cover the whole distance alone and on foot for eight dollars a month, but he very sensibly used the Pacific Fur Company's barge to convey the heavy burden of plants and necessary materials along the great rivers. Nuttall joined the Astorians, a group on a westward expedition funded by John Jacob Astor, and left St. Louis on April 21, 1811. Washington Irving wrote a stirring account of the "Overland Astorians," as they were known.

After years of literally wandering in the wilderness, Nuttall accepted a post at Harvard College in 1823. He had published *The Genera of North American Plants* in 1818, a book intended to supplement Frederick Pursh's two-volume *Flora Amer-*

icae Septentrionalis (1814). The latter book in turn had been inspired by Michaux (see next section).[88,89]

Nuttall wrote in English, a welcome innovation for Americans without a strong grounding in the classics. He proposed 60 new genera. The reaction was mixed. Botanists tended to approach taxonomy from two viewpoints. One group, the "splitters," wanted to create new genera and species very liberally. The other group, "the lumpers," wanted to condense the genera and be parsimonious with new taxa.

Constantine Samuel Rafinesque, a contemporary botanist and demon for minutiae, berated Nuttall for leaving out many plants and species and incorrectly naming others, but Augustin Pyramus de Candolle (1778–1841), a leading authority in Switzerland, considered Nuttall had done his work well and adopted the new classification. De Candolle has stood the scrutiny of time very well. He and his son Alphonse (1806–1893) left a body of work on taxonomy that is still useful. Both men held the chair of botany in Geneva, the son succeeding the father.

Nuttall also enlarged Michaux's work and caused it to be translated into English. He collected a total of about 5,000 botanical specimens in his journeys across North America as well as mineralogical and ornithological material.

Asa Gray, the dean of nineteenth-century American botanists, said of Nuttall "no botanist has visited so large a portion of the United States, nor made such an amount of observations in field and forest. Probably few naturalists have ever excelled him in aptitude for such observations, in quickness of eye, tact, in discrimination and tenacity of memory."[87] This praise follows a period in which Gray had nothing good to say about Nuttall.

Nuttall's uncle died in 1837, leaving Thomas the family property. The terms of the bequest stipulated that Nuttall spend at least six months of the year at home in Yorkshire.

Nuttall returned to England very reluctantly in 1841, purely from a sense of duty. His last years in his sister's home were quite unhappy, though not because of poverty or want. She appears to have been rather bossy.

Richard Henry Dana had been a student of Nuttall's at Harvard and was amazed to come across him on the beach at San Diego. Dana calls him "Old Curious" in the book *Two Years before the Mast.*[90]

FREDERICK PURSH

Frederick Pursh (1774–1820), a German, was the first scientist to write about the botany of the entire continent of North America, including the Pacific Northwest. He could do this because he started out as Benjamin Barton's assistant and Lewis and Clark deposited their collections with Barton.

Pursh came from Saxony and had been working as a gardener in Philadelphia. He was obviously somewhat superior, and several wealthy Philadelphians allowed him to study in their libraries and improve himself. When Barton asked for an assistant, Pursh was a good candidate. His only problem was a tendency to drink too much.

Barton seemed to be too involved with many more pressing issues to get busy and catalog the extraordinary treasures Lewis and Clark had found. Pursh saw their possibilities and gradually removed them from Barton's premises. In 1811, he sailed for England. By the time Barton decided to start working on them, Pursh had taken almost all of them to England and later Germany. Did he purloin them?[21]

Scholars have questioned Pursh's bona fides in removing the plants. He spent a long time in London using Joseph Banks' and Sir James Smith's libraries to document his analyses. Whatever the verdict, he temporarily overcame his alcoholism and produced his *Flora Americae Septentrionalis.*

Pursh, safe in London with his "bootlegged" specimens, named two genera *Lewisia* and *Clarkia* in honor of the leaders of the overland journey. Eventually his alcoholism caught up with him. He went to Canada to take another position but died destitute in Montreal.

AMERICAN EXPLORATION CONTINUED

As more of the continent came under the control of the United States and the wars that had impeded its smooth development ended, the country began to look around in the world. In 1836, Congress authorized a large American expedition for the first time, the United States Exploring Expedition, to explore the South Seas under the command of Captain Charles Wilkes.[91] The U.S. government hoped to expand and consolidate trade rather than discover new territory for colonies. It authorized the use of six vessels, rather than the modest two of the European expeditions, and sent 346 men. Among those were nine scientists and artists.

The ships did not leave for more than two years. The West Coast of North America was a significant part of the study, as well as many Pacific islands. Captain Wilkes was charged with covering the Pacific Ocean from north to south. The voyage lasted from 1838 to 1841. Wilkes brought back enormous numbers of South American seeds and plants from that expedition when he finally returned in 1842. This voyage established the foundations of modern science in the United States.

At about the same time, John Frémont made some very respectable plant collections in the Far West while on his surveying trips. Karl Andreas Geyer, the Saxon botanist attached to the Nicollet expedition to the Upper Mississippi region, had taught him how to dry and preserve specimens.

No one institution was large enough to handle the massive amounts of material that poured in. A considerable portion of the material was sent to private landowners to grow in their gardens. Many of these seeds and plants failed to survive, largely as a result of ignorance of the cultural needs of such different plant life. The huge losses were a serious blow to science.

All overworked, Torrey, Eaton, and Gray made major efforts to cope, but some of the reports did not appear for more than 20 years. The Lewis and Clark material had hardly been absorbed before all these exciting new specimens were received. The combination of all this with the unfortunate neglect of the plants collected on the Wilkes expedition led to the creation of the Smithsonian Institution, funded by the fortune that James Smithson, an Englishman, had left to the American nation. It came in very handy at that time.

Smithson was very unusual in that he studied chemistry at Oxford in the latter part of the eighteenth century and believed that science was the route to advancing civilization. The future lay in the New World. Even though he never traveled to America, he left his fortune to the new country. The U.S. Congress was very suspicious of an Englishman's motives in doing that and squabbled about it for many years. Eventually they passed an enabling act and moved forward.

The biggest difference between these explorations and those that had gone before was that the products of their travels remained in the United States. In spite of that, the immense delay in publishing the scientific aspects of the work probably hampered the release of ornamental plants to the nurseries and thus the public. They still relied on imports from England and Europe for novelties. Some of Torrey's comments on the collection of American plants came out as late as 1874.

With the end of the Mexican-American War, Congress appointed a Boundary Commission in 1846. Three botanists were attached to the commission and brought back very handsome plants. The California gold rush erupted during this period. It, too, had botanical and horticultural repercussions. One was almost immediate. The other took place more than a century later.

The contemporary event was the episode of William Lobb and the giant redwood trees, *Sequoiadendron giganteum* (see Chapter 12 for more about William Lobb).[92] When the gold rush started, the Veitch Nursery in Exeter had sent Lobb to the West Coast of North America to look for plants the nursery could adapt and sell. He was excellent at his job and had also found numerous valuable specimens in South America, including begonias.

When Lobb saw the first specimen of the giant tree, he realized that here was a unique opportunity. He rushed to the Calaveras Grove and gathered enough material to make a good identification and start cultivation if necessary. He sent it all back to his employer, James Veitch, in secret. The tree was named *Wellingtonia* in England and occasionally is still referred to by that name.

Lobb used very sharp practice. Alfred Kellogg had shown him the specimen in good faith as one botanist to another. Lobb rewarded this generosity by grabbing the rights to naming the tree and giving Veitch a head start in propagating it.

Lobb also sent the seeds of other wonderful trees and shrubs to Veitch from the West Coast (Table 3). Some of them were new to Europe. Others had been seen and noted by earlier explorers, but the seed had not been collected in adequate amounts for commercial exploitation.

The delayed repercussion of the gold rush was the rediscovery of numerous old roses growing untended in cemeteries and abandoned gardens in the gold country starting in the 1970s.[93] (Old roses were also found in Texas and Bermuda.) Con-

Table 3. Some Plants Collected by William Lobb in California and Oregon[1]

Species Name	Common Name	New to England
Abies bracteata	Santa Lucia fir	
A. concolor	Colorado white fir	New
A. concolor subsp. *lowiana*	Pacific white fir	New
A. grandis	Giant fir	
A. magnifica	California red fir	New
A. procera	Noble fir	
A. venusta		
Aesculus californica	Californian buckeye	
Ceanothus lobbianus	Natural hybrid	
C. veitchianus	Natural hybrid	
Delphinium cardinale		
Fremontodendron californicum		
Juniperus californica		New
Pinus attenuata	Knobcone pine	
P. coulteri	Big cone pine	
P. lambertiana	Sugar pine	
P. monticola	Western white pine	
P. ponderosa	Western yellow pine	
P. radiata	Monterey pine	
Pseudotsuga menziesii	Douglas fir	
Rhododendron occidentale		
Sequoia sempervirens	Coastal redwood	
Thuja plicata	Western red cedar	New

1. Adapted from Sue Shephard, 2003, *Seeds of Fortune, a Gardening Dynasty*, London: Bloomsbury Publishing. Whenever possible, the nomenclature represents the current, accepted names for plant taxa.

trary to the popular idea, there were quite a number of women in the raw gold towns. Miners and their wives settled briefly in the foothills of the Sierra Nevada. Many women had taken their cherished plants with them from their original homes, including roses, and planted them around their new homes.

A year or two later they might have had to move on, but the roses survived. If a woman died, her husband or friends planted a slip of her favorite rose at the grave. These, too, frequently survived. Varieties of rose that had been all but forgotten emerged untouched.

Ethelyn Keays wrote a landmark book about forgotten roses in 1935, *Old Roses*.[94] Many other rose lovers rebelled against the modern hybrid roses that look very pretty but are difficult to keep healthy and often devoid of scent. These rosarians popularized the idea of restoring antique roses for their toughness and fragrance.

Part Two

ASIAN AND SOUTHERN ASCENDANCY

CHAPTER 8

INDIA, AFGHANISTAN, AND NEPAL

ANY ACCOUNT of the movement of ornamental plants from the East to Europe has to start with India. The extreme interest in plants from China, the Himalayan countries, and Japan has slightly obscured the importance of India as a significant source of new plants in the West. However, it has some of the greatest plant resources in the world. Many plants that actually originated in India had migrated or had been taken to other countries and had become established in antiquity before Western invaders reached the subcontinent. The magnolia is only one case in point.

In the fifth century B.C.E., the Persians invaded the Indus Valley. Fanciful "reports" of the country and its people circulated in the ancient world and came down to Western scholars through authors like Herodotus. Theophrastus described a banyan tree in the fourth century B.C.E., based on a story from a soldier who had seen one. Indian plants merited attention.

The Romans traded with India, buying spices, especially pepper, as well as textiles and precious woods. Pliny the Elder described Indian plants in his books on natural history. Presumably the merchandise traveled overland through central Asia. Alexander the Great marched his armies to India overland after some serious fighting in Egypt.

India teems with endemic plants at all altitudes in its many climates. Many of the plants were sent to England during the colonial period. Some of the best-known collectors were Dr. William Roxburgh (1751–1815) and, later, Dr. Nathaniel Wallich (1786–1854), who deluged London with live and dried specimens (see the sections "William Roxburgh" and "Nathaniel Wallich" in this chapter).

Sir Joseph Hooker (1817–1911), son of Sir William Hooker, laid claim to being the dean of the Indian flora with good reason. The number and quality of the rhododendrons he took back to England exceeded anything found by anyone else. Frank Kingdon Ward (1885–1958) spent much of his time in northern India and the adjacent territories during the 50 years he collected in Asia. He, too, sent back spectacular collections of plants, some of which succeeded in cultivation.

The British hold on India had been tightening since the charter signed by Elizabeth I in 1600. Portugal continued to occupy Goa but did not have a commanding presence in the subcontinent. There were also small French and Danish colonies, but England was by far the greatest foreign power.

Initially, the semiprivate East India Company only controlled commerce there. Military intervention then gave the company political power. It could prosecute wars, collect customs, excise taxes, and even issue currency. Judiciary services in the English style, with courts of law, bewigged judges, and barristers, completed the litany.

By 1757, the official rulers of India had been marginalized. Its very weak central governments were not unlike England at the time of King John, with constant challenges from powerful barons. Robert Clive and Warren Hastings made short work of the flimsy indigenous rule.* The country was divided between the larger English section and the small principalities still ruled by native despots. By 1794, the East India Company's behavior was so outrageous that the English government established a Board of Control to oversee its activities.

After the Sepoy Mutiny in 1857, India's first war of independence, this exceedingly lucrative business was finally absorbed by the Crown, turning Queen Victoria into the Empress of India. She did not look too closely at what was being done in her name.

The opium trade was controlled from India. Clive had seized the monopoly on growing poppies in India after the East India Company victories in 1757, just as he seized the monopoly on salt. Vast quantities of opium were being produced by the nineteenth century. It was smuggled into China in the holds of both "East Indiamen" and American ships.

Protecting the salt tax led to one of the more astonishing aspects of British rule in India, which has been almost totally forgotten. The price of salt with the tax put it out of the reach of those who needed it most, the poor laborers and coolies. A lack of salt in that hot climate rapidly leads to weakness and debility. Salt is essential to maintaining life itself.

As a result, smuggling salt became very common. To prevent this smuggling, the British Raj caused a hedge of mixed types of thorn tree to be planted, five yards deep, running for 2,300 miles along the borders with the native states, the Great Hedge of India.[95] Here was an instance of the punitive use of plants. The hedge was impenetrable. Official crossing points were cut into it at intervals.

The existence of the hedge was recorded in a report of the British Customs department in 1865. In 1990, when an English writer, Roy Moxham, found this report in a secondhand bookshop, all knowledge of the hedge had disappeared. He traveled to India three times in vain to try to find the hedge. Only on the fourth try did he find someone who even knew what he was talking about.

The colonial authorities established botanical gardens in many of the Indian states and principalities they controlled. Plants were more likely to survive the shorter journey within the East than the very long one back to England. That was one of the reasons the East India Company somewhat reluctantly went along with botanical gardens in their territory. In turn, the directors of the botanical gardens

*The genus *Clivia* is named for Robert Clive's granddaughter.

sent specimens on to London and other capitals. These gardens were primarily concerned with economic plants, but ornamental ones were included.

Early British travelers in China and Japan often sent new plants to these gardens. Dutch collectors used the Dutch East India Company's headquarters at Batavia (now Jakarta) on the island of Java in the same way. The Dutch company maintained experimental gardens in the Indies and other parts of the world where they dominated.

Quinine, derived from the bark of *Cinchona* species, was a key to controlling the malaria that afflicted and killed so many colonials. Dutch botanists had succeeded in smuggling enough live *Cinchona* plants out of Peru to start their plantations in Java, long before the English managed to do it. Richard Spruce, the great English plant explorer who spent 15 years in South America, was commanded to send *Cinchona* plants to India in the 1850s. He did this at great cost to his health, rushing across the Andes as fast as he could rather than planning a more measured journey by water. In his careful and thorough way, Spruce sent seeds and plants that grew successfully in India.

Thirty years later, in the 1880s, a resourceful clerk in the India Office, Clements Markham, laid out a larger plan of action and was allowed to put it into effect.[96] He was not above using some slightly unprincipled methods to attain his ends. Cloak and dagger had nothing on Markham, yet he capped his career by being knighted and becoming president of the Royal Geographical Society, a very respectable man indeed. Everything he had done had been checked first with the Hookers at Kew to make sure the horticultural aspects of his endeavors were sound.

These questionable exploits gave rise to enormous resentment in the nominally independent South American countries and support the contention of revisionist historians who consider it "botanical imperialism."[97] Stealing sources of unique raw materials undercut those countries' capacity to compete in world markets.

INDIAN FLORA

India is estimated to have more than 45,000 plant species. About 8,400 of the approximately 15,000 vascular plants are endemic, a huge number. Many of these are concentrated in the Western Ghats, the mountain chain running along the west coast of the subcontinent.

It is hard to understand why this region was neglected by the extremely conscientious European scientists who studied the south, the plains around the Ganges, and, later, the Himalayas so thoroughly. One reason may be that the East India Company had no interest in it at the time and therefore there were no funds. Native-born botanists have since made up the lack with very fine studies.

A few English residents in the 1820s and 1830s did collect and name some of the plants around Bombay. For example, William Sykes, a government statistician, reported on about a thousand plants he had found near Poona (now Pune). He employed a soldier in the local artillery, Bombardier Llewellyn L. Fidlor, to augment his own black-and-white sketches with very competent watercolors.

India's truly tropical plants could not survive except in hothouses, but there were many plants from higher altitudes that entered the English garden repertoire. The Himalayan rhododendrons are a case in point. A few were still too tender for the open air in England, but many flourished. The same was later found to be true of rhododendrons in North America.

Clusius was one of the first people to build a conservatory for tropical plants in Leiden, and other people had Indian flowers in Leiden at the same time. John Gerard mentioned the Indian tree of sorrow, *Nyctanthes arbor-tristis*, in the *Herball* of 1597.[19] The *Herball* was largely taken verbatim from Dodoens and L'Obel, but Gerard was an equal opportunity plagiarizer. A well-known Italian physician of the period, Castor Durante, published his *Herbario Nuevo* in 1585, with a picture of the "tree of sorrow." It is a night-flowering jasmine. Linnaeus established its credentials in *Species Plantarum* in 1753. Durante's work was extremely popular.

Black pepper, *Piper nigrum*, came from India and had been dispersed in ancient times to the East Indies and other islands. Pepper trees grew in Naples and Venice in the fourteenth century.[36]

The Dutch took the lead in requiring their colonies to furnish new specimens every year. Ceylon (now Sri Lanka) was quite cooperative, but the governors of Dutch colonies in South Asia and South Africa had to be reminded to comply with this ordinance. By 1680 the botanical garden at Leiden had 3,800 tropical plants growing there. India and Ceylon were well represented.

A Bengali banyan tree, *Ficus benghalensis*, grew in the hothouse at the Amsterdam botanical garden before 1685. *Nerium indicum* (now *N. oleander*) and *Cucumis* both survived in this environment.

Prince William of Orange and his wife Mary succeeded to the English throne after the death of James II in 1688. Mary was James II's daughter. They took the idea of heated glasshouses with them from the Netherlands. Indian plants had not been thriving in England, but once this practice spread, the change was immense.

Henry Compton and the Duchess of Beaufort both had hothouses and enviable collections of exotic plants. *Cinchona* and camphor trees flourished in the warm environment at Chelsea Physic Garden. Many other notable persons also constructed hothouses.

Peter Collinson liked to exchange his American treasures for Indian novelties. His friend Lord Petre could afford a very large "stove." Sago palm and the Indian *Ficus religiosa* were among the more conspicuous plants in the heated enclosure.

Even during the Dutch navigation wars with England for control of the rich shipping lanes to the East, Dutch and English horticulturists continued to exchange plants as if nothing were happening.

Philip Miller not only grew exotic plants in the Chelsea Physic Garden's hothouses, but he also gave away seeds and allowed others to have the same pleasure. His dictionary gave hints about successfully growing these tender plants. Jasmine from Malabar grew quite well in England but did not set fruit.

Miller was a skillful politician. He paid attention to the whims of his patrons. If a plant was out of fashion, he would not recommend it.

FLORISTIC INTERACTION BETWEEN CHINA AND INDIA

Many plants went to Europe directly from India. Others had been growing in China for a millennium and then were collected with the native Chinese specimens (see sidebar).

Chinese Plants Growing in the Royal Botanic Garden at Calcutta during the Tenure of William Roxburgh: Director from 1793 to 1814

(The information that follows is adapted from E. Bretschneider. 1898 [reissued 1981]. *History of European Botanical Discoveries in China.* Leipzig, DDR: Leipzig-Antiquariat. Whenever possible, the nomenclature represents the current, accepted names for plant taxa.)

Acalypha conferta
Agrostis maxima
Alpinia calcarata
Amaryllis radiata
Amyris punctata
Arum divaricatum
Bambusa nana
Bauhinia corymbosa
Bryonia tenella
Cactus chinensis
Callicarpa purpurea
Calycanthus praecox
Camunium chinense
Castanea pumila
Chloranthus inconspicuus
Citrus acida
C. inermis
Combretum chinense
Convolvulus bilobatus
Corchorus capsularis
Crinum sinicum
Crotolaria elliptica
Cymbidium tesselatum
Diospyros kaki
Dolichos obcordatus
D. phaseoloides
Dracaena ferrea
Fagara nitida
Fagraea fragrans
Flemingia prostrata
Fraxinus chinensis
Gaertnera obtusifolia
Hastingia coccinea
Helianthus annuus
Hemerocallis cordata
H. fulva
Hibiscus chinensis
H. mutabilis
Hovenia dulcis
Ipomoea pileata
Iris chinensis
Ixora coccinea
I. stricta
Jasminum paniculatum
J. pubescens
Juniperus aquatica
J. cernua
J. communis
J. dimorpha
Laurus dulcis
Lobelia radicans
Lycopodium aristatum
Melia azedarach
Menispermum hexagynum
Mespilus japonica
Mimosa fruticosa
Moraea chinensis
Morus atropurpurea
Murraya exotica
Nelumbium speciosum
Panicum lineare
Phaseolus minimus
Phillyrea paniculata
Poa cylindrica
Pyrus chinensis
Rhus succedanea
Rosa glanduligera
R. microphylla
R. triphylla
Saccharum sinense
Sapium sebiferum
Scirpus tuberosus
Scytalia longan
Serissa foetida
Serratula cinerea
Sida abutilon
S. chinensis
Sinapis brassicata
Solanum decemdentatum
Sterculia balanghas
Tagetes erecta
Ulmus virgata
Vangueria spinosa
Volkameria kaempferi
Ziziphus albens
Z. nitida

The magnolia is one example. It had originally moved from India to China a thousand years before and had adapted very well to the new country. English plant collectors in China did not know this history when they sent it to London.

The opposite was also true. Some of the plants in India had been brought from China centuries before. Buddhist monks traveled between the two countries and carried seeds with them. When the soil and the climate matched, the new plants thrived and entered the flora of the host country.

JOSEPH HOOKER

In 1848, Joseph Hooker, later Sir Joseph, was allowed to collect plants in the eastern region of Nepal for three months (see Chapter 5). This was a triumph. No European had been permitted to do this before. The adventure was unexpected. He had intended to go to Sikkim, but the authorities there would not let him enter. Stuck with nothing to do, he arranged the Nepalese trip. The Nepalese government was suspicious even of him and insisted he be accompanied by a platoon of soldiers. While in the Himalayas, Hooker was paid by the British government.

Getting in and out of Sikkim was also unpredictable. Years could go by without foreigners being able to travel and collect plants. Once Hooker managed to collect in Sikkim, he came back with an immense number of rhododendrons. He had discovered 28 new species. His base was in Darjeeling. Although very fit and more than willing to go out himself, he employed local men to do much of the fieldwork.

This had three advantages. Twenty native men scouring the countryside amplified his ability to find more plants. It gave him leisure to make good notes and draw the specimens carefully, but it was also more politic. The tribesmen attracted far less attention from superstitious villagers than a white man would have done. Even so, he had some close calls, for example, when by mistake he wandered into Tibet. That led to some very unpleasant events. It is said he narrowly escaped death.

After Hooker moved down to the humid forests of Bengal and began collecting there, he brought back over 3,000 flowering plants and hundreds of ferns. By the time he had finished with his exploration of India, he and his colleague in Kashmir, Thomas Thomson, had 150,000 specimens to document. Many of these plants went into the creation of *Flora Indica,* published in 1855.[98] At that time many small and incomplete collections of Indian flora were scattered about London. Taken individually, none was adequate to support a definitive study of the subject, but the cumulative effect was significant.

Thomson, an old friend of Hooker's from their student days at the University of Glasgow, was an employee of the East India Company. He asked the company to allow him time off at full salary to help Hooker prepare a flora that would tie everything together. The company refused but offered to give him a small grant to aid in its publication. Hooker managed to get a salary from the Admiralty for the three years he needed.

Botanists from all over England and Europe wanted to help with this enticing project. Wallich came out of retirement to do his bit. In the end the work was never completed, but the first volume laid the foundations for important concepts of phytogeography.

Thomson went on to become superintendent of the Royal Botanic Garden at Calcutta in 1855. Joseph Hooker returned to work at Kew. In 1863, the Hookers were able to convince the Colonial Office that scientists needed small floras suitable for field use. Sir William Hooker had already seen floras of the West Indies, South Africa, New Zealand (by his son), and Australia succeed.

Now he thought the time was ripe to produce a *Flora of British India.*[99] Sir William died before the work, written by his son, appeared. It ended up being seven volumes. The last one appeared in 1898. Joseph started out working with Thomson, but the latter became very dilatory and had to be replaced. Joseph's deadlines were very hard to meet.

Compared with the reach of the Horticultural Society, Kew only sent a few collectors into the tropics, but the one they did send, Joseph Hooker, was outstanding. Nepotism in this case was very constructive. The prodigious energy and application of the Victorian savants were fully realized in the Hooker family.

BOTANICAL GARDENS IN INDIA

The British East India Company paid lip service to the idea of botanical gardens as reservoirs of economic plants and places to experiment with new possibilities. In fact, this was always a rather peripheral concern, and as soon as money became tight, the gardens' funds were quickly cut.

Because the 28 gardens had been created over a long period of time during different administrations, there were at least three ministries apart from the India Office (British East India Company's governmental arm) that might be in charge of them. McCracken has noted that colonial botanical institutions governed by the Colonial Office in London rather than by the company itself fared much better. They were less subject to these arbitrary changes.[100]

Contrary to the public's understanding, Kew provided inspiration plus the scientific and technical basis for the colonial gardens but did not actually control them. Moral suasion was their sole weapon, but under the Hookers and Thiselton-Dyer, this was very effective. A curator who was underpaid or a superintendent who was being harassed by an unsympathetic governor could expect Kew to weigh in and make sure things were corrected.

A few brief sketches of representative Indian botanical gardens and their directors will give a flavor of how samples of the glorious tropical vegetation inspired observant Englishmen and reached "home."

Almost all of the collectors believed that the dreadful hardships they endured in collecting and maintaining the plants were worthwhile. Roxburgh and Wallich suffered from serious chronic illness all the time they lived in India. William Griffith died much too young. Only Joseph Hooker seemed immune. He lived into his 90s after years spent traveling in all the most inhospitable areas of the world.

WILLIAM ROXBURGH

The botanical garden at Calcutta is generally considered the finest in the system.[100] William Roxburgh (1751–1815) was the first superintendent of the Royal Botanic Garden at Calcutta. He had had the best training available at the time under Dr. John Hope at the Edinburgh Botanical Garden. Roxburgh went to India in 1776 and learned a great deal from the surgeon-botanist Johann Gerhard Koenig, whom he met in Madras. Koenig was a Dane of German extraction who had studied with Linnaeus.

Banks noted Roxburgh's appointment and encouraged him to send new plants back to England. He began to do this in 1793. Most of them were very tender and could only survive in a hothouse.

Lt. Colonel Robert Kyd had started the Royal Botanic Garden at Calcutta in 1787. He had been very distressed by the famine in 1769 and planted more than 300 varieties of fruit and spice trees. Kyd thought that the Persian date and sago would be valuable reserve foods. Another important goal was to grow teak for shipbuilding. Seizing the spice trade from the Dutch was also on his agenda.

After Kyd's death in 1793, Roxburgh, a Scot, was appointed to this position. Roxburgh had established a very fine reputation while working as surgeon-botanist in a provincial garrison. Both Sir Joseph Banks and Sir William Jones were impressed by his zeal and application. (Jones, a judge in the British courts in Calcutta, was the scholar who recognized that Sanskrit was the probable ancestor of all Indo-European languages.[101])

The ground chosen for the garden was not too fertile. Roxburgh worked very hard setting the beds out and getting plants to grow. One of his notable achievements was an avenue of mahogany trees, but he was most interested in systematic botany. He had to accommodate plants sent out to India from England as well as the local plants and imports from China. Oak, ash, chestnut, willow, poplar, and plums came from home. Banks had initially protested when Kyd asked for funds to develop three acres, but with this type of planting plan, the space was essential.

Roxburgh wrote *Flora Indica,* a description of all the Indian plants he knew, but died before the book was published. Volumes 1 and 2 appeared in 1820 and 1824, respectively. The reason for the delay between the two volumes was that his successor, Nathaniel Wallich, added more material.

Roxburgh was friendly with the missionary William Carey, a skilled linguist who reached Calcutta in 1783. Carey took on the responsibility of editing and publishing the first volume and participated in the second. He prepared the manuscript.

Carey's son Bruce later assisted one of Roxburgh's sons in editing and reissuing *Flora Indica* in 1832.[102] They removed all traces of Wallich's contributions in this edition.

According to the convention in botanical nomenclature, the name used by the first person to examine and register the plant must be retained. This means that the first edition of Roxburgh's *Flora* remains extremely important, even if there are errors. Roxburgh had to give up his post in 1813 because of ill health.

NATHANIEL WALLICH

Nathaniel Wallich (1786–1854) became the superintendent of the Royal Botanic Garden at Calcutta after Roxburgh died.[103] He was born in Copenhagen, the son of an emancipated Sephardic Jew, Wolff Wallich. In accordance with the old Jewish custom, the son was known as Nathan Wolff (or Nathan ben Wolff). He was graduated from the Danish Royal Academy of Surgeons in 1806, having studied botany under Martin Vahl.

No matter how gifted he was, a Jew had little hope of academic advancement in Denmark. Wallich joined the colonial service instead and was appointed surgeon to the Danish settlement at Serampore, India, arriving late in 1807.

What happened next might have come out of a Gilbert and Sullivan operetta. Denmark had been allied with France in the Napoleonic Wars. In retaliation England seized the Danish colonies after beating Napoleon. The new English commandant immediately put Wallich in prison. By 1809 the commandant recognized he was a scholar and not a military threat. He was released and sent to assist Roxburgh in Calcutta.

Wallich had already started to collect plants and animals and catalog them. The East India Company decided to make good use of his skill. He was also commissioned into the English medical service and practiced briefly.

He modified his name to Nathaniel Wallich, to conform with English practice. For two years he had to stay in Mauritius because of illness, but once back in Calcutta in 1814, he remained there for the next 30 years. These years were quite tumultuous.[104]

One of his first accomplishments was to found the Oriental Museum of the Asiatic Society, later to be called the Indian Museum, in Calcutta. Unlike other scholars who commissioned paintings by local artists, Wallich insisted that the painter sign his name to the work. He was not especially enamored of Indian people as such but did want to keep things orderly.

In 1817 he became superintendent of the Royal Botanic Garden and gave up his professional association with the museum. Wallich published two valuable books: *Tentamen Flora Nepalensis Illustratae*, 1824–1826, and *Plantae Asiaticae Rariores*, 1830–1832.[105,106]

While on leave in England from 1828 to 1832, he prepared a catalog with more than 8,000 entries, an enumeration of the plants he had found himself on a number of field expeditions. He had spent the whole of 1821 in Nepal and later visited the forests of western India and Burma at the behest of his employers. On all these journeys he was able to collect plants.

The company leased a house for him in London to work on the collection. He had traveled back with 30 crates of plants. They contained more than 20,000 specimens, many of which were duplicates. Botanical colleagues cheerfully assisted him in the enormous task. It gave them a chance to participate in the excitement and write a few papers of their own on the novelties. People like Professor John Lindley and the de Candolles of Geneva, father and son, gave him advice and support.

In 1824, during a long visit to Singapore because of more ill health, he had suggested that the governor, Sir Stamford Raffles, set up a botanical garden in Singapore. This was done, but the garden only lasted only for five years. Raffles himself and later his widow Sophia suffered great distress under the strictures of the East India Company's Court of Directors.

Wallich's touchy temperament always caused him difficulty. There is some evidence to show that he was a serious snob. He was affability itself with the governor of Bengal, Lord Amherst, but could be very nasty to people beneath him socially, unless there was something to be gained from being nice.

Two younger men who were sent to help him became implacable enemies, and even some of his admirers were concerned by his irritability and capacity to bear grudges. John McClelland pursued him through the pages of the Calcutta *Journal of Natural History*, and William Griffith sabotaged the Garden during one of Wallich's leaves of absence but not, it has been said, through malice.[107]

Griffith and Wallich locked horns during a collecting trip. It appears that Wallich resented the younger man's energy and skill and threw out some of his specimens in a very high-handed way. There are indications that Griffith, while brilliant, was also very difficult to deal with on a day-to-day basis. Everyone in the field assumed that Griffith would succeed Wallich as superintendent of the Garden, but Griffith caught a lethal infection and died tragically young.

Despite his reputation for irascibility, numerous reports attest to Wallich's patience and kindness. Wallich was hospitable and generous toward other new young collectors sent into a strange and alarming country. A perfect example was that of John Gibson, one of the Duke of Devonshire's gardeners.

Joseph Paxton, the duke's head gardener, sent Gibson to find rare orchids for the duke's collection. Gibson was a shy country youth who had never been out of Derbyshire before. No one thought of telling him that India was very hot, and he arrived in great distress. He had fair skin and no protection against the sun and heat. Country boys from the English counties knew nothing about impenetrable forests, venomous reptiles, huge rapids and floods, and hostile native people. Many disappeared without trace.

Instead of being elated, Gibson was overwhelmed by the immense responsibility of pleasing the august Duke of Devonshire and reached Calcutta in a deep depression. Paxton had seen to it that he had the latest and best equipment, but that only had the paradoxical effect of making him even more afraid of failure. He settled down when Wallich immediately took charge. Gibson listened quietly as Wallich explained what to do. Gibson had a very impressive career after that. At the risk of sounding cynical, it is possible that Wallich befriended Gibson in order to get into the duke's good graces.

One of the chief ways Wallich assisted Gibson and the others was in carefully packing the specimens they found, ensuring their survival. He had found that seeds survived very well when packed in brown sugar. By the time Wallich had almost finished sending plants to England, the Wardian case became available.

In 1845, J. O. Voigt, another surgeon to the Danish government at Serampore, brought out his own catalog of the plants in the Botanic Gardens at Calcutta and in Serampore.[108] The work was well received.

THE BOTANICAL GARDEN AT DAPURI

The botanical garden at Dapuri was founded by Sir John Malcolm, then governor of Bengal, on his own, without authorization. The Court of Directors of the British East India Company censured him for this piece of impertinence. He had bought a large bungalow in the cooler region of Dapuri and created the botanical garden on its property with some assistance from Wallich. Alexander Gibson, a Scottish surgeon (1800–1867), became the first director in 1838.

Malcolm used the land to demonstrate the value of growing various native crops for economic purposes. Ornamental plants crept in as an afterthought but were particularly cherished. As many as a thousand species of ornamental plants were grown at Dapuri at one time.

Gibson retired to Scotland in 1860 after making many contributions to Indian botany. He returned to Auchinreoch in Kincardineshire where he had been born. Eleven species were named for him.

Gibson and his successor Nicholas Dalzell prepared lists and catalogs, but the truly exotic and rare plants seen in the drawings are not found in these lists.

Dalzell had to close Dapuri when the company no longer supported it. He retired penniless, not unusual for the time when there were no pensions for persons at his level. His widow sold his herbarium to Kew and may have also sold the drawings. She had no other means of subsistence. This may have been the initial step leading to the drawings eventually landing in Edinburgh.

There were 28 botanical gardens in India, but Dapuri is special because of the beautiful colored drawings of its contents that have survived in Edinburgh. It is not known how the Royal Botanic Garden, Edinburgh, came to have the drawings.

H. J. Noltie discovered the drawings about 15 years ago and devoted himself to publishing them.[109] It is hard to believe that no one knows the name of the artist. The expense records of the time simply list a "painter" and small outlays for paper and colors. Noltie made extraordinary efforts to find the name but did not succeed. Artists were simply a variety of workman, like plumbers or carpenters.

The pictures recorded plants from many countries. This is another surprising thing about Dapuri. Gibson was very enterprising. He bought a few seeds and plants from nurseries but got most of them in one form of barter or another. The budget was always very strained.

It is known that Gibson received seeds from Swan River in western Australia in 1843. He had strong links to Kew and might have received American and African plants through that resource.

THE HIMALAYAS

The Himalayas were of interest because there were geographic and political issues associated with them. They separated India from central Asia with all its complex international intrigue. *Himalaya* is a Sanskrit word connoting a place where there is snow.

The mountains were eventually studied with exemplary fervor up until the outbreak of World War II. The results were dazzling. Sir Joseph Hooker set the standard. No one has ever really surpassed his collection of Himalayan rhododendrons.

The explanation for the immense variety of species presumably lies in the geologically recent formation of the mountains and the isolation of one valley from another.[20]

From the end of the nineteenth century, George Forrest, Frank Kingdon Ward, Heinrich Handel-Mazzetti, Reginald Farrer, and Ernest Wilson devoted themselves to the Himalayan regions. The American Joseph Rock worked there in the mid-twentieth century, and a few explorers such as Ludlow and Sherriff continued to travel for a short time after World War II. In their matter-of-fact way, all of these explorers described dreadful hardships but dismissed them as part of the job.

NEPAL

Nepal encompasses about a third of the Himalayan chain and has some of its highest mountains. The vegetation represents many floristic ranges, from sea level to high montane. The alternating monsoon and dry seasons have major effects on plant growth.

Nepal contains about 6,800 species of flowering plants, a veritable treasure trove, but reaching the country and staying within it safely has always been difficult, even now. Much of the diversity has vanished, and there are warnings about how the lack of a good conservation policy in Nepal and the untrammeled spread of tourism are speeding the loss.[110]

Countries high up in the Himalayas tended to be closed to the outside world and for a long time were very secretive. Nepal has not responded well to modern changes, particularly because of the continuing political instability. Bhutan is said to have done a better job.

Nepal, Tibet, Bhutan, and Sikkim are all endowed with huge reservoirs of extraordinary vascular plants. These have attracted travelers and explorers, but this has not been translated into tangible value for the inhabitants.

The climate is very hard to adjust to, food can be hard to find, there are numerous pests such as leeches, and the mountains themselves are treacherous for climbers. The northern parts of Burma, Assam, and India itself have similar characteristics. Geology and meteorology are not influenced by political boundaries. In spite of these drawbacks, plant collectors and explorers have always returned.

The European presence in Nepal began with the East India Company in 1793.

In 1801, Francis Buchanan, who had worked at the Royal Botanic Garden at Calcutta, collected almost 2,000 specimens in Nepal. Twenty years later, Wallich managed to spend a year there and collected almost all the orchids shown in John Lindley's book *Sertum Orchidaceum,* published in 1838.[111]

Collecting orchids was not all Wallich did. He presented his report to the governor-general when he returned to Calcutta. It contained 12 portfolios of descriptions of the Nepalese flora, 364 drawings, catalogs of wood specimens, and fruit and flowers preserved in alcohol.

Wallich understood how much English gardeners hankered after new plants. He sent *Edgeworthia gardneri* with its fragrant, tubular flowers and also thought that *Rhododendron arboreum* would do very well. These are both a little tender for much of England. His acquaintances laughed at his transports of joy upon returning from Nepal surrounded by baskets of plants and dried specimens. As Desmond says, he came back a "fulfilled and contented man."[107]

BHUTAN

Warren Hastings, the first governor-general of India, had heard there was a valuable turnip in Bhutan and thought it could be grown in England as cattle feed. When William Griffith finally found one in 1837, it was very disappointing. To him it seemed nothing more than a trumpery radish. No more was said about it.

Griffith spent about four months in Bhutan and left a collection of 1,200 species, which was neglected until very much later. This is the same man who bedeviled Wallich. He was very talented, worked very hard, and had a clear vision of botany in relation to geography long before this became fashionable.

In a delightful example of botanical coincidence, Thomas Nuttall's nephew Thomas J. Booth was so fascinated by the pictures of Himalayan plants in Joseph Hooker's *Rhododendrons of Sikkim-Himalaya* that he set off for India in 1849 after only the sketchiest preparation to get some of these plants for his garden in Lancashire. He was a complete amateur but persisted in spite of fearful obstacles.

By 1851 he had sent his uncle three consignments of plants, seed, and dried specimens from the eastern border of Bhutan. Nuttall described them and published his work in 1853. He or someone else named one pure white rhododendron *Rhododendron nuttallii.* As for Booth, he left England and disappeared for good after this astonishing burst of activity.

One assumes that Nuttall had bought the publication that inspired his nephew, though of course Booth might have bought it for himself. It says something for their financial condition that either of them could afford to buy a book of this type. It was the first section of a very long work and would have been prohibitively expensive.

In 1914 and 1915, one of A. K. Bulley's collectors from Bees Ltd., Roland Cooper, found the seeds of plants that the nursery had sought, such as 19 species of primula. *Primula eburnea* was his favorite. Most of his primulas did not survive in the English climate.

AFGHANISTAN

While Afghanistan is a sad spectacle of dust and ashes today, in the early nineteenth century it bloomed with myriad wild and cultivated plants. The first Englishman to penetrate Afghanistan was William Moorcroft (1767–1825), a veterinarian trained in France (at that time England had no schools of veterinary medicine) and employed by the British army in India.[112] The army sent him to Bihar in 1808. His task was to buy good horses for the military.

Using the excuse that there were fine Arabian horses in Afghanistan, Moorcroft set off on a splendid journey in 1811 and collected many plants along the way. He continued to travel and explore Afghanistan, keeping voluminous journals. On his second grand tour, Wallich sent two native plant collectors with him.

His enthusiasm left a legacy of botanical and geographic information. Fruit and orchards were his principal targets. Moorcroft sent plants and seed both to Robert Brown in London and to Wallich in Calcutta. His most important contribution was to grow oats successfully in India, something that had been said was impossible.

The army was not amused by his very loose interpretation of their instructions and made no effort to preserve his papers after his death. He died or was murdered in 1825 and was buried at Kunduz, a place tragically made famous in the Afghanistan war of 2003. The Royal Geographic Society brought some of his work out in 1830, and a poorly edited version of the complete material appeared in 1841. He finally received his due in 1985 when Garry Alder published Moorcroft's biography, *Beyond Bokhara*.[113]

CHANGE ON THE HORIZON: THE GROWING CRAZE FOR EXOTIC PLANTS

As the world opened up further, Sir Joseph Banks wanted to find plants and animals that would benefit England (see Chapter 5). It was his idea to send collectors in the royal name proactively and not just wait for the new finds to come to him.

The systematic collection of Indian plants as compared with their random arrival from time to time was set in motion by Sir Joseph. In 1787 he sent a Polish man, A. P. Hove, to bring back a collection for Kew. The East India Company was still not interested in this nonsense and made things so difficult for poor Mr. Hove that he returned with almost no plants at all.[107]

Things like that never affected Banks' zeal. He asked Christopher Smith, a botanist newly appointed to Calcutta by the East India Company, to send him plants, too. This worked better. The actual carriage of the specimens was done by Peter Good, a gardener who later traveled to Australia with Robert Brown.

As the new plants came into England, they were widely disseminated. Lord Bute grew Indian plants at his estate in Hampshire. He also made sure that Kew had the same specimens. The Royal Horticultural Society's library has Bute's collection of 775 botanical drawings, principally of Indian plants. They were done from living specimens at his estate and at Kew.

Officials of the East India Company had first choice of anything from the colonies. James Forbes brought seed collected in Gujarat with him when he retired to England. He filled his greenhouse with Indian fruit and ornamental plants such as crimson *Ipomoea* and *Hibiscus mutabilis.*

The craze for tropical plants was so pervasive that it even reached a point at which Banks felt that enough was enough. He began to long for hardy plants that grew outside on their own or in a simple greenhouse. This is an amusing paradox because it was his exertions that had fostered the craze for exotics in the first place.

Desmond gives a great deal of the credit for this success to skilled nurserymen.[107] Joachim Conrad Loddiges in Hackney, just east of London in those days, was generally considered to have had the finest collection of tropical plants. With his son George, he had started the *Botanical Cabinet* in 1817 to encourage sales of his collection. This was a magazine issued at varying intervals. Visiting this nursery helped young men to recognize the plants they had to procure when they went out to India to collect for their employers.

Later in the nineteenth century, professional collectors tramped all over India, sent by nurserymen like James Veitch. Thomas Lobb, William Lobb's brother, found new orchids in Assam and Burma and barberry in Nepal. *Lilium wallichianum* var. *neilgherrense* was truly a trophy find, as was *Hypericum oblongifolium* (now *H. hookerianum*) from Assam.

Lobb was very astute, like his brother. He made it seem as if he were not interested in the plants and that, anyway, there was nothing of any value to be found. (For more about Veitch and the Lobb brothers, see Chapter 13.)

Many individuals felt the excitement of Indian plants. In Liverpool William Roscoe (1753–1831), a prominent lawyer and writer who was very interested in horticulture, was able to get different species of ginger directly from the Royal Botanical Garden at Calcutta. Roxburgh, director of the garden, and later Wallich, who succeeded him, were very helpful and cooperative.

Wallich sent dozens of marvelous plants both to private gardeners and to botanical gardens. Himalayan plants from Nepal were his favorites. He considered the Kahili ginger plant, *Hedychium gardnerianum,* to be at the pinnacle of the genus (Figure 24). It grows well in San Francisco in spite of the cool grey weather.

Wallich also found *Abies spectabilis, Picea smithiana, Pinus gerardiana, P. wallichiana,* and possibly *P. nepalensis. Cotoneaster frigidus* made a deep impression, but *C. microphyllus* and *C. rotundifolius* were also enjoyed. Gardeners owe *Hypericum uralum* and *Spiraea bella* to Wallich.

Geranium lambertii, G. nepalense, and *G. wallichianum,* as well as *Bergenia ligulata,* were a few of the others. He seems to have been the first person to send *Rosa brunonii,* also *R. sericea* and *R. macrophylla.* In the modern period, Indian rose breeders Girija and Viru Viraraghavan have returned to these roses as sources of new hybrids that can tolerate the enormously varied Indian climate. They are reversing the steps that made roses survive harsh winters in northern countries and breeding heat and drought tolerance into them.

Wallich introduced another permanent favorite, the poppy. He found the first Himalayan poppy, *Meconopsis napaulensis.* Joseph Hooker introduced many species. The famous blue poppy, *M. betonicifolia,* was found a century later (Figure 25). Wallich's prize find, *Amherstia nobilis,* named for the Governor-General Lord William Amherst's wife, Lady Amherst, and her daughter Lady Sarah, could not be made to flower in England until about 20 years after it was sent.

Lady Amherst was more than just an elegant aristocrat.[107] She and her daughter were active plant seekers themselves, within the limits of the time. They sent the seeds of *Clematis montana* and *Anemone vitifolia* to England, where they germinated and grew successfully. Wallich found the *Amherstia* growing at a temple in Burma, but it has not been seen in the wild since 1865. Even now it is unclear where it actually originates. Wallich gave two plants to one of the Duke of Devonshire's gardeners, John Gibson, one for the duke and one for the Court of Directors of the East India Company.

In spite of the best efforts of Paxton, the tree never flowered at Chatsworth. Twelve years later a surgeon's wife near London, Louisa Lawrence, managed to coax her tree into bloom. The duke was not amused. It appears that Mrs. Lawrence provided her specimen with more humidity than the duke, and this was the secret. She had surrounded it with a canvas tent.

Among the many discoveries, the deodar cedar, *Cedrus deodara,* spread rapidly and widely throughout London and England.[107] Deodar seed had originally been imported by Leslie Melville in 1831. He gave some of his seed to Kew. The tree had great potential for timber as well as ornamental value. In 1841, the Commissioners of Woods and Forests requested the Court of Directors of the East India Company to import 500 pounds of seed to furnish both private and public English gardens with these trees.

The high rate of dissemination continued for many years. In 1863, more than 2,000 pounds of seed came from a single botanical garden in Saharanpur, India. This illustrates the economic system working at its peak. The price of a sapling fell dramatically, from £5 or £6 each to £1 per hundred.

The superintendent of the East India Company's garden at Saharanpur, John Forbes Royle, was very conscientious. He wrote a flora of the Himalayas in 1832 and continued to send the Horticultural Society many specimens. When Royle retired, Hugh Falconer took over, and between them they supplied two more *Berberis*; two *Deutzia*; *Elaeagnus umbellata*; a new variety of white jasmine, *Jasminum affine*; some *Philadelphus*; *Spiraea*; *Pyracantha*; and Himalayan lilac, *Syringa emodi.*[106]

They did not stop there. Gardeners may not know that *Androsace lanuginosa, Anemone rivularis, Aquilegia fragrans, Bergenia ciliata, Codonopsis, Cynoglossum grandiflorum, Filipendula vestita, Iris aurea* (now *I. crocea*) and *I. biglumis,* two *Potentillas, Primula denticulata, Roscoea vaccinifolium,* and *Impatiens glandulifera* all went to England as a result of Royle and Falconer's exploring.

Himalayan fir trees, spruce, alders, magnolias, cotoneasters, and sorbus all became increasingly common, but the rhododendron was the most popular of all. Thomas Hardwicke had found the beautiful red *Rhododendron arboreum* in 1796 (Figure 26). Sir James Smith used Hardwicke's notes when reporting on the new plant in his *Exotic Botany* in 1804. This was the first Indian rhododendron to appear in England.

Wallich sent more seed and was certain the genus would do very well in England. Joseph Hooker's later collection of 25 species fully vindicated this confidence. Some of the most striking species first appeared in his collection. Hooker introduced *Rhododendron falconeri, R. griffithianum, R. thomsonii, R. cinnabarinum,* and *R. campylocarpum,* among others. Each of these was an extraordinary parent for future hybridists. There was a gap of about 20 years from the time that Wallich was collecting in the Himalayas and the rise of the new breed of professional collector later in the nineteenth century. Joseph Hooker was still an academic collector. He visited the Himalayas in the late 1840s.

Nonprofessional collectors, often military officers, filled in. Lytton Strachey's father, Sir Richard Strachey, sent the first *Nomocharis* to England. These were the people Sir William Hooker called his "*corps botanique.*"

Gorer revives the memory of Edward Madden, an army officer who sent his treasures to the National Botanic Gardens at Glasnevin, near Dublin. Madden's notebooks have been preserved. He was a careful and astute observer. For example, he saw that *Rhododendron arboreum* seed at 7,000 feet ripened in December, "but *R. campanulatum* D. Don and *R. anthopogon* D. Don, which flower in May, June and July at 12,000 feet are ripe by the end of October (p. 119)."[114] He thought this meant that seed at the higher elevation had a longer time to take hold before the full severity of winter set in, a very logical conclusion.

RHODODENDRONS BECOME COMMONPLACE

Rhododendron is now a staple of suburban gardens, thoroughly entrenched. No one gives it a thought any more. Thousands of hybrids have allowed the owners of even a modest garden to grow rhododendrons in as many bright or delicate shades as they wish. These shrubs are also the workhorses of public parks, combining rich dark green foliage, handsome color, and fine massing quality. They are relatively easy to maintain.

Only a few species hailing from very high in the Himalayas can withstand the winters in the northeastern United States. Ernest Wilson, while he worked for the Arnold Arboretum, regretted that so many rhododendrons were not hardy in Massachusetts.

John McLaren (1846–1943), superintendent of the Golden Gate Park in San Francisco for 50 years, prized rhododendrons above all other plants. They had been part of his world in Edinburgh, while he was a gardener's apprentice at the Royal Botanical Gardens in the 1860s. New rhododendrons continued to be found during and after his apprenticeship and were still considered to be extremely exotic. McLaren joined more than one consortium of backers financing exploration for rhododendron in the Himalayas during his tenure at Golden Gate Park in San Francisco.

The Rhododendron Dell in the park displayed the results. It is a favorite corner of the park. McLaren thought he was planting in a very naturalistic manner, recreating the glens of Scotland.

Recently, a lot of the bushes, originally planted in the 1920s and 1930s, had to be dug up and replaced by new ones. There was a huge public outcry in 2003 when the now antique, woody, and sad-looking rhododendrons were pulled up, to be replaced by new ones. San Francisco's citizens had no idea how old the shrubs were and thought this was mindless bureaucratic vandalism. The park's workers did not quite have to resort to bulletproof vests, but it was a tense few days.

PRIMULAS

Primulas also came from Nepal, though they are more frequently found in western China. More of them were found in the twentieth century. Desmond comments that some of the plants found by Wallich and Buchanan in Nepal were not introduced into English gardens for more than 100 years.[107] Later explorers in the same regions brought them back again, and this time their worth was recognized.

The unlikely team of Ludlow and Sherriff was responsible for many species of primula.[115] Frank Ludlow (1886–1972) was a well-educated Englishman who went to India as a schoolmaster. His hobby was natural history, especially ornithology. In his vacations he explored widely and collected many specimens of birds and plants. George Sherriff (1898–1967) was a professional officer in the Indian Army.

Ludlow acted as the British representative for Tibet in Lhasa in 1944. He was succeeded by Sherriff right at the end of World War II. Both worked in India while it was still part of the British Empire. These two rather different men met in India while on leave and became friends. They decided to explore and collect together.

Together they made seven trips over the next 16 years, from 1933 to 1949, collecting plants, birds, and other animals. They started from Tibet and managed to avoid George Forrest's unpleasant fate. At the end of their active days it emerged that they had collected more new species of primula than anyone else, about 130 in all. Although neither was a trained taxonomist, both devoted a lot of their retirement to naming and describing the primulas they had found.

A Scotsman by birth, Sherriff retired to a farm in Scotland, and his material is now in the Royal Botanic Garden at Edinburgh. Ludlow worked in London. Their journeys were a little more comfortable and civilized than so many others because Sherriff was an excellent planner. He had gained logistical skill in the army and organized their plant hunting in the same efficient manner. Sherriff calculated precisely how much food they would need, and he set up way stations where vegetable seeds would be planted on the way out. On the way back the vegetables were ready to eat. He forgot nothing that would keep them and their porters healthy and happy as far as possible. The only things that evaded his thoughtful care were the weather and the pests. No one could do anything about the huge, bloodthirsty leeches.

His most imaginative action was to send the specimens back to England by air. For a time he and Ludlow had tried sending their plants back in or near a ship's refrigerator, but then he had the even better idea of shortening the journey by using planes to convey the plants. In 1938 this was done successfully for the first time.

Once India became independent in 1947, Ludlow and Sherriff were no longer so welcome and unconstrained, and they reluctantly gave up exploring after 1949. China and Tibet were now completely closed.

Although the Himalayan plants arrived in good condition, they proved very difficult to keep in flower. Time after time Sherriff reported sadly that his seedlings either died young or, if they matured, did not flower long enough to set seed. In the end very few imports survived to enter the standard English nursery repertoire. The British climate, even in southwest Scotland, simply could not supply the temperatures, the endless rain, and the mist of the Himalayan slopes where the plants originated and to which they were exquisitely adapted. (In the current period, a plantsman in Scotland, John Roy, is proud of the fact that he was able to bring some Himalayan primrose seeds, such as *Primula elizabethiae,* into flower when others were unable to do so.[116])

Ludlow and Sherriff epitomized the restrained, unemotional Englishman of legend. After spending many months together for almost 20 years, they still addressed each other only by their surnames. Sherriff had married the daughter of an English missionary while in the service. Mrs. Sherriff was very helpful to her husband, especially when they were representing Britain in Tibet and also out in the field. Ludlow never married.

CHAPTER 9

CHINA

IN THE MID-1840S the hitherto inaccessible Orient was very slowly starting to open up. Tantalizing glimpses of an exciting flora appeared through the barriers. Even while everyone was absorbed in the wonders from the New World, change was coming.

In spite of the enormous difficulties in dealing with the Chinese empire, Western countries persisted in trying to make contact and constantly gnawed away at the barriers. Organizations and publications were started specifically to keep up with the increasing flow of information.

There was great enthusiasm among the knowledgeable public. They all wanted to own these treasures. The societies started to send representatives for the purpose of finding more plants, even before foreigners could go more than a mile or two inland from the coast. The transition from being a forbidden country to one where it was possible to move about relatively freely took many decades. Somehow the Dutch merchants managed to deal with the restrictions, but English diplomats were rebuffed time and again.

EUROPEAN EXPERIENCE IN CHINA

RUSSIA AND CHINA

The Russians had a "backdoor" affiliation with China. Going overland through Siberia took them directly into the northern sector of China. The Russian embassy was founded in Peking (now Beijing) in the seventeenth century in spite of the Chinese resistance to foreigners.

The Russian government organized many expeditions with explorers, geographers, and scientists. They were always accompanied by military personnel. The physicians and scientists at the legation were usually military officers. Russian motives were secret, but they were very wary and suspicious about their neighbors' intentions.

Several great botanists worked out of this embassy.[117] Alexander von Bunge (1803–1890), Carl Maximowicz (1827–1891), and Grigori Nicolaevich Potanin (1835–1920) are only a few of the best-known scientists. They sent their discoveries to St. Petersburg and Moscow but also to Berlin and Paris. Maximowicz was trained by Bunge. He made a number of journeys to China and Japan but later was based in St. Petersburg.

Emil Bretschneider

The painstaking historical work of Emil Vasilievich Bretschneider (1833–1901) has provided historians with a unique and unprecedented base from which to understand how western botanical activities in China evolved (Figure 27). This is a good place at which to consider the man himself in light of his achievements.

He chronicled the experiences of European botanists in China. Now, a century after his labors, he himself is emerging as an important historical figure in this story. He was born in Latvia and educated in Russia. Bretschneider studied medicine at St. Petersburg and became the physician to the Russian legation at Peking in 1866. After retiring in 1884, he spent the rest of his life compiling and editing the two-volume *History of European Botanical Discoveries in China.*

Although he was far from the center of scientific action, he was deeply interested in Charles Darwin's and William Lyell's theories. He paid close attention to all that was churning in biology, geology, and the history and geography of central and eastern Asia during those stirring years. He corresponded widely all over Europe. Scientific journals came to him in his remote fastness. No detail was too small to escape his attention.

The Archimandrite Palladius, Piotr Ivanovich Kafarov, religious leader of the legation at Peking, stimulated Bretschneider's interest in regional historical geography. The legation owned a most amazing collection of books going back to the seventeenth century, when the first Russians crossed Siberia into China. The young physician started to study the history of Russians in central Asia and wrote several significant papers on sinology.

Bretschneider knew that biological argument depended on accurate facts. If the information about plants from China was muddled, there was no way it could be used to support arguments for or against any theories. He set out to bring order into this confused mass. From the mid-1880s, he published preliminary papers on the early botanists, but the two-volume history is his magnum opus. Not the least of its striking aspects is that he wrote it in English himself. It has not had to be translated.

Bretschneider occupied his time very constructively under somewhat difficult circumstances. Chinese protocol was forbidding. He seems to have been alone in Peking, a cold, dark, and windy city in winter and very hot in summer. Biographical sketches do not mention a wife or children. There was almost too much leisure time.

Bretschneider himself made some useful botanical discoveries in the hills around the city. He sent seeds to St. Petersburg and also to Sargent at the Arnold Arboretum. The Chinese seeds germinated well in Massachusetts, and the plants were very successful.

I find Emil Bretschneider slightly enigmatic but wholly admirable.

JESUIT MISSION IN PEKING

Beside the modest but direct communication between England and China, other European powers were active in that period. A Jesuit mission had succeeded in setting up shop in Peking at the close of the sixteenth century, a privilege denied to almost everyone else. Father Matteo Ricci (1552–1610) made many attempts to do this and finally succeeded in 1600. The mission thrived for over a century, served by priests of many nationalities. The Society of Jesus held scholarship in high esteem. Many of the missionaries were distinguished in their temporal fields. Some of them pursued botany. It was sad that Ricci only lived for 10 more years after his great achievement.

From about 1700 the Jesuits exported samples of Chinese vegetation, chiefly overland through Siberia. Russian caravans and diplomatic embassies to China acted as conduits. Once across the borders, the plants were widely distributed. Specimens were received in Italy and Paris, for example.

Tournefort described and drew *Caryophyllus sinensis supinus* in his *Memoirs of the Academy*, published in 1705.[107,118] The Abbé Bignon had received it in 1702 from China.

In 1723, the Italian scientist Michael Angelo Tilli described a Chinese plant growing in Florence, *Cucurbita sinensis fructu.*[107,117] Abbot Josephus Ignatius Cordero took the seeds to Italy from China.

Philip Miller reported on the Chinese plant *Rhus chinensis* in 1768 in the eighth edition of *The Gardeners Dictionary*. It had reached him from Paris and grew successfully in the Chelsea Physic Garden.

Père Pierre d'Incarville

An outstanding member of the mission was Père Pierre d'Incarville (1706–1757). He worked in Peking from 1740 and died there after 17 years of service. Bernard de Jussieu had been his teacher. D'Incarville sent his collections to de Jussieu, but he also shared seed with other European experts. The genus *Incarvillea*, growing in the plains and mountains around Peking, commemorates his name. It is an exquisite herbaceous plant with pink flowers. D'Incarville sent the dried index plant to Paris in 1743.

According to Bretschneider, my authority in these matters, he sent magnificent collections to de Jussieu, but they languished in the Jardin des Plantes without being analyzed rapidly. Only about 15 of them were commented on ("noticed" in the quaint expression of the day) before 1800.

Many flowers and shrubs now grown commonly in Europe and North America were first described by this Jesuit. *Ailanthus glandulosa*, *Sophora japonica*, *Polygonum tinctorium*, *Thuja orientalis*, *Koelreuteria paniculata*, *Ziziphus sinensis*, *Gleditsia sinensis*, and *Lycium chinense* can all be traced to his collections.

Ailanthus spread rapidly because d'Incarville sent seed to Philip Miller at the Chelsea Physic Garden and a Philip Carteret Webb in Surrey. The seeds grew into plants that were described in 1751. Miller was always generous and distributed seed widely.

At first he thought it was a Chinese lacquer tree and held on to this idea until the trees fruited.[98] They could not have foreseen that it would become a nuisance in North American cities.

D'Incarville left a handwritten catalog in French of more than 260 new plants he found in China. From various clues it seems it was prepared hastily in 1755. The plants were grouped by site of origin. Most of them came from the region around Peking, but a good number were found near Macao. Allowing for the outmoded nomenclature and, hence, identification, it is thought that most of d'Incarville's discoveries remain ensconced in today's Western gardens.

He gave his list to a Russian surgeon to take to St. Petersburg. D'Incarville corresponded actively with Russian scholars. He also received materials and letters from de Jussieu in Paris while in Peking.

D'Incarville sent seed as well as dried specimens to France. The royal gardeners were able to germinate many of these seeds and thus grow the plants he collected, but he himself was more interested in the botanical significance of his finds and less concerned about whether a particular plant would do well in European gardens.

THE NETHERLANDS AND CHINA

Most Europeans trying to live and trade in China and Japan were subject to ferocious frustration until the end of the eighteenth century. Handfuls of plants were smuggled out and reached Europe by circuitous routes, such as India or the Dutch East Indies, but it was a very risky procedure.

The Dutch started to explore the Chinese coastline even before the formation of the Dutch East India Company. For a time they were ensconced in Formosa, but in 1662 a powerful Chinese pirate expelled them. The company's work in Japan went better. They signed a treaty with the Japanese emperor in 1611 for land near Nagasaki.

Although their stay in China was brief, the Dutch are credited with introducing tea (*Camellia sinensis*) from China into Europe. In spite of popular myth, it is more likely that the tea plants sent to Leiden came from Japan than from China.

Chinese plants flourished in Dutch gardens very early. This fact was recorded by Jacob Breyne (1637–1716), a German merchant and amateur botanist, in his book *Prodromus fasciculi rariorum plantarum...in hortis celeberrimis Hollandiae observatarum*.[117] He enjoyed visiting private Dutch gardens in Amsterdam, Haarlem, Leiden, and other cities and did so repeatedly.

The two volumes were published in 1680 and 1689. He mentioned tea plants but was possibly the first in Europe to comment on *Chrysanthemum sinense* (*Chrysanthemum morifolium*) and *Azalea indica*. The azalea had come from India, and Linnaeus named it accordingly.

Artists painted extravagant still lifes, with vases of exotic flowers (Figure 28).

It was not by chance that this form of art peaked in Holland at the time. The rich and varied imports played a significant role. A newly developed bourgeois class demanded symbols of their achievement. Possessing one of these paintings conferred status. Paintings also publicized the new acquisitions for the nurserymen. Finally, less opulent families could enjoy the new flowers vicariously without the expense of a garden.

THE ENGLISH EXPERIENCE IN CHINA

English ships first went to China in 1602, but nothing more happened until 1637 when Charles I sent Captain John Weddell with a fleet to explore further. He used force to get some supplies, but no more progress was made for many years.

John Tradescant (the elder) listed 32 different species and varieties of rose in his collection. One of these was named "Mensalis," or monthly. It is extremely tempting to speculate that this was an early import from the abortive Chinese expeditions, but it is not really likely. Weddell had a lot more on his mind than flowers and spent very little time in China. It is more likely that this was a variety of *Rosa moschata*, the musk rose. If the weather is warm enough, this rose will bloom more than once a year, although only a few scanty flowers appear.[43]

The East India Company dispatched a ship laden with merchandise from their headquarters in Madras to Macao, but the crew did not manage to sell anything. In 1677 and again in 1681, the company sent more ships and finally managed to obtain a tenuous hold in Canton. A larger expedition in 1701 led to more trading at Chusan.

JAMES CUNNINGHAM

James Cunningham, listed as a surgeon but in fact a rather shadowy figure, was on the latter voyage. Even the omniscient and indefatigable Bretschneider could only find faint traces of his activities.[117] A Scot, Cunningham joined the East India Company and first went to China in 1698. He was generally considered to be the first English plant collector there until Bretschneider produced evidence to the contrary.

Cunningham was only active for about 10 years. He shared his botanical finds with Hans Sloane, James Petiver, and Leonard Plukenet (see Chapter 5).

In addition to a fairly extensive collection of cultivated Chinese plants from Chusan, he also sent a package of flower paintings by Chinese artists to England. Petiver commented on them several times in marginal notes.

Petiver's *Gazophylacium Naturae et Artis* (1702–1709) contained descriptions of about 100 exotic plants, 20 of which came from Cunningham's material.[119]

It is believed Cunningham died in the East on his final voyage in 1709. His last letters were addressed to Sloane and Petiver. He sent about 100 different specimens. Chinese clematis, aquilegia, aconite, alcea, euonymus, roses, jasmine, asters, cypress, fir, and forms of rhubarb are among Cunningham's plants that were still in the British Museum as late as the end of the nineteenth century.

BRITISH ATTEMPTS TO REACH TRADE AGREEMENTS

China continued to attract the West, and George III made several efforts to sign trade treaties without success. The large and costly expedition headed by Lord Macartney in 1792 was contemptuously spurned by the emperor of China just as previous and subsequent ones had been.[7] What prevented it from being a total loss was the assiduous plant collecting by Macartney's secretary, Sir George Staunton, and his son.

They managed to elude their minders in and around Peking and on the long road back to Canton. Bretschneider records their finds under four headings. The Peking collection contains 114 items. Between Peking and Jehol they found 60 specimens. On the road in Shan Tung and Kiang Nan they found 122 specimens. Finally, on the last leg of the journey, in the provinces of Kiang Si and Kuang Tung, they collected about 100 plants.

In all, Staunton brought back about 400 specimens. It was very lucky that these plants survived the voyage back to England. Staunton presented his collections to Sir Joseph Banks. Plants were shared with botanists such as de Candolle in Switzerland. Some of the items were duplicated. Some were not new. Some were studied carefully and properly named, but many were not. In spite of these disadvantages, a large number of Staunton's finds became garden plants. He included the first Macartney rose, *Rosa bracteata*, in this collection.

WILLIAM KERR

The unfortunate William Kerr was the Royal Botanic Garden's representative in China at a measly £100 per annum.[1] Thirty years before, Francis Masson had received an annual salary of £200 and had barely managed to cover his expenses.

Kerr was appointed in 1803 and left for China soon afterwards. For about three years he was assiduous and sent many plants back to Kew that he had bought in the nurseries around Canton (now Guangzhou). His output then slackened very badly. It is unclear why, but other expatriates reported to their friends in England that Kerr fell into bad habits, perhaps using opium excessively.

Banks did not completely lose faith in him. Kerr was appointed superintendent of the Colonial Botanic Gardens at Colombo, in Ceylon (modern Sri Lanka) on the basis of Banks' recommendation. Kerr died in Ceylon in 1814.

In spite of the bad reports about his behavior, it seems Kerr did better than was thought. Starting in 1804, Kerr arranged with various captains of East India Company vessels to carry nearly 30 new specimens to London. The genus *Kerria* is named for him.

William Townsend Aiton (1766–1849), William Aiton's son who was later known as "Old Jock," edited and updated his father's catalog, finally bringing out the two new volumes in 1813. He listed Kerr's contributions in *Hortus Kewensis*, second edition, 1810–1813.[120] The amendments noted 90 new Chinese plants (Table 4). Lonicera, magnolia, aster, camellia, citrus of several kinds, nandina, nymphaea, pittosporum, lilac, roses, and many others had reached England in the 20 years since the Staunton expedition. A total of a 157 Chinese species were now officially recognized in England.

Kerr bought all his plants at the Fa Tee gardens, individual gardens on small plots of land on the outskirts of Canton very largely devoted to potted plants. This enabled the owners to make the best use of very limited space. Other English visitors and collectors also patronized these nurseries. They were the only source of flowers available.

Hazel Le Rougetel, an English horticultural writer and rosarian, visited China in the late 1980s.[121] She was determined to find the Fa Tee gardens or at least some trace of them because of their importance in the flow of Asian plants to the West. China had changed profoundly since the early nineteenth century. Modern residents of Guangzhou had no idea of what she was talking about. After a great deal of effort on her part she was taken to a badly run-down section of the town and shown a few remnants of gateways and doors that still bore faint symbols for nurseries. This was what she had wanted to see, though it had all deteriorated almost to the point of oblivion. There is a photograph of the residual Fa Tee in her paper.

JOHN REEVES

The success of both professional and amateur plant collectors in China would come to depend on the quiet, systematic activities of an English tea merchant, John Reeves (1774–1856). As a young man in London, Reeves was so expert at his trade that the East India Company sent him to China in 1812 to be an inspector of tea. He lived part of the time in Macao, but when the tea was being prepared for export moved to Canton.

Being in China suited Reeves. He only went back to England twice in all his years of service before returning permanently to retire. It may be that he could enjoy higher status in the small expatriate community than at home. In England he was only the youngest son of a rather impoverished clergyman and had been obliged to go into trade. In China he was an important person.

Reeves sent his son, John Jr., to England to be educated, but the young John Reeves returned to live in China and carried on his father's avocation. Both the Reeves had a passion for gardens. In Canton John Reeves Sr. bought choice plants at the nurseries. Whenever a foreign explorer managed to bring plants back from the interior, Reeves would take care of them. He became a fellow of the Horticultural Society and earnestly aided the collectors in every way he could.

Once the first Anglo-Chinese treaty was signed he rapidly suggested that the Horticultural Society send a professional plant collector to take advantage of the new freedom of movement.

Table 4. *Hortus Kewensis II*, 1810–1813: Chinese Plants[1]

Plant Name	Date
Aerides odoratum	1800
Aster spp.	1804
Astragalus sinicus	1763
A. chinensis	1795
Azalea indica[2]	n.d.
Begonia discolor[3]	1804
Bidens chinensis	1801
Bignonia grandiflora	n.d.
Bletia hyacinthina	1802
Bocconia cordata	1795
Camellia sasanqua	1811
Carpesium abrotanoides	1768 (or earlier)
Chrysanthemum indicum	1764
Citrus nobilis	1805
Clerodendron fragrans	1790
Conyza chinensis	1796 (or earlier)
C. hirsuta	1767
Cookia punctata	1795
Corchorus japonicus	1804
Crataegus glabra	1804
C. indica	1806
Croton aromaticus	1793
Cuscuta chinensis	1803
Cymbidium sinense	1793
Desmanthus natans	1800
Dianthus japonicus	1804
Dimocarpus litchi	1786 (or earlier)
D. longan	1786 (or earlier)
Elaeagnus latifolia	1712 (or earlier)
Ficus pumila	1759
F. stipulata	1771
Gardenia micrantha	1806
G. radicans[4]	1804
G. spinosa	1806
Grislea tomentosa	1804
Hedysarum lagopodioides	1790
H. tomentosum	1782
Hoya carnosa	1802
Hydrangea hortensis	1788
Iris fimbriata	1792
Ischaemum aristatum	1803

(continued)

Table 4. *Hortus Kewensis II*, 1810–1813: Chinese Plants *(continued)*

Plant Name	Date
Jasminum hirsutum	1759
Juniperus chinensis	1804
Laurus glauca	1806
Ligustrum lucidum	1794
Lilium concolor	1806
L. japonicum	1804
L. tigrinum	1804
Limonia trifoliata	1799
Lonicera flexuosa	1806
L. japonica	1806
Magnolia conspicua	1789
M. fuscata	1797
M. obovata	1790
M. pumila	1786
M. tomentosa	1804
Manisuris granularis	1784
Mazus rugosus	1780 (or earlier)
Melastoma malabathricum	1795
Musa coccinea	1792
Mussaenda spp.	1805
Nandina domestica	1804
Nepenthes distillatoria	1789
Nymphaea pygmaea	1805
Paederia foetida	1806
Paeonia moutan	1787
Pergularia odoratissima	1784
Phaseolus radiatus	1732
Philydrum lanuginosum	1801
Pinus lanceolata	1804
Pittosporum tobira	1804
Polygonum chinense	1795
Rhus javanica	1799
Rosa banksiae	1807
R. bracteata	1795
R. indica	1789
R. multiflora	1804
R. semperflorens	1789
Sagittaria obtusifolia	1804
Sansevieria carnea	1792
Senecio divaricatus	1801
Syringa chinensis	1795 (or earlier)

(continued)

Table 4. *Hortus Kewensis II,* 1810–1813: Chinese Plants *(continued)*

Plant Name	Date
Taxus macrophylla	1804
T. nucifera	1764
Trapa bicornis	1790
Uvaria odorata	1800
Uvularia chinensis	1801
Vitex incisa	1758
V. ovata	1796

n.d., no date known.

1. Adapted from Emil Bretschneider, 1898 (reissued 1981), *History of European Botanical Discoveries in China,* Leipzig, DDR: Leipzig-Antiquariat. Whenever possible, the nomenclature represents the current, accepted names for plant taxa.

2. Original taxonomist: Linnaeus.

3. Original taxonomist: Dryander, 1791.

4. Original taxonomist: Thunberg, 1780.

Reeves' most important contribution was in nurturing the new discoveries. No one before him had done that so skillfully. He ensured that any plant found by the collectors was growing successfully in his garden before it was shipped to England (Table 5). This made an immense difference to the survival rate of the imports.

He usually planted the new specimens in pots and took good care of them. They might stay in his garden for months. The rooted plant was then enclosed in a protective case before taking its chances on the open deck of a ship. Before Nathaniel Ward invented the now ubiquitous glazed case, Reeves' efforts were commendable.

Reeves' other original idea was to have the flowers painted by an artist, recording them in their fresh state. He employed Chinese artists. They had an ancient tradition of painting flowers and shrubs with great skill. The results were not only scientifically valid but staggeringly beautiful, too.

OPENING UP THE ORIENT

The English tried again in 1816 to reach a trade agreement, sending Lord Amherst to China on another unsuccessful mission.[1] Dr. Clark Abel was his physician and collected a wide variety of plants in his turn. Unfortunately, most of these were lost in a shipwreck on the voyage home and never reached England. The *Abelia* was named for Abel. Amherst was later appointed governor of Bengal.

Perhaps the most significant aspect of this and the previous expedition was the fact that the diplomats managed to collect plants out in the countryside and were not confined to haunting nurseries and private gardens in the coastal rim. That freedom would not become general until after the Opium Wars, which forced China to allow trading with the West. The East India Company had been steadily

Table 5. Plants in John Reeves' Garden at Canton[1]

Plant Name	Date	Comment
Abelia uniflora	1824 (1830, Wallich)	Reeves in Fu Kien
Andromeda chinensis	1829	
Astranthus cochinchinensis	1824	Sent to HS: syn. *Homagium fagifolium* (?)
Azalea squamata	1846	Lindley
Blackwellia padiflora	1830	To HS
Bulbophyllum bicolor	n.d.	Orchid
Callicarpa reevesii	1830	From China to Calcutta
Camellia japonica	1829	
Capparis acuminata	1828	
Chloranthus monostachyus	n.d.	Sent by Reeves
Choripetalum obovatum		
Chrysanthemum (12 varieties)	1821	Capt. C. O. Mayne of HMS *Atlas*
Chrysanthemum	1826	Reeves
Cibotium barometz	1834 (or earlier)	
Clerodendron canescens	1804	
Croton lachnocarpus	n.d.	
Daphniphyllum calycinum	n.d.	
Dendrobium auriferum	n.d.	Orchid
D. catenatum	n.d.	Orchid
Dienia congesta	n.d.	Orchid
Elaeocarpus decipiens	n.d.	
Erycibe obtusifolia	n.d.	
Euonymus gracillimus	n.d.	
Eupatorium reevesii	1828	
Glycine sinensis	1818	to HS
Helicteres spicata	1831	
Ilex reevesiana	1851?	Fortune Gard. Chron. Not sent by Reeves (could be *I. fortunei*)
Lagerstroemia indica	1825	
Laurus aggregata	1824	Capt. Lindsay of HMS *Orwell* "Obscure plant" not in Hemsley
Liparis nervosa	n.d.	Orchid
Microdesmis caseariifolia	n.d.	
Ormosia pachycarpa	n.d.	Reeves, near Canton
Photinia prunifolia	1836	
Plectranthus amethystoides	n.d.	

(continued)

Table 5. Plants in John Reeves' Garden at Canton[1] *(continued)*

Plant Name	Date	Comment
Primula sinensis	1819	Capt. Rawes
Prunus pseudocerasus	1822	Reeves sent to HS
P. salicina	n.d.	"An obscure plant"
P. serrulata	1822	Variety of *P. pseudocerasus*
Psychotria reevesii	n.d.	To Calcutta Botanic Garden
P. reevesii	Unclear	Staunton saw it earlier
Reevesia thyrsoidea	1827	
Rumea chinensis	n.d.	To Calcutta Botanic Garden
Rubus parvifolius	1818	Sent by Reeves to HS
Salvia plebeia	1810	Staunton saw it earlier
Scutellaria rivularis	n.d.	
Stachys aspera var. *glabrata*	n.d.	
Strychnos paniculata	n.d.	
Vanda multiflora	1821	Orchid
Viburnum sempervirens	1853?	
Wisteria sinensis	n.d.	

HS, Horticultural Society, London; *n.d.*, no date known; *syn.*, synonym.

1. Adapted from Emil Bretschneider, 1898 (reissued 1981), *History of European Botanical Discoveries in China*, Leipzig, DDR: Leipzig-Antiquariat. Whenever possible, the nomenclature represents the current, accepted names for plant taxa.

increasing the amount of cheap opium flooding China and eventually forced the very weak regime to accept this poisonous material at the end of a gun.

In 1843 a preliminary treaty was signed. There was still considerable resistance, but in 1860 a permanent treaty was signed in Nanking and the situation settled down. The imperial powers set up a trade system within China and an Imperial Maritime Customs service to enforce compliance.

Once travel to the East became easier and more or less routine, the horticultural societies quickly seized the initiative. They mounted expeditions for the express purpose of finding new plants to enrich British gardens.

Still, even in 1846, a European in China was an object of derision and at considerable risk, as is obvious in Robert Fortune's account of his journeys in search of the tea plant.[122] On occasion he disguised himself as a Chinese in order to go safely from one town to another, but trusted to his gun to get him out of ticklish situations when necessary.

He recounts incidents and situations that would have been hilarious if they had not been so perilous. The few people who knew him failed to recognize "Sing Wah" as he slunk by in his Chinese garments.

Fortune was the Horticultural Society's first plant collector of this new era, the forerunner of a heroic age of plant collecting.

ROBERT FORTUNE

A rugged Scotsman, Robert Fortune (1812–1880) was first employed by the Royal Horticultural Society as a groundsman (Figure 29). His gifts were recognized early. For example, he did extremely well in the examinations that were instituted at the society shortly before he started to work there.

Fortune was chosen to go to China very soon after the first treaty was signed. He initially found many new plants in the nurseries of the treaty ports but later explored the interior successfully.

Fortune landed in Hong Kong on his first trip and had only contempt for its bleak and windswept appearance. He searched diligently for plants in Hong Kong and the nearby Ning Po, chafing all the time to move on. In his book *Three Years' Wanderings in China*, he confidently predicted that nothing could be made of such a barren place as Hong Kong.

Fortune managed to reach Shanghai, Soochow, and Hangchow among other large cities, often traveling by water. The need for disguise actually arose because his clothes were stolen one night while he slept on a barge. His servant had to rig him up in local garments, then made it all look more convincing by shaving his head and somehow creating a pigtail. Fortune let it be known that "Sing Wah" came from a remote part of the country. This explained his poor speech and clumsiness at table. He was able to participate in a mandarin's banquet without being unmasked.

Although he would have much preferred to hunt plants in the wild, he found many plants new to him in the Shanghai nurseries and in the gardens of prominent citizens.

Cemeteries were a source of remarkable trees. Fortune found a mourning cypress, *Cupressus funebris*, on one of his journeys farther inland. In an odd way the funerary practices of the Chinese resembled those of the Scottish Presbyterians, and he responded to this tradition.

The list of his introductions is massive (Table 6). Some of the best known are the balloon flower, *Platycodon grandiflorus*; Chinese anemone, *Anemone hupehensis*; 'Fortune's Double Yellow' rose, *Rosa odorata* var. *pseudoindica*; and bleeding heart, *Dicentra spectabilis*.

Spongberg comments on golden larch, *Pseudolarix amabilis*, which Fortune found in diminutive form, or *peng-jing* (similar to Japanese bonsai).[62] Although this tree grows slowly, it attains enormous height. It seems an odd choice to be made into a miniature but that presumably was the challenge. Fortune sent it to England, as he did all his other discoveries. From there, it subsequently reached Cambridge and the Arnold Arboretum.

Cryptomeria japonica seed went directly to both England and the United States. Fortune sent a shipment to England from Shanghai in 1844. One shrub that Fortune sent only to England did get to the United States quite quickly. The pearlbush, *Exochorda racemosa*, was imported by a nurseryman in Morrisville, Pennsylvania, Mahlon Moon. *(text continues on page 136)*

Table 6. Plants Collected in China by Robert Fortune[1]

Plant Name	Common Name	Date Imported	Source of Introduction
	Guava	n.d.	
	Pomegranate	n.d.	
	Melon nos	n.d.	
	Watermelon	n.d.	
Abelia rupestris		1843	Lindley, J. Hort. Soc. 1846 I, p. 63
A. uniflora		Unclear	Paxton's Fl. Gard. 1852 II, p. 145, fig. 208
Abies jezoensis		1850	
A. kaempferi		1853	
Acer	Maple	1848	
Aconitum autumnale		1846	Lindley, J. Hort. Soc. 1847, p. 77
A. chinense		1833	Siebold (Japan), Paxton's Mag. Bot. V 1838
Actinidia chinensis		1847	Planchon, Lond. Bot.VI, p. 303
Adamia versicolor		1844	Fortune, J. Hort. Soc. 1846 I, p. 298
Aglaia odorata		n.d.	
Akebia quinata (as *Aquebia*)		1845	Fortune, Siebold & Zuccarini, 1836, Fl. Jap. I, p. 143 tab. 77
Amaranthus oleraceus		1845	
Amelanchier racemosa		1845	Bot. Reg. 1847, 38 note
Anemone japonica		1835	Fortune, Lindley, Siebold & Zuccarini, 1835; Fl. Jap. I, p. 15 pl. 5; Bot. Reg. 1845, tab. 66
Apios fortunei		1873	
Arachis hypogaea	Peanut	n.d.	
Aralia papyrifera	Rice-paper plant	1854?	
Aristolochia sinarum		1859	
Artemisia indica	Mosquito tobacco	Unclear	
Arundina chinensis		1844	Fortune, J. Hort. Soc. 1846 I, p. 215
Aster turbinatus		Unclear	
Aucuba japonica		1861	
Averrhoa carambola	Chinese gooseberry	n.d.	
Azalea crispiflora		1853	
A. indica		Unclear	
A. narcissiflora		1853	Planchon, Rev. Hort. III, p. 66
A. obtusa		1846	Lindley, J. Hort. Soc. 1846 I, p. 152
A. ovata		1843	Lindley, J. Hort. Soc. 1846 I, p. 149

(continued)

Table 6. Plants Collected in China by Robert Fortune[1] *(continued)*

Plant Name	Common Name	Date Imported	Source of Introduction
A. ramentacea		1846	Lindley, J. Hort. Soc. 1849 IV, p. 291
A. sinensis	Yellow azalea	1845	
A. squamata		1844	Lindley, J. Hort. Soc. I, p. 152
A. vittata		1844	Planchon, Rev. Hort. 1853 VII, t. 887
Bambusa	Bamboo	n.d.	
Barthea chinensis		Unclear, before 1856	
Berberis consanguinea		1848	J. Hort. Soc. 1852, p. 225
B. fortunei		1846	Lindley, J. Hort. Soc. 1846 I, pp. 231, 300
B. trifurca		1849/50	Paxton's Fl. Gard. 1852, p. 57, fig. 258
Brassica	White cabbage	n.d.	
B. chinensis	Oil plant	1846	
Broussonetia papyrifera	Japanese paper	Unclear	
Buddleja lindleyana (as *Buddleia*)		1843	
Buxus fortunei		Unclear	
Caloscordum exsertum		Unclear	Herbert, Lindley, Bot. Reg. 1847, subt. 5
Calystegia pubescens var. *flore pleno*		1844	Lindley, J. Hort. Soc. 1846 I, pp. 70, 221
Camellia japonica		1846	J. Hort. Soc. 1846 I, p. 223
C. reticulata		(1820)	Bot. Mag. 1857, tab. 4976
C. sasanqua	Yellow camellia	1850	
Campanula grandiflora		1843	
C. nobilis		1846	Lindley, J. Hort. Soc. 1846 I, p. 232
Campsis fortunei		Unclear	Seemann, J. Bot. 1867, p. 373
Canarium album	Chinese olive	1845	
Cannabis gigantea		1846	Crevost
Carthamus tinctorius	Safflower (scarlet dye)	1855	
Castanea	Chestnut	1853	Fortune, Gard. Chron. 1860, p. 170
Cephalotaxus fortunei		1848	Hooker, Bot. Mag. 1850, pl. 4499
Cercis chinensis	Judas-tree	1854	

(continued)

Table 6. Plants Collected in China by Robert Fortune[1] *(continued)*

Plant Name	Common Name	Date Imported	Source of Introduction
Chamaerops excelsa?	Palm (used for ropes, etc.)	1843	
Chimonanthus fragrans		Unclear	Fortune (?date) found it wild in Che Kiang
Chionanthus retusus		1845	Paxton's Fl. Gard. 1852 III, p. 86
Chirita sinensis		1844	Bot. Reg. 1844, t. 59
Chloranthus inconspicuus		Unclear	
Chrysanthemum		1843	
Cinnamomum camphora	Camphor	n.d.	
Citrus	Fingered citron	1844	
C. aurantium	Mandarin orange	n.d.	
C. decumana	Pumelo	n.d.	
C. japonica	Kumquat	n.d.	
Clematis fortunei		1862	T. Moore, Gard. Chron. 1863, pp. 460, 676
C. lanuginosa		1850	Lindley, Paxton's Fl. Gard. 1852 III, p. 107
Clerodendron cyrtophyllum		Unclear	
C. foetidum		1850	
C. fortunei		Unclear	Hemsley, Index Fl. Sin. II, p. 259
Cleyera fortunei		Unclear	Gard. Chron. 1895 I, p. 10
Convolvulus reptans		Unclear	
Cookia punctata		Unclear	
Corchorus	Jute	n.d.	
Corydalis sheareri		n.d.	
Corylopsis pauciflora		Unclear	
C. spicata		Unclear	
Cryptomeria japonica	Japanese cedar	1844	Siebold & Zuccarini, Fl. Jap. 1836 I, p. 43
Cunninghamia sinensis	Lance-leaved pine	1849	
Cupressus funebris		1846?	Lindley, Gard. Chron. 1849, p. 243
Cyrtandra fortunei		Unclear	
Cyrtomium fortunei		1866?	
Daphne fortunei		1843	J. Hort. Soc. 1846 I, p. 147

(continued)

Table 6. Plants Collected in China by Robert Fortune[1] *(continued)*

Plant Name	Common Name	Date Imported	Source of Introduction
Deutzia		1861	
D. crenata (flore pleno)		1863?	Illustr. Hort. 1864 XI, p. 389
Dicentra spectabilis		1846	
Dielytra spectabilis			Gen. Hist. 1 1831, p. 140.
Dioscorea batatas	Yam	n.d.	
Dryandra cordata	Varnish tree	Unclear	
Edgeworthia chrysantha		1845	Lindley, J. Hort. Soc. 1846 I, p. 148
Elaeaodendron fortunei		1863	Turczaninow, Bull. Mosc. 1863
Enkianthus reticulatus		Unclear	
Eriobotrya japonica	Loquat	n.d.	
Euonymus radicans (variegated form)		1860	Japan Gard. Chron. 1863, p. 1016
Eupatorium fortunei		Unclear	
Euryale ferox		1855	
Fagopyrum esculentum	Buckwheat	n.d.	
Farfugium grande		1855	Lindley, Gard. Chron. 1857 IV, p. 215
Ficus nitida	Chinese banyan tree	1843?	
Flacourtia chinensis		n.d.	
Fontanesia fortunei		1846?	Carrière, Rev. Hort. 1859, p. 43, fig. 9
Forsythia fortunei		1861	Lindley, Gard. Chron. 1864, p. 412
F. viridissima		1844	Lindley, J. Hort. Soc. 1846 I, pp. 215, 226
Fortunaea chinensis		1844	Lindley, J. Hort. Soc. 1846 I, pp. 150, 221
Fragaria indica		n.d.	
Fraxinus	Wax insect tree	n.d.	
F. retusa		Unclear	Champion, Kew J. Bot. IV 1852, p. 330
Fritillaria verticillata		Unclear	
Funkia fortunei		Unclear	Baker, Gard. Chron. 1876 II, p. 36
Gardenia florida var. *fortuniana*		1844	Lindley, J. Hort. Soc. 1846 I, pp. 221, 226
Gentiana fortunei		1848	Hooker, Bot. Mag. 1854, tab. 4776
Glochidion fortunei		Unclear	

(continued)

Table 6. Plants Collected in China by Robert Fortune[1] *(continued)*

Plant Name	**Common Name**	**Date Imported**	**Source of Introduction**
Glossaspis antennifera		1852?	
Glyptostrobus heterophyllus	Water pine	Unclear	
Gossypium	Cotton	n.d.	
Gymnocladus chinensis	Soap bean tree	1853	
Halesia fortunei		Unclear	
Helianthus	Sunflower	Unclear	
Hibiscus syriacus var. *chinensis*		1844	Lindley, J. Hort. Soc. 1853 VIII
Hordeum	Barley	n.d.	
Hydrangea		1844	J. Hort. Soc 1846 I, p. 221
H. chinensis		Unclear	
H. hortensia		Unclear	
Ilex cornuta	Holly	1843	Lindey, Paxton's Fl. Gard 1850 I, p. 43
I. leptacantha	Holly	n.d.	Paxton's Fl. Gard. 1852 III, p. 72
I. microcarpa	Holly	n.d.	Paxton's Fl. Gard. 1850 I, p. 43 fig. 28
I. reevesiana	Holly	1848	
I. terago		1852?	Gard. Chron. 1859, pp. 508, 972
Indigofera decora		1843	
Ipomoea batatas	Yam	n.d.	
Iris tectorum		Unclear	
Isatis indigotica	Blue dye	n.d.	
Ixora coccinea		Unclear	
Jasminum nudiflorum		1844	Lindley, J. Hort. Soc. 1846, p. 153
J. paniculatum		Unclear	
Juncus effusus	Matting fiber	n.d.	
Juniperus sphaerica		n.d.	Lindley, Paxton's Fl. Gard. 1850 I, p. 56 fig. 35
Lamium chinense		1848?	
Lespedeza ciliata		Unclear	Bentham, Hooker's J. Bot. Kew Gard. 1852 IV, p. 48
Ligustrum coriaceum		Unclear	
L. lucidum		1856	
L. sinense		1855	
Lilium auratum		1860	Lindley, Gard. Chron. 1862, p. 644
L. brownii		1854	

(continued)

Table 6. Plants Collected in China by Robert Fortune[1] *(continued)*

Plant Name	Common Name	Date Imported	Source of Introduction
L. fortunei		1860?	Lindley, Gard. Chron. 1862, p. 212
L. sinicum		n.d.	Lindley, Paxton's Fl. Gard. 1851, p. 384
Linum stelleroides		1848	Hooker's J. Bot. Kew Gard., p. 178
Lonicera aureo-reticulata		n.d.	
L. fragrantissima		1852?	Paxton's Fl. Gard. 1852 III, fig. 268
L. maackii		1843	(Fortune)
L. standishii		1843?	Hook. f., Bot. Mag. 1868, tab. 5700
Lycopodium cernuum		1843	Fortune, J. Hort. Soc. 1846 I, p. 221; Bentham, Fl. Hongkongensis, p. 437
Lycoris radiata		Unclear	
L. straminea		1845	Lindley, J. Hort. Soc. 1848 III, p. 71
Lysimachia candida		1846	Lindley, J. Hort. Soc. 1846 I, p. 301
L. klattiana		n.d.	Hance, J. Bot. 1878, p. 236
Lysionotus pauciflorus		Unclear	
Malus	Apple (edible)	n.d.	
Malus	Crabapple	1754	
Mangifera indica	Mango	n.d.	
Mastacanthus sinensis		1844	Bot. Reg. 1846, tab. 2
Medicago denticulata	Trefoil	1844	
Monochasma savatieri		Unclear	
Morus	Mulberry tree	1843	
Murraya exotica		Unclear	
Musa textilis		1845	
Myrica nagi		1853	
Nandina domestica		1849	Fortune, T. 122 (1849)
Narcissus	Jonquil	n.d.	
Nelumbium speciosum	Lotus	1844	
Nephelium litchi	Lychee	1845	
Nicotiana	Tobacco	n.d.	
Olea fragrans		1844?	
Oryza	Rice	n.d.	
Osmanthus aquifolium		1860	
Paederia foetida		1843	

(continued)

Table 6. Plants Collected in China by Robert Fortune[1] *(continued)*

Plant Name	Common Name	Date Imported	Source of Introduction
Paeonia moutan	Tree peony	1846 (1786)	
Pholidota chinensis		1846?	J. Hort. Soc. 1847 II, p. 308
Photinia fortuneana		1843	Amoy, China
P. glabra		n.d.	
Pinus bungeana		1857?	
P. sinensis	Chinese pine	Unclear	
Pittosporum glabratum		1846	J. Hort. Soc. 1846 I, p. 230
Platanthera galeandra		Unclear	
P. mandarinorum		Unclear	
Plumbago larpentae		1846	
Polycoelium chinense		1847	
Polypodium fortunei		1857?	
Populus alba	Poplar	n.d.	
Premna microphylla		1863?	
Primula japonica		1870	
Prunus	Plum (edible)	n.d.	
Prunus	Flowering peach, double white	1843	J. Hort. Soc. 1847 II, p. 311
Prunus	Flowering peach, double crimson	1843/45	J. Hort. Soc. 1848 III, pp. 246, 313
P. persica	Peach	1844	
P. sinensis (undescribed alba variety)	Flowering plum, double flowered	1846	Lindley, J. Hort. Soc. 1847 II, p. 158
P. triloba	Flowering plum	1853–1855	Lindley, Gard. Chron. 1857, p. 268
Pterostigma grandiflorum		1843	Lindley, J. Hort. Soc. 1846 I, p. 66
Pyrus	Pear (edible)	n.d.	
Quercus bambusifolia		1854	Fortune, Gard. Chron. 1860, p. 170
Q. chinensis		1861	
Q. inversa		1849	Paxton's Fl. Gard. 1850 I, p. 58
Q. sclerophylla		1849?	
Rhamnus	Green indigo	1857	
Rhaphiolepis indica		1846	Lindley, J. Hort. Soc. 1847 II, p. 306
Rhododendron fortunei		1855	Lindley, Gard. Chron. 1859, p. 868
Rhus	Varnish tree	n.d.	

(continued)

Table 6. Plants Collected in China by Robert Fortune[1] *(continued)*

Plant Name	Common Name	Date Imported	Source of Introduction
R. microlepis		Unclear	
R. succedanea	Japan waxtree	n.d.	
Rhynchospermum jasminoides		1844	Lindley, J. Hort Soc. 1846 I, p. 74
Ribes fasciculatum		1843	Siebold & Zuccarini, Fl. Jap. 1843, p. 189
Rosa	Rose (non-perfumed)	1843	
Rosa	Five-coloured rose	1844	
R. anemoniflora		1844	Lindley, J. Hort. Soc. 1837 II, p. 318
R. banksiae		n.d.	Lindley, Paxton's Fl. Gard. 1851 II, p. 71
R. rugosa		n.d.	J. Hort. Soc. 1848 III, p. 313
R. sinica	Fortune's double yellow rose		Lindley, J. Hort. Soc. 1851 VI, p. 52
Ruellia indigotica	Blue dye	1853	
Sabia japonica		1862 (?pre)	
Salisburia adiantifolia	Maidenhair tree	Unclear	
Salix babylonica	Weeping willow	n.d.	
Salvia fortunei		1848?	
Saururopsis chinensis		1848	
Saxifraga fortunei		n.d.	Bot. Mag. 1863 t. 5377
Sciadopitys verticillata	Umbrella pine	n.d.	Siebold & Zuccarini, 1870
Scirpus tuberosus		Unclear	
Sesamum orientale	Sesame seed	n.d.	
Silene fortunei		n.d.	Sem. Rar. Hort. Patav. Coll. 1847 ex Linnaea XXIV, p. 181
Skimmia japonica		1852/3	
S. reevesiana		1850	
Soja hispida	Soybean	n.d.	
Solanum melongena	"Egg-apple" (?Eggplant)	n.d.	
Solopia chinensis		n.d.	
Sophora japonica		1853	
Spathoglottis fortunei		1844	Bot. Reg. 1845, tab. 19
Spiraea callosa		ca. 1852	Lindley, Paxton's Fl. Gard. 1852 II, p. 113, fig. 191
S. palmata		Unclear	Bot. Mag. 1868, t. 5726
S. prunifolia (flore pleno)		1844	Siebold & Zuccarini, Fl. Jap. 1840 I, p. 131

(continued)

Table 6. Plants Collected in China by Robert Fortune[1] *(continued)*

Plant Name	Common Name	Date Imported	Source of Introduction
S. pubescens		Unclear	Lindley, J. Hort. Soc. 1847, p. 157
Spiranthes stylites	Orchid	1856?	Lindley, J. Hort. Soc. 1857 I, p. 178
Statice fortunei		1843	
Sterculia platanifolia		1845	Gard. Chron. 1863, p. 1040
Stillingia sebifera	Tallow tree	1848	
Styrax fortunei		n.d.	Hance, J. Bot. 1882, p. 36
Symplocos congesta		Unclear	
Syringa oblata		Unclear	Lindley, J. Hort. Soc. 1859, p. 868
Taxus cuspidata		1856 (or before)	Siebold & Zuccarini, Fl. Jap. 1846
Teucrium fortunei		Unclear	
T. nepetifolium		1848?	
Thalictrum fortunei		n.d.	Moore, J. Bot. 16(185), p. 130
Thea sinensis	Tea	1843/1853?	
Thuja orientalis (arborvitae)		1867?	
Thujopsis standishii		n.d.	Pinetum Suppl. 1862, p. 100
Torenia concolor		1844	Lindley, Bot. Reg. 1846, tab. 62
Torreya grandis		1855	Gard. Chron. 1857, pp. 788, 1860
Tricyrtis hirta		1862?	Bot. Mag. 1863, tab. 5355
Triticum	Wheat	n.d.	
Urtica nivea		n.d.	
Vernonia fortunei		1852	Sch. Bip., Flora 35, 1852
Viburnum dilatatum		1846	Lindley, J. Hort. Soc. 1848, p. 247
V. hanceanum		1843	
V. macrocephalum		n.d.	Fortune, J. Hort. Soc. 1847 II, p. 244
V. plicatum		1846	Siebold & Zuccarini, Fl. Jap. 1839 I, p. 81, t. 38
Vitis	Grape	n.d.	
Weigela rosea		1844	Lindley, J. Hort. Soc. 1846 I, pp. 65, 189, 215
Wikstroemia nutans		Unclear	
Wisteria sinensis		1844	J. Hort. Soc. 1846 I, p. 223
Youngia dentata	Chinese lettuce	1846	
Zingiber (as *Zinziber*)	Ginger		
Ziziphus vulgaris	Chinese date/jujube	1845	

n.d., no date known.

1. Adapted from Emil Bretschneider, 1898 (reissued 1981), *History of European Botanical Discoveries in China*, Leipzig, DDR: Leipzig-Antiquariat. Whenever possible, the nomenclature represents the current, accepted names for plant taxa. Source information is as given in Bretschneider.

The East India Company employed Fortune next. The Horticultural Society had not been too generous with money, and he may have decided to go with a higher bidder. The society had paid him the same £100 per annum, which explorers 40 years before had received and which was truly a niggardly sum. His second and third tours were to find valuable economic plants, particularly tea. The huge tea industry in India and Ceylon (Sri Lanka) was based on his introductions. More recently, it has been noted that the original smuggled tea plants failed and were replaced by plants endemic to Assam.[92]

In 1858, the United States government requested him to go to China for the fourth time, to study plant life and find new plants for them. The principal quest was to import tea plants, returning to the fruitful areas he had visited years before. Fortune dispatched the tea plants, this time to the Patent Office in Washington D.C., but tea was never established as an American crop. That was the last time he visited China.

He returned to Scotland and remained active until late in life, even though he had retired from professional work. For a time he sold imported seed, such as that of the lacebark pine, *Pinus bungeana*, which he had found on his last journey. His family burned all his diaries and correspondence after he died, making it hard to fathom his personality. From peripheral sources it seems he was dour and unsmiling but exceedingly competent at his work.

CHARLES MARIES

Charles Maries (1851–1902) traveled in China and Japan for James Veitch. His first journey was in 1877. Part of the time he worked in Hong Kong and its environs, the rest he spent in Japan. Maries was very disciplined and conscientious and found dozens of important plants. On at least one occasion when his collections were lost at sea, he returned to the source and found everything again. His contributions include *Abies mariesii*, *Daphne genkwa*, *Actinidia kolomikta*, *Platycodon grandiflorus*, and *Lilium formosanum*.

Even though he had many difficult journeys in the wilder parts of Japan, he always felt more comfortable there. Maries found China very hostile and intimidating. On seeing the nature of the upper Yangtze River, above the Ichang Gorge, Maries threw up his hands and said that everything worth having in China had already been collected. He turned back. His outlook was probably affected by the fact that he had been robbed and badly beaten by bandits en route.

The Veitches were very annoyed with him, and it colored their attitude to his earlier fine results. He has had a "bad press" with very little reason. Maries left Veitch and found success in India, as director of two private botanical gardens owned by maharajahs. Sir Joseph Hooker found him the posts.

The next Veitch "traveler," Ernest Wilson, did not make Maries' mistake. He pushed and pulled and got his boats up the river past the gorges and triumphantly sent back an entire new flora.

ERNEST H. WILSON

Ernest Wilson's triumph in finding *Davidia involucrata* and Veitch's in propagating it successfully for sale by 1903 raised the threshold for exotic flora. One dove tree does not constitute a deluge, but when it came to deciduous and evergreen trees, rhododendrons, azaleas, camellias, chrysanthemums, peonies, primulas, lilies, ornamental fruit trees, and roses, there was indeed a flood.

England, primarily, but also the rest of Great Britain and the Continent, revelled in this glorious cargo. Plants long grown in Great Britain and the by-now-familiar American plants, no matter how worthy, could not compete. The dove tree somehow symbolized all that was rare and different.

With each generation, the range was greater, the capacity to send plants in good condition improved, and the results ever more spectacular. Fifty years after Fortune's success, Sir Harry Veitch sent Ernest H. Wilson (1876–1930) to China in 1899, to bring back the elusive dove tree, *Davidia involucrata*, described a generation before by Père Armand David.[92]

No one had seen a living specimen of the dove tree, but the dried fragments were so enchanting that Veitch bet on this as a commercial winner. Wilson's instructions were to focus solely on that tree and not to waste effort or resources picking up other, less spectacular plants. He was told to seek out Dr. Augustine Henry and learn where the dove tree was to be found. The doctor gave him a sketch map and helped him navigate some of the difficulties in dealing with the Chinese.

Amazingly, Wilson found the very specimen that Augustine Henry had seen, but only the stump remained. A local resident had cut the tree down to build a house. Wilson wasted very little time on feeling sorry for himself but got busy and found another dove tree a short distance away. He procured its seeds safely.

These trees were not precisely the same species of *Davidia* as Père David's. Wilson finally sent *D. involucrata* seeds to Veitch at a later date. In spite of his very clear instructions not to mess about with other plants, he decided to make use of this opportunity and collect in the area. He returned to England with hundreds of fresh and dried specimens. Veitch could hardly complain.

Wilson was born in the charming Cotswold town of Chipping Camden. He may be its most famous citizen. Some years ago the town created a memorial garden to commemorate his name. He was very ambitious and not only became a gardener but went on to obtain a certificate in botany, intending to become a teacher if he could not find anything more exciting to do.

While working in a subordinate position at Kew, he was noticed by the director, Sir William Thiselton-Dyer. Veitch needed a likely lad to go to China for them, and Thiselton-Dyer recommended Wilson. He fit the bill. They took six months to train him properly and then dispatched him to China via Boston and the Arnold Arboretum. Charles Sprague Sargent could be relied upon to give him the final polish.

Wilson traveled to China twice for Veitch with staggering results. Sargent then prevailed upon him to switch to the arboretum's employ. It was an opportunity he could not overlook. He had married and had one child. His wife and daughter went with him on a few of the journeys.

The Arnold Arboretum published *Plantae Wilsonianae* in 1913, the results of devoted work by their taxonomists such as Alfred Rehder.[123] The literally thousands of species and numerous genera Wilson discovered and collected all over Asia are described in this work. The book also contains some plants not collected by Wilson but which it made sense to include.

Wilson was eventually promoted to the top position at the arboretum but preferred to be known as the "Keeper of the Arboretum" rather than director.

The extra education he had obtained as a young man stood him in good stead. He wrote several excellent books about his travels and the plants he had found.[124,125] It was a tragedy that he died at the relatively early age of 57 in a car crash in Worcester, Massachusetts, after visiting his daughter and her husband in upstate New York. His daughter arranged to have her parents buried in Montreal. Canada was still part of the British Empire and hence symbolically England in her eyes.

Wilson was proudest of the regal lily he found, *Lilium regale*. During that expedition he was almost killed by an avalanche. The mules that survived had to walk over his prostrate body to regroup before he could be safely strapped to a makeshift stretcher and moved down the mountain. His leg was broken by the force of the avalanche and ended up permanently damaged. The missionary surgeons thought it might have to be amputated, but he prevailed on them not to do that. Wilson always limped afterwards. He called it his "lily limp."

BEES LTD. AND A. K. BULLEY

So far the emphasis has been on southern England, but there were ambitious nurserymen all over the country. An outstanding example was Arthur Kilpin Bulley (1861–1942) of Liverpool. He started out as a cotton broker but turned to horticulture. He founded Ness Gardens and the nursery, Bees Ltd.*

In those prosperous days, families in the north of England and its principal seaport benefited from the British domination of the textile industries. With wealth came refinement. Even if the husbands were rough and ready, their wives laid down the law about things like music and flowers. Nurseries entered a heyday. Bulley was ready.

The cotton trade had indirectly gotten him started. Moving textiles across Asia required him to deal with customs officials. Augustine Henry was one of them in Yunnan. Henry sent Bulley seed but also incited him to send plant collectors on his own. The atmosphere was ripe for something like this as the Royal Geographic Society published the results of its explorations of this same region in 1899.

Bulley turned to Sir Isaac Bayley Balfour, professor of botany at the University of Edinburgh, for advice. Balfour suggested George Forrest.

---

*Plants termed *Beesia* are named for Bulley and his collectors.

GEORGE FORREST

George Forrest (1873–1932) was yet another of those poor Scottish boys with good brains and an endless capacity for hard work. After a sound education at Kilmarnock Academy in Ayrshire, he earned a living as a pharmacist's assistant. This enabled him to learn something about botany. Pharmacy was still based largely on botanical extracts. The first completely synthetic drug, aspirin (acetyl salicylic acid), was not synthesized until 1899.

After briefly trying his luck in Australia, he returned to Scotland and applied to Balfour for a post at the Royal Botanic Gardens in Edinburgh. The only one that was available was in the herbarium, but this turned out to be an advantage. Two years of working with methodically arranged plants from all over the world provided him with an invaluable education.

When Bulley appeared in 1905, Forrest was psychologically prepared. In all, he made eight trips to China, the first few for Bulley, later ones for private syndicates, always basing himself at Yunnan. The mountains and gorges were the source of the three great rivers in China and the neighboring countries of Burma (now Myanmar), Tibet, and Indochina (Vietnam): the Mekong, the Salween, and the Irrawaddy.

Forrest believed in organization and delegation. It was essential because of the terms of his later contracts. The areas specified were so vast that no one man could have covered them alone. There were six more expeditions, many of which were instigated by J. C. Williams of Caerhays in Wales. He created syndicates to pay for the journeys and benefit from the results.

Forrest had teams of native people collecting on a massive scale, usually Mosso tribesmen. During his first excursion, between 1904 and 1906, he sent Bulley seeds of 95 species of rhododendron. In fact, he sent so much material back to England that a lot of plants were simply shunted to one side out of desperation. Some of them were never developed. Others were evaluated years after he sent them. This delay was of course not good for his backers, who hoped for rapid commercial gain.

One of his favorite plants was the rhododendron. More than 300 species are attributed to him.[126] In spite of these mammoth results, he always felt that there were many more species left to be discovered.

Forrest became unwittingly embroiled in religious and political upheavals in China and Tibet. In 1905 he was exploring the west bank of the upper Mekong with three French missionaries. The monks from the local lamasery were angry with the Christians for taking their parishioners away. They swarmed into the mission and began murdering the foreigners.

Forrest managed to escape into the wild surrounding country, but the lamas chased after him. For nine days he staggered from one swamp to another, starving and parched, with torn and bleeding limbs. He said that he reached such a low point that he did not care whether he lived or not. Finally he found refuge and was nursed back to health.

Forrest died unexpectedly in the field, just before he was due to return after his last trip. From the reports that have survived, it sounds as though he had an acute intracerebral hemorrhage, dying within a few minutes of calling out for help. Writing many years later, his son George Forrest Jr., a gardener at the Savill Garden, Windsor, still recalled the shock and distress the family felt when they heard about the father's death.[127]

REGINALD FARRER

Reginald Farrer (1880–1920) was the son of a wealthy Yorkshire dalesman and a cousin of the Sitwells through their mothers. In spite of his undoubted gifts, he was slightly marginal in smart society because of his cleft palate, odd speaking voice, and sarcastic, brittle conversation.

Sir Osbert Sitwell described him in his memoir *Left Hand, Right Hand*.[128] The contemplation of plants in nature offered solace. He became passionate about rock gardens and alpine plants in particular.

From that passion to deciding to explore for plants on his own was not a profound leap. He had enough money to indulge himself in this whim. One of the very last Veitch collectors, William Purdom, became available as a professional assistant. Sir Harry had not been too pleased with him, but Farrer found him ideal for what he needed.

Purdom's collecting had not been voluminous, but he had brought back seed of the lacebark pine, *Pinus bungeana*; several fine clematis; primulas; and a horse chestnut that Sargent wanted. With Farrer he concentrated more on alpine plants. They found many gentians, including *Gentiana farreri*.

Eastern religion attracted Farrer, and he converted to Buddhism. His style of writing was flowing and rhapsodic, unlike the more prosaic work of Wilson and the others. *On the Eaves of the World* is Farrer's best-known work.[129] Although he traveled in comfort and was less subject to the harsh Chinese conditions, he caught pneumonia while collecting and died at the age of 40. Purdom took Farrer's plants and possessions back to England. He, too, died within a couple of years.

FRANCIS ("FRANK") KINGDON WARD

Francis K. Ward (1885–1958) was also known as Frank Ward and Frank Kingdon-Ward (Figure 30). His father was a mycologist and the professor of botany at Cambridge University, so his choice of career was hardly surprising. There were always explorers and geographers in his father's study while he was a boy. He absorbed their stories and longed to follow suit.

As soon as Ward had finished his degree at Cambridge, he took almost the first job he was offered, to be a schoolmaster in Shanghai. He hardly cared what he did as long as he was in Asia. While he was on leave, he was able to collect plants in Indonesia and get started on the career he really wanted.

Right about that time Bulley needed to replace Forrest. Very unhappy with Bulley's stinginess and lack of understanding, Forrest signed on with a syndicate of rhododendron fanatics headed by J. C. Williams. Bulley asked Balfour for help again. Balfour knew about Ward and suggested Bulley try him. Forrest was, of course, not pleased about someone collecting in his "patch" and behaved very stiffly toward Ward when he reached western China in 1913.

There may also have been class antagonism. Ward was everything Forrest was not. Even though Forrest had been the one to break away and resign from Bulley's employ, it is understandable that he felt threatened by how quickly he could be replaced.

Ward pursued his calling with profound intensity. Time and again he went off into wild and dangerous parts of Asia.[130] Ward suffered from acrophobia. Jean Ward, his second wife, latterly Mrs. Rasmussen, was a botanist herself and traveled with him in the later years. She described his ashen face as he steeled himself to climb out over a chasm to reach an attractive plant. He never gave in to his fears, but pushed himself in an almost inhuman manner.

Ward had very broad interests, rather like Spruce. He was almost more interested in the geography of the regions he explored than in the plants. He wrote 23 books, some of them truly geographical in nature. Ward wrote very well, in a dry and skeptical tone, completely unsentimental and yet obviously full of wonder.

The books read as well today as they must have done when they were first published. His term for his occupation was "geographical botanist," very apt. He won awards from the Royal Geographical Society as well as the Royal Horticultural Society. Ward traveled so continuously that he stated he was of "no fixed abode" in a friend's guest book.

The great mystery of the Tsangpo River in Tibet and its gorges absorbed him deeply. His book *The Riddle of the Tsangpo Gorges* left a question mark for 70 years. Ward had not been able to find the source of the rivers.

In 1993, a band of vigorous explorers determined to follow the river to its source and complete his unfinished search. Kenneth Cox, a rhododendron expert and grandson of the E. H. M. Cox who had collected in China with Reginald Farrer, assembled a party of like-minded people and led the ultimately successful search. Even with all the equipment and resources of modern exploration, it was a profoundly difficult journey.

Ward brought back plants that established him as a serious collector. Forrest may have scoffed at this soft and inexperienced young upstart but did not live to see how Ward developed. Combing Ward's books, one finds dozens of species. The Tsangpo volume lists over 200 of them alone. *The Land of the Blue Poppy* tells about *Meconopsis betonicifolia*, a flower that created a sensation at the Chelsea Flower Show in 1927.[131] His finds included numerous rhododendrons, with some dwarf species from 16,000 feet up; primulas; lilies; viburnums; jasmines; irises; and other *Meconopsis* species. This is still only a partial list. Like Wilson, he was attached to the lilies he found.

During and after World War II, he and his wife led daring expeditions in the Himalayas, assisting the American and British forces. He was too old to be in the front lines, but his knowledge of the mountains and back country was invaluable. Many American planes had crashed, and in the late 1940s, the U.S. Air Force wanted to collect whatever remained of them. Ward noted that almost all the crashes had been a result of flying in unsuitable weather.

His remarks about the "tenderfooted" American search and rescue teams were very gentle. Ward recognized they could not possibly keep up with him and his wife in that extraordinary terrain. The photographs show a man seemingly made of whipcord. He and his wife Jean survived a 9.2 earthquake in 1951, up in the Himalayas. The mountain split open and simply swallowed some of their men and tackle.

After the war he spent some time in London but did not really enjoy it. Sitting at a desk did not suit him. When he had to stay in London between trips, he planned further expeditions. He could not just sit at a desk and consider his time well spent, although he wrote many excellent books between travels. In 1958, at the age of 73, he was arranging to go plant hunting in Vietnam when he suddenly died.

With the death of Ward, an era closed. There are still excellent explorers and collectors, but one man alone could not handle what is needed now. Exploring and collecting have to be an organized group activity depending on teamwork.

CHINESE FLORA

EARLIEST CHINESE SPECIMENS

Plants from Asia arrived sporadically in Europe in the seventeenth century. James Petiver and John Ray both described dried plants collected from the offshore Chinese island of Amoy before 1688. *Ricinus chinensis sebifera* came from China "some years ago."[119] John Ray noted three ferns from China in 1688, received from Petiver.[39]

The Chelsea Physic Garden had at least one Chinese plant, noted as *Shunda chinensis pannosos minore*, in 1691, according to Petiver. "Shunda" is the Malabar name for *Solanum*. The plant may have been *Solanum chinense* or the similar, rejected name *S. indicum*. Table 7 shows the Chinese plants that the Chelsea Physic Garden contained in the eighteenth century.

EARLY CHINESE PLANTS IN PARIS

The Jardin des Plantes was well stocked with Chinese plants by the end of the eighteenth century (Table 8). Its herbarium was also overflowing. Comte Jean-Baptiste Lamarck (1744–1829), René Louiche Desfontaines (1750–1833), and Jean Louis Marie Poiret (ca. 1755–1834), all botanists at the garden, did the laboratory work.

In most cases they did not comment on who had collected the specimens, but they did record the specific contributions of Père d'Incarville, Pierre Poivre (for whom pepper is named in French), and Pierre Sonnerat. The latter three men had all traveled in China themselves. Eighty-four Chinese plants were noted.

France had a series of departmental botanical gardens in addition to the Jardin des Plantes, some of which rivaled the parent institution.

Table 7. Chinese Plants in the Chelsea Physic Garden[1]

Plant Name	Date	Comments
Alcea rosea	1768[2]	Actually arrived years earlier
Amaranthus cruentus	Unclear	
Ammannia baccifera	1768	
Artemisia minima	1768 (or before)	
Aster chinensis	1736	
Astragalus chinensis	1763	
Brassica violacea	1767	
Carpesium abrotanoides	1768 (or before)	
Celosia coccinea	1768	
Conyza hirsuta	1767	
C. patula	Unclear	
Datura ferox	1731	
Dianthus chinensis	1705	French missionaries to Paris: flowers single until 1719; double flowers later
Ficus pumila	1759	
Hibiscus manihot	1712	
Hypericum monogynum	1753	Duke of Northumberland shared his seeds
Ixia chinensis	1759	
Lycium chinense	1768	
L. halimifolium	1755	
Malva chinensis	1747	
Matricaria indica	1768	Ning Po
Raphanus chinensis	1768	
Rheum	1734	Seeds from Boerhaave
Rhus chinensis	1737?	Killed 1740 by severe winter
Rosa spp.	1759	
Siegesbeckia orientalis	1768 (or before)	
Sinapis juncea	1731	
Smilax china	1759	
Thuja orientalis	ca. 1752	
Trichosanthes anguina	1755	
Urena lobata	1731	
Urtica nivea	1750	
Vitex chinensis	Unclear	Obtained from France (French missionaries)
V. negundo	1737?	Killed 1740 by severe winter
Xanthium chinense	1768	

1. Adapted from Emil Bretschneider, 1898 (reissued 1981), *History of European Botanical Discoveries in China*, Leipzig, DDR: Leipzig-Antiquariat. Whenever possible, the nomenclature represents the current, accepted names for plant taxa.

2. Listed in Philip Miller's dictionary.

EARLY CHINESE PLANTS IN OTHER EUROPEAN CITIES

Chinese plants were coming to the fore in the principal capitals of Europe in the eighteenth century. There is considerable resemblance between the principal collections of the gardens in these cities.

A few Chinese plants were grown at Schönbrunn in Vienna during this same epoch. The natural destination for exotic plants was the botanical garden that had been created for just this purpose. In 1755, the Emperor Franz Josef had appointed Nikolaus von Jacquin, a Dutch physician, to be in charge of his gardens. The latter acquired some plants from China in 1759.

In the Netherlands, Leiden's historic botanical garden dominated the scene and had numerous imports. St. Petersburg and Moscow both received Chinese specimens from their embassy in Peking. Peter the Great had included a physic garden for his military apothecaries when building St. Petersburg.

Prussia had its botanical garden in Berlin. Its successor, the Berlin-Dahlem Botanical Garden, is still justly famous as a center for Eastern and tropical plants.

Banks intended Kew to be the core of a great web of colonial botanical gardens, all feeding into the center (see Chapter 8). William Aiton (1731–1793), the gardener in charge at Kew, published *Hortus Kewensis* in 1789, with the assistance of the Swedish naturalists Daniel Solander and Jonas Dryander who worked for Sir Joseph.[132] It was a listing of all the plants in the Royal Botanic Garden at Kew. They were classified in the book according to Linnaean principles. Sixty-seven Chinese plants were recorded (Table 9).

CHINOISERIE

Camellias, azalea, peonies, and chrysanthemums caused great excitement in the West. All aspects of daily life were affected. Printed cloth called "chintz," decorated with huge pictures of the new flowers, became the rage. Delicate Chinese porcelain with floral motifs abounded. Gardens were designed in the "Chinese" manner. The inexpensive, everyday pottery known as willow ware, used by poorer people, had pictures of Chinese villages on it. Chinoiserie reigned in England and on the Continent for years.

REPEATING "MONTHLY" ROSES

Over the course of the eighteenth century, more and more exotic flowers appeared. A sense of what was happening can be seen in the story of early Chinese roses. Similar series of events occurred with all the major ornamental plants—the camellias, chrysanthemums, rhododendrons, azaleas, and so forth. For each one there was a champion, someone who made it his goal in life to obtain all varieties, improve on the originals, and do more than anyone else with a particular flower.

(text continues on page 150)

Table 8. Early Chinese Plants in the Jardin des Plantes[1]

Plant Name	Date	Synonym	Scientific Introduction
Adenanthera pavonina	1786		
Ailanthus glandulosa	1789?		D'Incarville; Desfontaines; Lamarck
Alcea chinensis	1783		
Aletris chinensis	1783		
Amomum zingiber	1783		
Amygdalus pumila	1804	*Prunus sinensis; P. japonica*	
Artemisia capillifolia	1783		
A. minima	1783		
A. sinensis	1804		
Aster chinensis	1783		
Bidens chinensis	1783		
Brassica chinensis	1770		Linnaeus; Aiton Hortus (Kew 1789)
Calycanthus praecox	1783		Lamarck; Desfontaines
Camellia japonica	1783		Linnaeus; Lamarck
Caragana chamlagu	1783		(Kew 1776)
Cassia spp.	1783		
C. pumila	Unclear		
Cedrela sinensis	Unclear		D'Incarville; Adrien de Jussieu
Chenopodium purpurascens	1776		
Chimonanthus fragrans	Unclear		Lindley
Citrus sinensis	1804		
Cocculus spp.	Unclear		D'Incarville
Convallaria japonica	1796		
Convolvulus chinensis	1789		
Conyza spp.	1786		
Cuscuta chinensis	1786		
Cycas revoluta	Unclear		D'Incarville; Lamarck; Thunberg
Cydonia sinensis	1812	*Pyrus cathayensis*	
Cyperus pangorei	Unclear		
Dalechampia parvifolia	Unclear		D'Incarville; Lamarck
Daphne sinensis	1789		
Datura ferox	1806		
Dianthus chinensis	1705		Linnaeus
Dolichos sinensis	1776		Thouin
Eugenia michelii	1789		

(continued)

Table 8. Early Chinese Plants in the Jardin des Plantes[1] *(continued)*

Plant Name	Date	Synonym	Scientific Introduction
Fagara avicennae	Unclear	*Zanthoxylum avicennae*	Lamarck
Ficus hirta	Unclear		D'Incarville; Vahl
F. pertusa	1786		
Fumaria caudata	1789	*Corydalis longiflora* or *C. bungeana*[2]	Lamarck
F. spectabilis	Unclear	*Dicentra spectabilis*	D'Incarville; Lamarck
Geranium sibiricum	Unclear		D'Incarville; Lamarck
Ginkgo biloba	1786		
Gleditsia macrantha	1804		
G. sinensis	1786		
Grewia nitida	Unclear		D'Incarville; A. L. de Jussieu
Hedyotis paniculata	1789		
Hemerocallis plantaginea	1789		
Hydrangea hortensis	1790		(England 1789)
Hypericum chinense	1796		Lamarck
H. erectum	1789		D'Incarville; Linnaeus; Lamarck (Peking)
Hyssopus lophanthus	1789		
Incarvillea sinensis	1789		D'Incarville; Lamarck (Sole) specimen
Ionidium heterophyllum	Unclear		D'Incarville; Ventenat
Justicia picta	1783		
Kalanchoe spathulata	1804		
Koelreuteria paniculata	1804		
Lagerstroemia chinensis	1789		Lamarck
Latania chinensis	1804		
Limonia trifoliata	1787		(Kew)
Lychnis grandiflora	1789		Lamarck
Lycium chinense	1789		
Magnolia discolor			Desfontaines (from London)
M. obovata	Unclear		Thunberg (to Paris from London)
M. pumila			Desfontaines (from London)
M. purpurea			Desfontaines (from London)
Malva verticillata	1789		Linnaeus; Lamarck
Matricaria indica	1789	*Chrysanthemum indicum*	
Mentha perilloides	1796	*Perilla ocymoides*	
Mespilus japonica	1804		
Mogorium multiflorum	1796		

(continued)

Table 8. Early Chinese Plants in the Jardin des Plantes[1] *(continued)*

Plant Name	Date	Synonym	Scientific Introduction
Moraea chinensis	1796	*Ixia chinensis*	Thunberg
Myrtus spp.	1776		(England)
Nauclea orientalis	1796		
Olea fragrans	1796		
Phaseolus radiatus	1789		
Polypodium dichotomum	1894?		
Pyrus spectabilis	1780		(Kew)
Raphanus sativus	Unclear		Linnaeus
Rhamnus theezans	1804		
Rheum palmatum	1804		
Rosa semperflorens	1794		Curtis's Bot. Mag.
R. ternata	Unclear	*R. sinica; R. laevigata*	
Salsola spp.	1806		
Saxifraga sarmentosa	1804		
Senecio pseudochina	Unclear		
Serissa foetida	1806	*Lycium foetidum*	
Sinapis brassicata	Unclear		
S. chinensis	1782		Thouin (per Aiton)
S. juncea	Unclear		Linnaeus
Sterculia platanifolia	1788		
Syringa villosa	Unclear		D'Incarville; Vahl
Taenitis chinensis	Unclear		D'Incarville
Terminalia vernix	1783		Maluku Islands
Thea bohea	Unclear	*T. viridis*	
Thuja orientalis	1806		
Trichosanthes anguina	1804		
Urena lobata	1723		(England)
Urtica nivea	1786		
Viola spp.	Unclear	*Ionidium heterophyllum*	D'Incarville; Poiret
Zanthoxylum trifoliatum	1786		
Ziziphus sinensis	1789	*Z. vulgaris*	

1. Adapted from Emil Bretschneider, 1898 (reissued 1981), *History of European Botanical Discoveries in China*, Leipzig, DDR: Leipzig-Antiquariat. Whenever possible, the nomenclature represents the current, accepted names for plant taxa.

2. Probable synonym.

Table 9. Chinese Plants in the Royal Botanic Gardens, Kew, before 1789[1]

Plant Name	Date	Reference[2]	Source
Acorus gramineus	1786	A. I 474	Allan Cooper RN
Ailanthus glandulosa	1757	A. III 448	
Alcea rosea	1597?, 1753		Gerard's Herball
Althaea rosea	1753?	A. II 446	
Amaranthus spp.	1728	A. III 350	
A. tristis	1759	A. III 348	P. Miller
Amaryllis aurea	1777	A. I 419	Fothergill
Annona hexapetala	1758	A. II 253	Hugh, Duke of N'land
Artemisia minima	1778	A. III 173	Thouin
Aster chinensis	1731	A. III 208	
Brassica chinensis	1770	A. II 402	Mr. Richard intro.
Camellia japonica	1739 or earlier	A. II 460	Cultivated by Lord Petre
Celosia argentea	1714	A. I 287	Duchess of Beaufort
C. coccinea	1597	A. I 287	Gerard's Herball
Chenopodium atriplicis	1780	A. I 311	Thouin
Chloranthus inconspicuus	1781	A. I 160	James Lind
Citrus sinensis	1595		Francis Carew
C. sinensis	1629	A. III 101	
Clerodendron fortunatum	1784	A. II 364	Lee and Kennedy
Conyza patula	1758 or earlier	A. III 184	P. Miller
Croton sebiferum	1755	A. III 375	Hugh, Duke of N'land
Cucumis acutangulus	1692	A. III 383	Bishop Compton
Cycas revoluta	ca. 1758	A. III 475	Richard Warner (from Capt. Hutchinson)
Daphne odora	1781	A. II 26	B. Torin
Datura ferox	1731	A. I 238	P. Miller
Dianthus chinensis	1719 or earlier	A. II 90	
Dolichos sinensis	1776	A. III 31	Thouin
Dracaena ferrea	1771	A. I 454	B. Torin
Epidendrum ensifolium	1780 or earlier	A. III 499	Fothergill
Gardenia florida	1754	A. I 293	
Gleditsia horrida	1700		(N. America)
G. horrida	1774	A. III 444	(China)
Hibiscus manihot	1712	A. II 457	(Chelsea Physic)
Hypericum monogynum	1753	A. III 108	Hugh, Duke of N'land
Hyssopus lophanthus	1759	A. II 284	

(continued)

Table 9. Chinese Plants in the Royal Botanic Gardens, Kew, before 1789[1] *(continued)*

Plant Name	Date	Reference[2]	Source
Ixia chinensis	1759	A. I 62	
Koelreuteria paniculata	1763	A. II 7	George, Earl of Coventry
Lagerstroemia indica	1759	A. II 230	Hugh, Duke of N'land
Limodorum tankervilleae	1788	A. III 302	
Lychnis coronata	1774	A. II 117	Fothergill
Lycium barbarum	1709	A. I 257	Duchess of Beaufort
Malva verticillata	1683	A. II 449	James Sutherland
Morus alba	1596	A. III 342	John Gerard
Myrtus tomentosa	1776	A. II 159	Mrs. Norman
Nicotiana fruticosa	1699 or earlier	A. I 241	Duchess of Beaufort (dowager)
Olea fragrans	1771	A. I 14	B. Torin
Panax spp.	1773	A. III 448	Fothergill
Polygonum tinctorium	1776	A. II 31	John Blake
Pteris serrulata	ca. 1770	A. III 459	J. Gordon
Pyrus spectabilis	1780	A. II 175	Fothergill
Rhapis flabelliformis	ca. 1774	A. III 473	J. Gordon
Rheum compactum	1779	A. II 42	Thunberg
R. palmatum	1768	A. II 41	P. Miller
R. undulatum	1759	A. II 41	
Rhus semialata	1780	A. III 489	Banks
R. succedanea	1773	A. I 363	
Robinia chamlagu	1773	A. III 54	
Rosa sinica[3]	1759	A. II 203	
Saxifraga sarmentosa	1771	A. II 79	B. Torin
Siegesbeckia orientalis	1730	A. III 244	
Sinapis chinensis	1782	A. II 403	Thouin
S. juncea	1710	A. III 404	Petiver? (1731 P. Miller)
Smilax china	1759	A. III 402	
Sophora japonica	1753		J. Gordon
Sterculia platanifolia	1757	A. II 378	
Thea bohea	1768	A. II 230	John Ellis
Thuja orientalis	1752	A. III 371	P. Miller
Trichosanthes anguina	1755	A. III 379	P. Miller
Urena lobata	1732		James Sherard
Xanthium orientale	1713	A. III 343	Bishop Compton

A, Aiton's *Hortus Kewensis* 1789; *N'land,* Northumberland; *RN,* Royal Navy.

1. Adapted from Emil Bretschneider, 1898 (reissued 1981), *History of European Botanical Discoveries in China,* Leipzig, DDR: Leipzig-Antiquariat. Whenever possible, the nomenclature represents the current, accepted names for plant taxa.

2. As given in Bretschneider. Format for references is title, volume number, page number.

3. Different rose from Linnaeus' "R. sinica" (*R. indica*).

The challenges brought out the obsessional and compulsive nature of plant fanciers. It was not unlike the old tulip madness but without the financial schemes.

The first repeating, "monthly" roses were a revelation. Pehr Osbeck (1723–1805), a pupil of Linnaeus, traveled to China in 1751 and found a small pink rose in the Custom House at Canton. He took it back with him to Sweden in 1752. Osbeck was very diligent, commenting on over 200 plants and describing at least 11 totally new ones. He is said to have given all his Chinese plants to Linnaeus, but the latter described only 37 of them. Neither Osbeck nor Linnaeus dreamed of what this rose could do.

It was unknown in England until Sir George Staunton sent a specimen in 1792. He too found it in Canton. Staunton had accompanied Lord Macartney to Peking when the latter tried to get the emperor of China to allow trading with England. The mission failed.

It was not a very exciting rose, with small pink flowers and only a slight perfume, yet these little unassuming flowers broke a great logjam in horticulture. At first only a few people paid any attention to it. One was James Colville, who propagated it at his nursery in the King's Road, Chelsea.

Most native European roses bloomed once and then did nothing until the following year. 'Parson's Pink China,' as it came to be known (or 'Old Blush'), blossomed all summer long. It was a variety of *Rosa chinensis.* Mr. Parson owned a garden in Rickmansworth, at that time a country village in Middlesex, and so it was named for him.[13] Breeding the repeating feature of the Chinese rose with other desirable characteristics would create hitherto unimaginable roses. Now perhaps some progress could be made.

This was what, in fact, professional and amateur breeders proceeded to do for the next 50 years and beyond. Intricate crosses and backcrosses continue to be done today, constantly refining a flower that is already enchanting. For a time the breeders concentrated on perfecting the appearance only, ignoring the fragrance. Today's insistence on cherishing the old, premodern roses is restoring the fragrance.

Three other Chinese roses played major roles in the development of modern hybrid roses. Taken together, the four of them are known colloquially as the "four studs." The other three were 'Slater's Crimson China,' 'Hume's Blush Tea-scented China,' and 'Parks' Yellow Tea-scented China.' Between them, these four studs incorporated the valuable qualities of repeated blooming, rich color, climbing, and perfume. They were all hybrids of ancient lineage in China.

The "five ancestors" were fundamental to the process: *Rosa gallica, R. moschata, R. damascena, R. alba,* and *R. centifolia.* Only the species rose *R. gallica* is native to Europe. It is thought to have been taken to England in the eleventh century, perhaps by Benedictine monks.

The musk rose, *Rosa moschata,* originated in the Himalayas but reached the Middle East very early in recorded history, probably along the Silk Road. It is both a climbing rose and a repeating rose under warm enough conditions and has a striking fragrance. Attar of roses is distilled from the musk rose. All three characteristics were important for breeding.

Rosa damascena, *R. alba*, and *R. centifolia* are ancient hybrids. The species name *damascena* reflects the plant's origin from Syria. It, too, is highly fragrant and is a rich source of attar of roses. *Rosa alba* was known to the Romans. *Rosa centifolia* also came from the East but was introduced to Europe via Austria.

Gilbert Slater of Low Leyton in Essex was connected with the East India Company and had imported the 'Crimson China' rose two years before 'Old Blush' arrived in England. This rose had even smaller flowers than the other but bloomed almost continuously. It was nicknamed "semperflorens."

The Horticultural Society sent a gardener, John Parks, to collect plants in China. In 1823, he returned to England with his 'Yellow Tea-scented China,' which was a climber. John Reeves in Canton had helped him to keep his plants alive and ensure their survival. Reeves also employed a Chinese artist to paint the flowers, and this yellow rose was among them. The painting is still in the Lindley Library of the Royal Horticultural Society in London.

Sir Abraham Hume (1749–1838) and his wife Lady Amelia were both ardent horticulturists at their estate in Hertfordshire. Sir Abraham introduced many of the first chrysanthemums and peonies in the first 20 years of the nineteenth century.

Lady Amelia died in 1809 at the age of 57. She had studied with Sir James Smith and was serious about botany. Smith named a new Australian genus *Humea* in her honor. Sir Abraham introduced his rose in 1809, perhaps in her memory.

The sources for such introductions were usually obliging sea captains willing to find the plants and be responsible for their survival on the return voyage, a very tall order at the time. The indispensable Bretschneider lists many of the captains and their ships. These were almost always "East Indiamen."

C. C. Hurst

Reciting this seemingly straightforward narrative is a result of complicated detective work done in the 1930s by Dr. Charles C. Hurst, a botanical geneticist.[133] He was able to piece together what happened from old paintings, catalogs, and nursery records, bolstered by examining the modern plants' chromosomes with the fairly crude techniques of the time. Hurst did similar work on orchids with Robert Allen Rolfe, coauthoring the *Orchid Stud Book* in 1909.

The only original rose of the quartet to survive in Europe was 'Parson's Pink China.' Hume's 'Old Blush' and 'Parks' Yellow Tea-scented China' rose could not tolerate the English climate and had to be confined to greenhouses. With changes in fashion, they were supplanted by newer hybrids and disappeared. Any China roses planted outdoors in milder sections of the country succumbed to the exceptional frosts in 1874 and 1895.

A specimen of 'Slater's Crimson China' rose was discovered in Bermuda. The island is a valuable source of old roses because of its isolation. Hurst thus had two of the four "studs" for his tests. His findings seem plausible, and no one has come up with better explanations.

Dissemination

Seeds of 'Parson's Pink China' and 'Slater's Crimson China' both reached Paris in spite of the French Revolution.[13] Thory and Redouté grew them and many resulting hybrids for the Empress Josephine at Malmaison in 1798.

In 1802 'Parson's Pink China' was imported into the United States by Champneys of South Carolina. This is only one of the exotic roses that reached North America very early, where it became the parent of 'Champneys' Pink Cluster' and the Noisette roses (see Chapter 6).

VERIFICATION OF SOURCES

All the Asian plants of the period came from private gardens or nurseries in Chinese or Japanese towns. The Chinese had been cultivating them for centuries. These were the only sources of plants to which Europeans had even limited access. There was no way they could go out into the wild. Giving a foreigner a map was punishable by death in Japan.

Finding a specimen of an important plant in the wild was a goal of many plant collectors throughout the nineteenth century. It would establish the heredity and origins of the hybrids previously obtained. One or two explorers succeeded, but it was still uncertain in fact whether these were the same wild species that had been brought into cultivation.

Roses were a priority. They exercise a powerful hold on all gardeners' imaginations. In 1821 Captain Richard Rawes, captain of HMS *Warren Hastings*, was said to have seen the wild tea rose, *Rosa gigantea*, and the Banksia rose, *R. banksiae*, in Yunnan province.[13] This was really very far afield for the day. It is possible that this is an apocryphal story.

Bretschneider records the captain's introductions in detail. Rawes' cousin, Thomas Carey of Bromley, asked him to bring plants from China. Between 1816 and 1824, Rawes imported white chrysanthemums, wisteria, white camellia, sasanqua camellia, moutan peonies, primula, and *Camellia japonica*. There are no roses in this list. Perhaps he had indeed simply seen them and noted them, but in 1821, British sea captains were not likely to be as far interior as Yunnan.

Sixty years later, in 1888, Dr. Augustine Henry (1857–1930), an Irish physician and an official in the Maritime Imperial Customs, reported seeing wild *Rosa chinensis* in the Hubei mountains, beyond Ichang (Figure 31).[92] Henry had a lot of time on his hands and is said to have complained of terminal boredom. Taking up botany helped him overcome this. The pursuit was so absorbing that it changed his life. He found prodigious numbers of new plants in western China above Ichang, creating a major sensation at Kew when he sent them a packet of the specimens.

Dr. Henry eventually left China and returned to Ireland. He did not retire but transformed himself into a skilled forester. With H. J. Elwes, he developed the school of forestry at Cambridge and subsequently accepted the new chair in forestry at Dublin University. The ravages of uncontrolled deforestation in China had made him very concerned about the plight of British trees and forests and its connection with rural poverty. In this account of unusual men, Henry holds a very high place.

The story of this wild rose is credible. Henry spent seven years in Ichang. The Customs service extended this far inland because seagoing vessels could reach the interior along the great rivers. Only the gorges beyond Ichang stopped them. Later collectors were halted in their tracks by the forbidding changes in the Yangtze River.

A hundred years after Henry, in 1984, the Japanese botanist Mikinori Ogisu, a pupil of Graham Thomas, found several wild China roses in southern Sichuan. He photographed them and was able to study them more carefully than his predecessors.

Roger Phillips and Martyn Rix, contemporary experts in plant history, traveled to China in 1989, seeking the wild forms of old roses, using Ogisu's work as a point of departure.[13] They are the authors of several distinguished books about roses and other botanical topics. Within a few days of their arrival they came across wild *Rosa gigantea* and *R. banksiae* along the side of the road in Kunming. Vast numbers of pink *R. multiflora* were also seen.

Many hybrid rambler roses have *Rosa multiflora* genes. The standard form in cultivation is cream. It has been so successful in cultivation that it has naturalized itself in the hedgerows of eastern North America.

Phillips and Rix found not only wild forms but also hybrids that appear to have been created within the central and western regions of China rather than in Guangdong (formerly known as Canton district). They had at least some answers to their questions about the origins of roses received in the West.

CHAPTER 10

JAPAN

ENGLAND did not engage with Japan as directly as it had with China and India. The Dutch reached Japan early, soon after the Portuguese, and kept a fierce hold on whatever privileges were extended. After the Dutch, the Americans dealt with Japan, not the English.

In 1854, the American Commodore Perry battered his way into the heart of Japan and imposed the signing of a treaty. All pretense of keeping the foreigners out collapsed. The treaty gave the United States a firm hold in Japanese trade. Behind the soldiers came traders, and behind the traders came administrators and missionaries, both in China and Japan.

Great Britain came onto the scene much later. One of the few English collectors in Japan was John Gould Veitch, who was responsible for bringing "Boston" ivy to England. Ivy gives the impression of age to a building, but the commonest variety only arrived in Britain from Japan close to 1870.

DUTCH COLLECTORS

ENGELBERT KAEMPFER

The Dutch East India Company always sent a physician with their trading representatives in Japan. Engelbert Kaempfer (1651–1716) was the first to go. Kaempfer was born in the German province of Westphalia and studied medicine in Germany. He joined a Swedish expedition to Persia, traveling from Moscow in 1683. Kaempfer stayed in Isfahan for some time but then signed up with the Dutch East India Company as a ship's surgeon on a voyage to Batavia (Jakarta). The company sent him to Deshima, their island off the coast of Japan, in 1690. He served as physician to the Dutch governor.

Kaempfer was curious to see as many plants as he could. Knowing he could not move about, he bribed and cajoled local Japanese people to collect plants for him. Plying them with drink made them careless and a little more amenable. They were terrified of the regime. Once convinced, they were meticulous in their collecting and brought the specimens to him in good condition, according to William T. Stearn.[134]

Kaempfer was the first person to describe Japanese plants in Latin for the West and to provide good illustrations of them. Section 5 of *Amoenitatum Exoticarum (Fasciculus V)* was devoted to plants.[134] It was supposed to be part of a much larger work. Right at the end of his life, in 1712, Section 5 was published.

It created a sensation. For the first time Europeans read about aucuba, skimmia, hydrangea, chinonanthus, ginkgo, azaleas, tree peonies, two kinds of magnolia, some prunus, and *Lilium tigrinum* (*L. lancifolium*). When Carl Thunberg published *Flora Japonica* in 1784, a much larger and more detailed work, it did not create the same stir.[135] It was as if Kaempfer had stolen all the thunder in advance.

The rest of the book he wrote about Japan was not published in his lifetime. Sir Hans Sloane bought the manuscript of a *History of Japan*, a very detailed treatise on Japan written in German.[136] He had it translated into English and published it in two volumes in 1727. Fifty years later a German version was published, retranslated from the English. The actual German manuscript has never been published.

Kaempfer returned to Amsterdam in 1693. He could not find work in the Netherlands and so went home to Westphalia, where he is said to have made a particularly bad marriage. He became Count de Lippe's physician. Medical practice kept him so busy he could not organize his papers and complete his writing

He had quickly described many species but could not gather more specimens to verify his conclusions. He could not have known which plants were endemic to Japan or that some plants had only recently reached Japan from China.

The camellias impressed him. He noted 23 varieties of *Camellia sasanqua*. Kaempfer liked hostas, but he only saw two of the almost 40 species in the country. The two he described were later named *Hosta sieboldiana* and *H. lancifolia*.

CARL THUNBERG

Little if anything is known of Kaempfer's immediate successors. The next serious physician-plantsman to go to Japan for the Dutch East India Company seems to have been Carl P. Thunberg (1743–1828), a Swede.

Thunberg had studied botany under Linnaeus at the Uppsala University Botanical Garden. He was devoted to his work but had a peculiar personality that caused trouble from time to time. He was apparently rather brash, irresponsible, and eager for the limelight. People could get very angry with him, but their anger did not last because they felt he was really like a child in many ways.

His brief tour of duty in Japan, as physician to the Dutch governor, from 1775 to 1776, followed four years in South Africa, where he worked briefly with Francis Masson from Kew. In spite of the advantages of combining resources, Masson could not cope with Thunberg and preferred to collect on his own.

The pattern of life for the Dutch in Japan did not change much in the 85 years between Kaempfer and Thunberg. The company's ships reached Deshima in August and departed in November. A new governor arrived each year.

Every spring the governor and his senior officials had to go to Edo (Tokyo) and pay homage to the emperor. Provincial and other native noblemen also had to do this. In Kaempfer's day the Europeans were treated like performing animals, obliged to sing and dance and display European manners for the amusement of the court.

Thunberg escaped this torment. There had been a partial change in outlook. The Dutch had begun to make the Japanese understand the benefit of European science and skills. Astrologers and local doctors spontaneously heaped Thunberg with gifts, such as Japanese herbals. After he left Japan, some of them sent him seeds and plants.

It was also a little easier for Thunberg to look for plants on the way to Edo and back again than it had been for Kaempfer. He could take short side trips in the countryside without being "recaptured" by the guards assigned to the party. Thunberg could also buy plants in a Japanese nursery, something Kaempfer was not allowed to do.

He laid out a small botanical garden on Deshima to grow his plants until they could go to Europe. At least one of the plants was forbidden to leave Japan, but he managed to send two specimens of *Cycas revoluta* anyway. *Thujopsis dolabrata* was another plant he noted. Thunberg ended up as professor of botany at Uppsala, a most distinguished position.

PHILIPP VON SIEBOLD

Philipp Franz von Siebold (1796–1866), a German physician from Würzburg, was a very different type of man from either Kaempfer, an inquisitive scholar, or Thunberg, a somewhat naive botanist. Medicine was a means to an end for him and botany only a profitable sideline. He had grandiose visions of becoming wealthy and famous. He achieved little of either in his lifetime, but has since been recognized as a major figure in the change of European and thus American flora.

Siebold joined the Dutch East India Company soon after qualifying in medicine at Würzburg in 1820 and went to Batavia in 1822. He was good at cataract surgery, still a risky and primitive procedure in those days. The company needed to renegotiate their treaty with the Japanese and sent Siebold with the embassy to Edo. The directors of the company thought he would impress the court with his Western medical skill. They were correct.

Many of the same footling rules were still in place by the time Siebold got to Japan, but he turned the whole thing upside down. His aggressive self-confidence, bolstered by successful surgery on grateful patients, allowed him to learn Japanese, buy a house, and almost marry a very fine Japanese woman. That was not allowed, but he lived with her monogamously and had a daughter who became a noted midwife in Japan years after he left the country. Grateful nobles gave him paintings, carvings, and other treasures to decorate his house outside Nagasaki.

With all these wonderful advantages, Siebold still managed to cause so much trouble for himself that the company had to remove him after little more than three years in Japan. The emperor had put him in prison for possessing a map of Japan. One trembles to think what happened to the poor court astrologer who gave it to him.

He was permanently barred from returning. In 1858 he went there again after some relaxation of the rules, but the authorities were not pleased. The company was warned to get him out of the country and had to resort to stratagems to do so. He would not leave voluntarily.

With this background, it is hard to realize how effectively he assembled magnificent collections of plants. His motive was to sell them in Holland for very high prices. Friends and volunteers sought out rare and special plants. In 1829 the first consignments of plants left to go to the Netherlands.

In 1830 his employers permitted Siebold to return to the Netherlands via Batavia after his release from prison. He packed more than 480 plants to take with him on the voyage. On his way back to Holland, he stopped in Belgium. A violent revolution was in progress. Fortunately, four of his *Camellia japonica* escaped destruction. His other precious imports were *Lilium speciosum* and four species of *Hosta.*

He ended his career as a nurseryman in Leiden, but his arrogant and paranoid behavior continued to embroil him in controversy and affected his livelihood. Siebold alienated his most important patrons. The king of Holland awarded him various honors, and he assumed that these distinctions meant he could ride roughshod over all his competitors.

During these difficult times, Siebold imported numerous shipments of exotic plants and wrote several long and short works on the geography, ethnology, language, and natural history of Japan. *Flora Japonica*, written with J. G. Zuccarini, was issued between 1835 and 1842.[78] Siebold and Zuccarini illustrated it with colored plates made from drawings of the plants by Japanese artists. The latter were not named. *Flora Japonica* remains the primary reference for Japanese plants found in the initital stages of contact.

Siebold's nursery imported camellia, hydrangea, bamboo, azalea, and lilies in profusion. It continued in business after his death in 1866. Siebold received a specimen of *Hydrangea macrophylla* 'Otsaka' from a Buddhist priest before he left Japan for the last time (Figure 32). This was said to be one of the ancestors of the Hortensia series of cultivars, but this is probably not true.

All these exciting plants eventually went directly from Japan to the United States, but it was still a long time before they ceased arriving via Europe. The next collectors in this narrative were from America and short-circuited this route.

AMERICAN COLLECTORS

Dr. George Rogers Hall (1820–1899) was a graduate of Trinity College in Hartford and the Harvard University medical school. He was moderately interested in plants, but he did not go to Asia as a collector. That was a later by-product of his sojourn there.

The business opportunities opening up in Asia were what attracted him. At first he built a hospital in Shanghai and ran it with some modest success but soon found that he could make more money in commercial ventures.

Hall traded in fine art and curios from Japan and eventually moved to Yokahama. While there, he began to collect Japanese plants in a careful and methodical way. He also housed some of Robert Fortune's Japanese specimens temporarily until they could be planted in their Wardian cases. A few of Hall's plants came from Siebold's collections.

Once Hall had made his fortune, he returned to the United States with his collection, giving Japanese seeds and plants to Parson's Nursery on Long Island and to collectors in New England.

Many now very familiar plants entered the United States with Hall. The kobus magnolia, *Magnolia kobus*, and the star magnolia, *M. stellata*, are hardy in most of the country. Hall packed seeds of the umbrella pine, the zelkova, and Japanese maples (*Acer palmatum*). Hall did not discover these plants, but he did make it possible for them to be grown widely. This is a very early example of new plants going directly to America and not coming via Europe.

Rosa wichuraiana was also sent directly to America from Japan. The breeders immediately began to cross it with the tea rose to try to make it flower longer. The species rose and the hybrids then trickled back across the Atlantic, arriving in England at the end of the nineteenth century.

Another source of exotic Japanese plants in the United States was Thomas Hogg (1820–1892), a Scotsman sent to Japan by Abraham Lincoln in 1862 to be the U.S. consul. Hogg's father had started a nursery in New York a generation earlier. After the father's death, Hogg and his brother took it over.

Hogg sent Japanese plants to his brother to give his family business an edge. In his desire to make sure the plants were more widely distributed, he also supplied Parson's Nursery in Flushing with seed. The Japanese stewartia, *Stewartia pseudocamellia*, is one of the best-known plants from this period.

Charles Sprague Sargent visited Japan 30 years later, in 1892, and collected extensively. One of his best coups was to attract Ernest Wilson away from Veitch and send him on collecting trips for the Arnold Arboretum. Sargent also corresponded with Emil Bretschneider in Peking and carefully nurtured the seeds Bretschneider sent to Boston.

He tried to get Augustine Henry in western China to join his team, too, but it never happened. Despite his very best efforts, they could not come to terms. Henry visited the United States once, briefly traveling to Oregon and Montana to see the forests.[137] He did not like America and only really ever wanted to live in Ireland.

CHAPTER 11

AUSTRALIA, NEW ZEALAND, AND THE SOUTHERN HEMISPHERE

AT ABOUT the same time as all the events in the transatlantic and Eastern arena were occurring, the first explorers of the Southern Hemisphere were finding plants quite unlike anything they had ever seen. Proteas, banksias, acacias, and eucalyptus emerged from vast and terrifying new lands. Australia, South Africa, and New Zealand were botanical treasure houses.

THE PACIFIC ISLANDS

The earliest expeditions to the Southern Hemisphere were instigated and paid for by the French kings. Royal expeditions supplied material for analysis. Louis XV sent Louis-Antoine de Bougainville to circumnavigate the known world. Although this voyage was primarily to establish political claims, Bougainville took Philibert Commerson (1727–1773) with him to study the natural history of the Pacific and its islands as they moved from place to place. The voyage lasted from 1766 to 1769.

Commerson was one of the most diligent plant collectors ever known, sending back more than 30,000 specimens. Some of the Tahitian plants assisted Antoine-Laurent de Jussieu, Bernard's nephew, in defining his natural classification system. Commerson is also remembered for the odd way in which his valet turned out to be a woman in disguise.

The French king sent another scientific expedition to the Pacific in 1786. Jean-François de Galaup, Comte de La Pérouse, commanded a small squadron of two ships completely fitted out as scientific laboratories. Politics were still important, but natural history came a very close second. The La Pérouse expedition had a professional gardener, Jean-Nicholas Collignon, on board to care for all the plants.

La Pérouse was dispatched by Louis XVI to evaluate the natural history of the Pacific coasts and find new edible plants and anything else of value. He recorded all his experiences meticulously in notebooks. He left under royal command and was away during the revolution.

Fortunately, La Pérouse sent each set of notebooks back to France as he went along. When he put into different ports, he gave them to a ship's captain with instructions to take them to Paris. His caution was well justified. He, his men, and their ships disappeared somewhere in the South Pacific.

In spite of the extraordinary turmoil in French politics, La Pérouse's failure to return was noticed. The French government began to look into why he had not returned. The story did not end for almost 40 years. A British sea captain, Peter Dillon, came across the wreckage of one of his ships in the Solomon Islands in 1827. It appeared that La Pérouse had been murdered, possibly by pirates. The French government rewarded Dillon with the Croix de Légion d'Honneur and a pension.

AUSTRALIA

At first much of the exploration of Australia and New Zealand was done from the sea. Brief landings on the coast or on islands were the only way the earliest collectors could work. Once settlement began, it was theoretically possible to travel overland although it was still excruciatingly difficult. Thousands of miles separated the two coasts.

WILLIAM DAMPIER

The unusually well-educated pirate William Dampier (1651–1715) made the first collection of Australian plants while sailing down the west coast of the unknown land. For a time it was known as New Holland because Abel Tasman had named it for his native land.

In spite of his being an outlaw, Dampier moved in elegant English circles. The ultimate accolade is that his portrait hangs in the National Portrait Gallery in London (Figure 33). He had just avoided being hanged. The title of a recent biography is *A Pirate of Exquisite Mind.*[2]

He wrote several books about his travels, and the University of Oxford was glad to have his herbarium specimens (Figures 34–36). They are still there.[138,139]

CAPTAIN COOK AND JOSEPH BANKS

The world was opening up. England had to match the French and other dominant European countries in the new territories. The admiralty, acting for the British government and the king, commanded Captain James Cook to find routes to the southern seas and make useful astronomical observations while doing so. These results were crucial for future navigation in strange waters. Sir Joseph Banks boldly attached himself to Cook. He was extremely curious about the natural phenomena they might encounter in distant lands. Cook and Banks were amazed by the new plants they saw when they landed in the southeast region of Australia on

28 April 1770. As a result they named this stretch of shore "Botany Bay." The site was commemorated by a plaque in 1821.

Banks had a grand vision of the world's flora. He avidly wanted to obtain as much of it as he could for Kew. He suggested that colonies would prosper at the eastern end of Australia. He persisted in this view and dispatched other collectors while waiting for events to catch up with his imagination.

SWAN RIVER COLONY

After Dampier's journey down the west coast of Australia, very little happened until almost a century later, when the entrance to the Swan River was found and a small settlement sprang up near it. James Drummond (1784–1863) and Georgiana Malloy (1805–1843) emigrated from England to the Swan River country for different reasons and were botanical pioneers. Both found exotic plants and sent them to England.

James Drummond

James Drummond was a Scotsman who lived and worked in Ireland in the early nineteenth century. He was in charge of the British government's botanical garden in Cork, enjoying the work of collecting plants and studying botany. It was a shock when the job ceased to exist in about 1827 because of severe budget cuts.

In 1828, Drummond was offered the post of official naturalist in the new Swan River Colony. The only reason the government even considered such an expenditure was because the French were showing too much interest in the southern colonies. Once he arrived in western Australia, even this post did not work out, and he was left to make a living as best he could in a remote and forlorn wilderness.

The family turned to farming. Drummond began long exploratory trips to collect plants and send them to Kew. Sir William Hooker was willing to pay or to get paying customers for Drummond's wonderful plants. Between 1837 and 1852 Drummond sent thousands of specimens to London.

Georgiana Molloy

Because he was a very poor man and had great difficulty in supporting his family, Drummond appeared to be extremely mercenary to his aristocratic customers. He became a tiresome nuisance. The customers grew to prefer Georgiana Molloy, a woman in a class much closer to their own and one who did not importune them for money.[140,141]

Her husband, Captain John Molloy, was assigned to a tour of duty in western Australia soon after their marriage. They stayed there for many years, until Mrs. Molloy died soon after having her eighth child. Childbirth in those days was always extremely dangerous. The primitive conditions Molloy endured only increased the hazard.

In spite of heavy family responsibilities, she found time to explore the area near her house and collect the unfamiliar plants. She used the same agent as Drummond, James Mangles. Among her contributions were several acacias, kangaroo paws (*Anigozanthos* species), clematis, casuarinas, hakeas, hibbertias, kennedia, and lechenaultia.

According to Lines, Molloy's biographer, no complete list of her specimens exists. There were hundreds of them in at least a hundred species. Only the tall boronia, *Boronia molloyae*, was named for her.

It is estimated that there are about 9,000 species in the botanical province she and Drummond inhabited. Approximately 8,000 are endemic and unique to the region.

OTHER PIONEER COLLECTORS

David Burton

David Burton was probably the grandson of James Lee, of Lee and Kennedy at the Vineyard Nursery in London. Lee sent him to explore for plants in Australia, but after being there for only six months he died in a shooting accident, in 1790. Despite his short stay, he managed to find a number of important plants.[1]

George Caley

George Caley (1770?–1829) was in a class by himself when it came to terrible behavior, lack of manners, and inability to deal with other people in an acceptable way. The degree of these failings and the fact that they affected all aspects of his life suggest that he may have had undiagnosed mental disease. In the end, he turned everyone against him except Banks.

His father was a horse dealer, and Caley started out as a stable boy in Yorkshire. His interest in botany began when he tried to find medicinal plants to dose the horses in his care.

Banks paid for Caley's passage to Australia to collect plants and also paid his salary. The reward for his investment in Caley was seeds of over 170 species of plants, including *Epacris purpurascens* and *Jacksonia scoparia.*

Banks finally had to admit that Caley could no longer handle all his projects as he slowed down in old age. He became superfluous, and Banks regretfully stopped paying him. Caley returned to England only briefly, to give evidence in Captain Bligh's favor after yet another mutiny, this time in Australia. He soon sailed away again.

Caley ended as director of the St. Vincent Botanic Garden in the West Indies. *Banksia caleyi*, *Grevillea caleyi*, and *Acacia caleyi* are named for him. His colleagues, who could probably have cheerfully murdered him, all agreed he did wonderful work.

Allan Cunningham

Allan Cunningham (1791–1839) stayed in Australia for 15 years.[142,143] Much of that time was spent at sea because of the great obstacles to exploring by land. The governor of New South Wales at the time of his first journey did not help him, making things even more difficult.

Cunningham's father had wanted him to become a lawyer, but the son preferred horticulture. He initially worked for the younger Aiton at Kew, editing *Hortus Kewensis.* The Kew authorities sent Cunningham to South America in 1814 with James Bowie. They stayed in Rio de Janeiro for two years. Brazil was very xenophobic in that period and would not allow visitors to travel widely. Bowie and Cunningham managed to reach São Paulo, where they found gloxinia (*Sinningia speciosa*) and *Jacaranda ovalifolia.*

In 1816, Aiton instructed Cunningham to go to Australia and sent Bowie to South Africa. Cunningham's explorations were valuable to the development of Australia quite apart from the plants he discovered. Many plants were described and named long after he found them. On one journey, he brought back *Acacia pendula, A. spectabilis, Grevillea acanthifolia, Hovea celsii,* and *Pandorea pandorana.* He was a field man, not a desk scholar. Others labeled his discoveries and gave them accurate names.

One of those was George Bentham (1800–1884). He prepared *Flora Australiensis* while sitting quietly in an office in Kew, using many of Cunningham's specimens for this work.[144] Bentham was a very retiring man who did not put himself forward but who was the backbone of British botany for decades (see Chapter 12). When Robert Fortune sent back the treasures of his China expedition, Bentham wrote *Flora Hongkongensis.*[145]

Cunningham sailed along the Australian coast with Lieutenant Phillip Parker King RN between 1817 and 1822; he managed to collect up to 400 specimens on his brief periods ashore.

He also explored the interior of eastern Australia and would have done more but for his health. Cunningham returned to England in 1831 and stayed there six years before going south again.

Cunningham was appointed to be superintendent of the Royal Botanic Gardens, Sydney, but resigned a short time later to explore and collect plants in New Zealand, another piece of pioneering work.[146]

John Bidwill

The flora of Australia continued to emerge piecemeal, as did the geography. John Bidwill (1815–1853), a former employee of Lucombe and Pince in Exeter, Veitch's rivals, arrived in Australia in 1839. He penetrated a little way inland and found the "bunya bunya" tree, *Araucaria bidwillii.*[147] He had emigrated to take up a land grant and work for a firm of merchants, but there was a long delay before he could get started. To fill in the time, he sailed to New Zealand and was the first European to collect plants there.

Pince had taught Bidwill how to graft and hybridize. Bidwill became well known for these skills and eventually was appointed to be the director of the botanical garden at Sydney. He died of exposure in the Australian bush quite soon after assuming this position.

BOTANY AND HORTICULTURE IN EARLY MELBOURNE

The dominant figure in nineteenth-century Australian botany was Ferdinand von Mueller (1825–1896). The Australian flora was collected and identified at first by European scholars, mostly British, French, and a few other nationalities including German.

Mueller emigrated from Germany in an effort to avoid the ravages of the tuberculosis that had afflicted his family. He came to personify Australian scholarship. He had had a thorough training in the sciences before leaving Germany and was well suited for what lay ahead.

Mueller became director of the Royal Botanic Gardens, Melbourne, and quickly transferred his allegiance to the British Empire and crown. The queen knighted him for his services to British science. He had added the aristocratic "von" himself at some point in his career.[144]

Mueller was a pure scholar in the German tradition, utterly indifferent to public relations or the superficial needs of the citizens of Melbourne. To him a botanical garden was a laboratory, with plants in scientific order regardless of appearance or accessibility. The idea that the garden should be attractive and give pleasure to the public was derisory. In a well-run establishment, the professor makes the rules.

His attitudes were not unique. Other directors of botanical gardens shared this view. Joseph Hooker fought a long and exhausting battle with his masters over this matter.

There was so much public outcry in Melbourne over Mueller's disregard for the public's wishes that the authorities relieved Mueller of his duties running the garden. They appointed one of his protégés, William Guilfoyle, in his place.

Guilfoyle (1840–1912) had trained with Mueller and knew the garden well. While running the family business, he had also done extensive landscape design and construction. Guilfoyle's father Michael had emigrated from England to Australia in 1853 and started a nursery in Redfern. His sons learned the trade from him but complained that he was a strict disciplinarian, a martinet who made their lives very difficult.

Guilfoyle immediately tore the botanical garden apart and rebuilt it as a beautiful park. Mueller sulked for the next 30 years. He continued to work in his laboratory but never set foot in the garden again. He went to his laboratory across the street every day without ever looking back.

As in so many other similar situations, if Mueller had had a tactful manner and the ability to recognize someone else's point of view, he might never have been banished from the premises. He was strictly correct, but in a large, burgeoning city, the public desire for a park in which to relax had to be accommodated.

TRANSPACIFIC TRADE

There was extensive transpacific trade very soon after the California gold rush created a vast market. The Australian gold rush, about two years later in 1851, also played its part. This timing was not by chance. A miner who had been in California noticed that the terrain in Ballarat was very similar to the kind he had seen in California and began to prospect.

There has been a lot of disagreement about the origin of eucalyptus trees in California, but it is now accepted that the first contingent of Australian gold seekers took the seed with them in 1849.[148]

NEW ZEALAND

The biota of New Zealand is utterly different from that of Australia. The islands were separated from the great southern continent known as Gondwana many years before Australia. It has more affinity for coastal South America and the Pacific islands than for the continent 1,500 miles to its west. Species of beech, *Nothofagus*, form forests over the southern part of New Zealand's South Island, not unlike those of Chile.

The absence of endemic mammalian species other than bats helped to protect the original flora from excessive grazing or destruction. Daisies that grow into trees, carrots with leaves like swords, gigantic buttercups, and yellow forget-me-nots constitute some of the amazing sights that greeted later collectors.

The dominant human population was of rather more recent vintage. Pacific islanders, forerunners of the Maori people, reached New Zealand in the thirteenth century, giving them only about 600 years to become acquainted with the territories they fought so hard to maintain.

They made tremendous use of the trees and shrubs they found in the wild. Tall Kauri trees, *Agathis australis*, were valuable for boat building and spears. The wild trees have since been almost completely decimated because of their value.[149] The Maori grew *Clianthus puniceus* quite soon after they reached New Zealand, but it was not found in the wild until Allan Cunningham saw it on his second trip.

Explorers visited the shores and offshore islands of the New Zealand coast following Cook's initial voyages. George Forster, another young botanist of the period, went with Cook in 1772. On that voyage, Cook named a small island for the Duke of Norfolk. This is the source of the Norfolk Island pine, *Araucaria heterophylla*.

French vessels were not far behind Cook. The well-known scientist Labillardière was sent to see what had become of La Pérouse in 1791. He found the

Phormium, New Zealand flax, noted by Forster, and collected enough seed to grow it on his return to Paris. It was the first New Zealand fiber to be grown in Europe. There was no trace of La Pérouse.

The French government gave Jules Dumont d'Urville the task of finding La Pérouse in 1826. He was a skillful and conscientious explorer. He noticed that the native flora of New Zealand was already corrupted by foreign imports and insisted that the original plants be carefully described. Dumont d'Urville made two journeys to New Zealand and took back many plants on both occasions.

He was enlightened in another way. In his opinion, the names used by the native inhabitants were the appropriate ones. He made every effort to understand their language and keep the terms pure. At a time when Europeans assumed automatically that they were much cleverer and superior to any indigenous peoples, that was a remarkable attitude.

During his voyage, Dumont d'Urville heard of Captain Peter Dillon's discovery of the La Pérouse expedition. He reached the Solomon Islands in 1828 and recovered additional remains from the wreckage.

Archibald Menzies also reached New Zealand with Captain George Vancouver in 1791. All these naturalists did the best they could with lightning-fast forays from the big ship across the surf onto the beach and into the adjacent forest. They could not stay more than a few hours at most because the captains were always afraid of either being ambushed or having problems with the tides.

When European settlers arrived in the early nineteenth century, the native flora overwhelmed them. Allan Cunningham traveled to New Zealand from Australia in 1826.[7] It was almost impossible to penetrate very deeply into the interior, but he managed to go the 60 miles needed to see the opposite shore. The Kauri forests were still pristine at the time.

Cunningham returned to New Zealand in 1838, working in the Bay of Islands. He was already sick with the tuberculosis that killed him a couple of years later, but he inspired the young missionary William Colenso (1811–1899) to take up plant collecting (Figure 37).

Cunningham's brother Richard also collected in New Zealand, finding a climbing fuchsia, *Fuchsia procumbens*, and veronicas, which are now called *Hebe*.

Other very early explorers were John Bidwill, who went there in 1839, and Dr. Ernest Dieffenbach, who arrived soon afterwards. Alice Coats notes that Bidwill's rough ways and complete insensitivity to Maori beliefs and customs made Dieffenbach's stay very difficult.[1] Bidwill had left a trail of hatred and resentment behind him.

Joseph Hooker collected plants there in 1841. He prepared *Flora Novae Zelandiae* as part of his series of Southern Hemisphere floras. When he visited Paris in the later 1840s, he was a little piqued to see how much the French had achieved before him. This is slightly surprising since Dr. Etienne Raoul, another fine French botanical explorer, had shared his finds with Sir William Hooker at Kew when he returned to France. Joseph must have been aware of this.

William Colenso, a printer and lay deacon of the British and Foreign Bible Society, was the first European collector to cross the fearsome Ruahine Range.[150] Colenso was serious about plants and corresponded with both Hookers, father and son. His interest, already apparent before he left Cornwall to become a missionary, was further stimulated by meeting Hooker, Darwin, and Dr. Andrew Sinclair, another Scottish ship's surgeon who visited New Zealand in 1841. Sinclair also collected plants assiduously.

Colenso made many journeys through frightful obstacles, sometimes not eating for two or three days in a row because of the poverty of the land he traversed. He passed through endless fernbrakes, which he hated, but sent hundreds of new plants to the Hookers at regular intervals. The fern spores made him sneeze violently.

He seldom collected seed, so the value of his work for commercial horticulture was indirect. *Fuchsia colensoi*, *Phormium colensoi* (later *P. cookianum*), and *Pittosporum colensoi* all commemorate his name. He lived to be almost 90, breaking the rules of respectability by fathering a child with a native woman, but ending up as the member of Parliament for the modern New Zealand city of Napier.

When the citizens of Dunedin in Otago decided to create a botanical garden in 1853, the first in the country, there was no mention of Colenso. His work had not yet penetrated the national consciousness.[151]

The chief plants of New Zealand origin in modern European and American gardens are species of *Phormium*, *Hebe*, and *Olearia*.

SOUTH AFRICA

After his triumphal return from Australia, Banks suggested to George III that South Africa might also be a source of valuable new plants.[1] The king, in one of his lucid periods, agreed. Kew appointed one of the young gardeners, a Scotsman from Aberdeen, Francis Masson, to go out to South Africa.

FRANCIS MASSON

Ultimately, Francis Masson (1741–1805) spent 25 years collecting plants in different parts of the world (Figure 38). He was given £200 a year, quite a lot of money in those days. Sixty years later, the Horticultural Society of London would give its collectors only £100 a year, making it very hard for them to manage.[65]

Masson first went to the Cape of Good Hope in 1772 and stayed for three years. For a time he worked in tandem with Carl Thunberg, who had formerly been posted to Japan by the Dutch East India Company. Banks had arranged for Thunberg to collaborate with him, but they were antithetical in personality. Masson did much better on his own, even though it made sense to share the expenses of the journeys.

From what has come down over the years, Thunberg was egotistical, bombastic, and rash but had an almost childlike quality that prevented people from getting too angry with him. Masson had many of the social and personality traits of the Scottish, reserved, measured, exceedingly reliable, and not flashy at all.

Masson returned to the Cape in 1786, when the political situation had changed. In the interval he had worked in the West Indies and Atlantic islands. The Dutch were no longer happy to cooperate with the English in South Africa, and technically he was not welcome. In fact, the governor of Cape Province treated him very courteously. He stayed until 1795, when things became too dangerous and he asked to be recalled.

Masson found pelargoniums, numerous ericas, ixia, nerine, gladiolus, stapelia, *Mesembryanthemum*, oxalis, diosma, and many aloes, all of which, or their descendants, remain among the most popular and treasured ornamental plants today. He sent more than 500 species back to Kew. He managed to pack the bulbs, seeds, and other plant material well enough to survive the long sea voyages.

Banks recognized the value of what Masson had done. He wrote enthusiastically to the king about the discoveries, saying that ". . . His Majesty's appointment of Mr. Masson is to be accounted among the few Royal bounties which have not in any way been misapplied."[1]

Eighteenth-century naturalists did not have the advantage of modern scientific concepts, but in choosing to explore southern Africa, Banks had stumbled onto one of the most prolific endemic centers of floral life in the world. Square mile for square mile, there are more endemic plants in the three principal zones of the Cape than anywhere else.

It is sad to report that Masson was badly treated by the authorities at Kew after his tremendous accomplishments in South Africa. He was a poor man and could not live without work. They gave him no pension. The younger James Lee, son of the Lee of Lee and Kennedy, was outraged by Kew's mistreatment of him after years of service. "(He) deserves to have some memorial given of his extreme modesty, good temper, generosity and usefulness."[1]

After struggling financially, Masson petitioned Kew to find him some means of subsistence. Their response was to send him to Québec. He was no longer young, he was not in robust health, and the Canadian winters very rapidly led to his death.

Even when *in extremis* he was diffident and, deeply aware of his lowly status, he was concerned about troubling anyone. Anna Saltmarsh wrote a paper about his dying days, based on a brief surviving note from Canada.[152]

Only one genus has been named *Massonia* to commemorate his contributions. He was a lowly gardener, not trained in botany or horticulture and so not considered important enough. He published one book, *Stapeliae Novae*, in two parts, in 1796 and 1797. It was illustrated with many of his own drawings (Figure 39).

OTHER COLLECTORS

Beside Thunberg and Masson, other explorers scoured the Cape. Shortly after Masson's period, James Niven was sent to explore the Cape by George Hibbert, a wealthy amateur. He stayed there for four years, from 1798 to 1803. When he came back with a good collection, he was immediately conscripted into the British army and thrust back into South Africa. He died in 1827.

Britain was at war with the Dutch, trying to annex the Cape. This was to forestall the French from seizing this essential region. Niven spoke some Dutch and a little Zulu and so was invaluable as a scout. In spite of being preoccupied with military matters, he managed to send back at least 30 new ericas.

William J. Burchell (1781–1863), an eccentric but highly productive merchant and traveler throughout southern Africa, collected plants as part of his exploration of the country. He studied its people, geography, flora, and fauna and traded with the native people from a sturdy old wagon in which he also lived. His father was a nurseryman in Fulham, South London.

The family supported William, and he returned with more than 40,000 specimens, 2,000 species of seeds, and almost 300 bulbs after a five-year sojourn, from 1810 to 1815. Burchell's collection was notable for its *Mesembryanthemum* species. He generously shared many of them with the leading authority of the day, A. H. Haworth. The other seeds and bulbs were all planted in his father's garden and entered commerce that way.

Banks also sent James Bowie (ca. 1789–1869) to collect in South Africa. Bowie sent back the glorious *Clivia*, though in fact, Burchell had taken it back before Bowie did, not knowing what he had.

After Banks died in 1820, Bowie's masters summoned him home. There was no more money to pay such men. Life in London did not suit him, however, and he went back to Africa to try his hand as a freelance collector. Bowie's character let him down, and he did not prosper. Eventually another English settler took pity on him and employed him as a gardener near Cape Town.

There were other British collectors, such as the Irishman William Henry Harvey and Thomas Cooper from Surrey. The Austrian emperor Franz Joseph sent two men from Vienna who stayed more than 10 years.

CHAPTER 12

MEXICO AND CENTRAL AND SOUTH AMERICA

MEXICO

Mexico is an important source of plants in the modern gardening repertoire. Because of its proximity to the United States and long familiarity, its significant role as a donor of floral material tends to get shunted aside. The next few paragraphs provide only some of the more salient features of this story.

The country has an unprecedented range of floras because of its broad variations in geography and climate zones. Deserts, tropical jungle, and high-altitude pine forest are just a few of its extremes. The Sonoran Desert extends up into the southwestern United States. There are more species of oak and pine tree in Mexico than anywhere else. Pine trees from the higher altitudes grow very well in the cooler regions of the United States and Europe.

Many plants continue to be found and examined for horticultural possibilities in the United States. Drought-tolerant trees and shrubs, such as *Plumbago scandens*, white desert; *Prunus mexicana*; and *Cercis canadensis* var. *mexicana*, are among their number.

The Horticultural Society sent Karl Theodor Hartweg (1812–1871) to explore in Mexico for three years between 1836 and 1839. He then moved on to Guatemala, followed by Ecuador, Jamaica, California, and finally Madeira by 1854. Hartweg was a German gardener who worked conscientiously for the British but crowned his career by going back to Germany and becoming inspector of the Imperial Gardens in Swetzingen.[65]

John Lindley, key taxonomist at the Horticultural Society, named a genus *Hartwegia* in his honor. This genus has been subsumed by another one, and the name is no longer in use. For a period between 1839 and 1857, George Bentham issued *Plantae Hartwegianae*, codifying Hartweg's collection and augmenting it with plants "beautifully dried" by G. J. Graham, another collector who was most conscientious but not widely known today.[153]

The most handsome and popular plant from Mexico is probably the dahlia. Hartweg and Graham brought back almost all the other annuals that are widely enjoyed today, such as cosmos, zinnias, lobelias, calceolarias, marigolds, penstemons, buddleia, salvia, clematis, gentians, and ageratum. Many popular succulents also originated in Mexico, such as *Agave*, *Yucca*, and *Hesperaloe*.

The European acquaintance with Mexican plants goes back to the sixteenth century, but it is possible that the native people had specialized botanical gardens long before that. Cortés encountered the astonishing "floating gardens" in Tenochtitlan, the great city on the lake he attacked.[154]

Fray Bernardino de Sahagún, a Spanish Franciscan, wrote a history of Mexico in the 1530s that was suppressed by the church and not published until 1830.[155] Sahagún wrote in Nahuatl, hoping that the native people would be able to understand his book, but the church was alarmed by his radical ideas. They wanted to suppress anything to do with the pagan period before Catholicism became the official religion.

Book XI of the 12 books is devoted to plants: "En que se trata de todas las hierbas."[156] Sahagún divided this book into different sections dealing with trees, herbs, and other types of plant. He was most interested in anything with medicinal uses.

In another epoch, 1570, Philip II of Spain sent his personal physician Francisco Hernández to Mexico as Protomédico of the Indies. Dr. Hernández was passionate about plants and worked himself beyond endurance studying them all.

He, too, used the native Nahuatl plant names.[157] A dahlia was "acocoxochitl," as can be seen from one of his woodcuts. There are very faint hints that Hernández might have been a crypto-Jew. Even though Hernández was never officially persecuted by the Inquisition, Philip kept him cooling his heels in Mexico for five years after he was supposed to return and actively prevented his book from being published. Hernández was said to be "impious" because of his giving names to new discoveries. Naming was considered to be a form of creation.[158]

After seven years, he had to return to Spain for his health and died in 1578. He left a huge body of work, all unpublished in his lifetime. Some of it was completed by a medical colleague. Hernández described 3,000 previously unknown plants. Six volumes were finally published, including 10 folios of paintings by native Mexican artists. The manuscripts were all deposited in the Escorial but were lost in a fire in 1671.

The Malaspina expedition was a major landmark in the New World 200 years later.[159] Captain Alejandro Malaspina (1754–1810) was an aristocratic Italian seaman and scholar who visited the west coasts of Mexico and California in 1789 soon after the Franciscan missions were founded. He and his colleagues collected plants and animals, but they were not primarily botanists.

One botanist, Thaddeus Haenke, was supposed to go with them but missed the ship when it left Cádiz and had to catch up with them belatedly. Haenke was a Bohemian. He collected plants between Acapulco and Mexico City. He is also known for his portrait of Malaspina at sea.

José Mariano Mociño (1757–1819) was an important native-born Mexican botanist. He assisted the Spaniards Vicente de Cervantes and Martín Sessé y Lacasta in a major botanical exploration launched by the king of Spain in 1788. Curiously, on the one hand we are told that the Spanish were indifferent to the flora of their home country (see Chapter 4), and yet on the other hand they put a great deal of stock into the flora of the New World. The Real Expedición Botánica lasted for about 15 years until 1804.

Mociño then traveled to Spain with Sessé but became a refugee in France during the Peninsular War. He gave his manuscripts and drawings to Augustin Pyramus de Candolle in Montpellier before he died in 1819. Part of them was published years later in 1886 as *Plantae Novae Hispaniae. Flora Mexicana* appeared in 1891.[160]

During this phase the local scientists overlapped with European collectors, both scientific and commercial. Much later, in the 1930s, an American artist with a great interest in nature, Cecile Hulse Matschat, spent seven years in Mexico while her husband carried out large engineering projects there. She wrote a very valuable book showing that, despite the rich history of plant hunting in Mexico, dozens, maybe hundreds, of plants that would do well in American gardens had not become well known.[154] Mrs. Matschat thought this was partly because the data are very fragmented. The relevant information is scattered among numerous journals and inaccessible to nonspecialists.

CENTRAL AMERICA

Once known as the Spanish Main, Central America includes the islands and the countries of the Caribbean Sea. Spain dominated this region rather as the Netherlands dominated the East Indies for years.

The Caribbean islands allowed pirates to flourish. They could hide in dozens of small channels and harbors. Because of their proximity to the mainland of the Americas, these islands had rich and complex floras modified by geological isolation and the formation of new land.

Cromwell's annexation of Jamaica in 1655 was an important coup for Britain. Sir Hans Sloane (1660–1753) began the modern scientific collection of plants while serving as the Duke of Albemarle's private physician between 1687 and 1688. An English gardener, Thomas Willisel, had started to collect there shortly before Sloane arrived and had been most assiduous but died only a year after reaching the island.[1]

A number of other collectors worked in Jamaica, partially stimulated by Sloane. The best known was Charles Plumier (1646–1704), for whom the frangipani tree (*Plumeria rubra*) is named. He named more than 50 genera, often commemorating well-known scholars, such as *Fuchsia* (Fuchs) and *Lobelia* (L'Obel).

SOUTH AMERICA

The section of the Americas between the Tropic of Cancer and the Tropic of Capricorn contains almost 30% of the world's known genera and species of flowering plants. Every day, vast quantities of this diverse flora are lost because of human activity without ever being identified or evaluated for medical or economic use. Despairing comments on this sad state of affairs have been appearing since the mid-nineteenth century (see Chapter 13).

Charles Marie de La Condamine (1701–1774) is important to historians of science. He was the first European to cross Brazil, traveling from the Andes to the sea coast along the Amazon River between 1743 and 1744. The chaotic politics of the region had kept all foreigners out before that time. The marquis did not explore North America. He confined his exploration to the south. Louis XIV had sent him to the equator to measure the length of an arc of one degree of meridian. Peru was the sole place at which the equator was accessible at the time.[161]

The definitive study of the flora of Brazil is agreed to be that of Carl Friedrich Philipp von Martius (1794–1868), a German botanist who devoted his entire career to that subject. He founded *Flora Brasiliensis* and, before his death, brought out 46 of the eventual 130 fascicles. The series finally came to an end in 1906.[162] Von Martius traveled in Brazil from 1817 to 1821, only four years. He brought back more than 12,000 herbarium specimens. Much of what he described may no longer be extant.

Certain other names are indelibly associated with exploration and plant discovery in South America. Alexander von Humboldt (1769–1859) is perhaps the very best known, Aimé Bonpland (1773–1858), Humboldt's companion, much less so. Robert Schomburgk (1804–1865) is hardly known at all.

Two Englishmen with totally different goals have to be included in this list, Richard Spruce and William Lobb, one a pure scientist, the other an employee of a nursery and charged with finding commercially useful plants. Both discharged their duties in an exemplary fashion but are known only to a small group of scientists and interested persons. Modern British and European gardens would not look the way they do today if William Lobb had not traveled widely in South America.

Horticulturally, South America can be divided into the Amazon Basin, the Argentine, and the Andes.[1] It is a useful way of looking at things. Lobb did not find plants suitable for English gardens in Brazil, the Amazon Basin. He found them in the Andean region, Peru, and Chile.

Bonpland collected plants in the Argentine very soon after that country liberated itself from Spain. A much humbler explorer, the Scottish landscaper James Tweedie (1775–1862), put Argentinian plants into British gardens. He emigrated to the Argentine from Scotland in 1825 and sent the plants he collected to Sir William Hooker at the University of Glasgow. Tweedie introduced *Petunia violacea*, one of the two principal ancestors of all modern petunias. Some gardeners in California may wish he had not sent pampas grass (*Cortaderia selloana*), but among his other finds were *Triteleia uniflora*, *Verbena platensis*, and *Cyphomandra betacea* (now *Solanum betaceum*), the tree tomato.[1]

In the modern era, Richard Evans Schultes (1915–2001) remains unique in his accomplishments and renown. Schultes was seeking economic plants but came across many orchids in the Amazonian forest.

Spanish scientists traveled more often to Central and South America than to

North America. Botanical gardens in Madrid and the other principal Iberian cities contained plants from the tropical and semitropical New World. King Carlos III had founded the Royal Botanic Garden in Madrid in 1774 to accommodate his possessions in New Spain.

European scientists from the other countries also went to South America and brought back many samples of the unusual flora. Before the Allied bombing in World War II, the Berlin-Dahlem Botanical Garden contained an exceptional collection of Brazilian plants, perhaps because of von Martius' work.[163]

With the expansion of the nursery trade in Europe, particularly in Britain, commercial collectors began to replace purely scientific travelers in the second half of the nineteenth century. Orchids were the driver.

ALEXANDER VON HUMBOLDT AND AIMÉ BONPLAND

Alexander von Humboldt set a standard for accuracy, comprehensiveness, and a sweeping outlook that informed the rest of nineteenth-century exploration and natural history (Figure 40). Like some other very influential scientists, he did not spend a great deal of his life in the wild, although he traveled in both South and North America. He was deeply influenced and inspired by La Condamine, but the latter never traveled in North America. It was the breadth of Humboldt's observations, the conclusions he drew from them, and the use he made of the notes he had taken that are his legacy. He enriched geography, geology, anthropology, oceanography, cartography, natural science, and meteorology both theoretically and practically. Even his errors were fruitful in some ways.

Humboldt and his brother, Wilhelm, had the advantages of wealth and privilege in growing up, but he made exceptional use of them. Their father died when Alexander and his brother were quite young. He was devoted to his mother, but once she died, he felt free to pursue his desire to travel.

Bonpland was a French physician and botanist who had tried to be a steady practitioner of medicine but had chafed continuously at this humdrum existence and longed only to collect plants. Humboldt and Bonpland met each other in the most casual way, collecting their mail from the same source. Their skills and personalities were complementary, making them lifelong friends.

The English blockade of Mediterranean and Atlantic ports in 1798 sealed off all the usual exits, but Humboldt and Bonpland climbed over the Pyrenees and left from Spain in a neutral vessel in 1799. The king of Spain had given Humboldt a broad laissez-passer document that stood him in good stead wherever he went.

The ship deposited them in Venezuela. Humboldt was intent on finding the source of the Orinoco River. Bonpland zealously collected plants as they traveled. Electric eels and a tree that gave "milk" were some of the first marvels they found. Bonpland was so diligent he collected more than 16,000 specimens. In spite of his attempts to preserve everything chemically, at least 40% of the specimens were ruined by insects, mold, and damp.

Humboldt and Bonpland were very concerned about the fate of their collections. They divided them in three and sent each portion by a different route. In this way they managed at least to get part of the material back to Europe safely. La Pérouse had fortunately adopted a similar technique in light of his ultimate fate.

The two men traveled through Venezuela, Ecuador, Peru, Cuba, and Mexico. Fifty years later, Richard Spruce met villagers in the Venezuelan uplands who remembered Humboldt, a "little man with grey eyes."[162]

In all, Humboldt was away for five years. Bonpland never returned to Europe but eked out an existence in Uruguay on the pension Humboldt obtained for him. Humboldt published *Plantae Equinoctiales* in 1808, a work based on English and German models such as James Edward Smith, Carl Willdenow, and Martin Vahl (Nathaniel Wallich's teacher).[164]

This book gives some idea of the richness of Humboldt and Bonpland's collections. Humboldt was very much impressed by the variety of palm trees, enormous number of grasses, and subtropical cryptogams. A bald recital of the numbers tells much of the story. He listed more than 150 melastomes, 480 eupatoria, 86 molina, 52 calceolaria, 58 psychotria, 40 lobelia, 40 plants in the Ranunculaceae family, and more than 400 grasses.

RICHARD SPRUCE

In 1849 a young Yorkshire schoolmaster sailed from Liverpool to Pará (now Belém) in Brazil. He traveled extensively in several contiguous countries, in some cases beyond Humboldt and Bonpland's farthest reach. Richard Spruce (1817–1893) had served his own apprenticeship, alone, methodically collecting, labeling, and examining plants in ever-widening circles from his birthplace to larger swaths of Yorkshire and the north of England. His particular area of interest was bryophytes (liverworts, hornworts, and mosses).

Spruce's father was the schoolmaster at the Howard estate, seat of the dukes of Norfolk, and Richard was the only child of his first wife. The school was supported by the family at Castle Howard. His father taught Richard at home and sent him out to earn a living as a schoolmaster, too, at a boarding school in York. He was just 17.

Richard Spruce's skill in botany grew from his combing the surrounding country minutely whenever he had time off. At first he merely wanted to collect samples of plants he had read about, but he gradually found more and more new ones that no one had noted previously. He was not content simply to list them and move on, but spent hours poring over them to verify their anatomy and learn everything he could about them.

Well-known botanists grew accustomed to receiving packages from Spruce with new mosses, asking searching and pertinent questions. When a collector was needed to go to the Pyrenees, officials at Kew turned to him. They actually wanted someone to go to Spain, but as noted elsewhere in this work, Spain was too dangerous for a lone Englishman to wander around in isolated places. The French Pyrenees had to do. Spruce spent 1845 and 1846 in France.

He had not liked teaching, and the school where he taught was going out of business. It seemed obvious to accept the timely offer from Kew.

His work in the Pyrenees made Spruce the obvious choice to go to South America, to collect on a serious scale. He ultimately sent back more than 7,000 species from South America, working from 1849 to 1864. His "agent" was George Bentham at Kew. Bentham supported Spruce financially as well as academically but never visited Asia or the Southern Hemisphere

Bentham was a remarkable man. In his teens he came across the work of the de Candolles and devoted his life to botany. Not only did he compile the great floras of Hong Kong and Australia, but he also made an immense contribution to taxonomy in *Genera Plantarum.*[165] He gave a definitive description of all the known seed plants in this three-volume work. The whole family was talented. Bentham's uncle was the utilitarian philosopher Jeremy Bentham.

Bentham coordinated all Spruce's activities, making sure that plants went to the people who were supposed to get them. He also collated the specimens and named many of them himself. Bentham had brought out *Flora Hongkongensis* with Fortune's plants in the same way. The work was so careful and accurate that it is hard to believe that Bentham, who grew up in France, never left England again after returning there in early adulthood.

Spruce brought back multiples of everything. He was expected to recoup the expense of going to South America by selling these sets of plants to wealthy private collectors. Sir William Hooker, Bentham, and a few other botanists laid out money to help him buy a ticket and have something with which to pay for travel, food and lodging, porters, and other menial expenses at his destination.

Besides these rather tenuous resources, Spruce had no other money. At one point in his *Notes*, he plaintively remarks that if only he had some social standing, he could get a position as British consul in South America and thus be assured of a living. As a poor and very obscure person, this possibility was out of his reach and he knew it.

George Bentham was an indefatigable workhorse but always gave Spruce immense credit for exquisitely neat labeling, minute and careful descriptions, and accurate recording of where he had found the plants. Bentham helped to sort out the duplicates and prepare the sets of plants for sale.

Not too many of Spruce's plants can survive outdoors in northern climates. He found many orchids. Richard Evans Schultes had been stirred by the story of a blue orchid, *Acacallis cyanea* (Figure 41).[166] It had not been seen in the wild since Spruce first collected it a century before. By a wonderful chance Schultes found this orchid quite early in his travels and was delighted. Margaret Mee (1909–1988), a renowned English painter of tropical flowers, painted it when she was drawing and painting from nature in Brazil at the age of 75.[167]

Much of the information about Spruce's life is found in Alfred Russel Wallace's introduction to *Notes of a Botanist on the Amazon and Andes.*[162] Wallace (1823–1913) was the man who almost upstaged Charles Darwin with his theory of evolution. In 1858 Wallace sent a copy of his manuscript on natural selection to Charles

Darwin. Darwin's friends recognized the significance of the work, as did Darwin himself. Charles Darwin had been hesitating to publish his very radical theory, but once he saw that he might lose priority, he immediately changed his mind.

Both papers were read at the same meeting of the Linnean Society that year. Neither man was present. Wallace had such a high opinion of Darwin and such a low opinion of his own importance that he considered this arrangement perfectly fair.

Wallace edited Spruce's notes and prepared the manuscript for publication after Spruce died. The notes were hard to decipher, in spite of Spruce's usually careful methods.

Modern scholars have studied the original documents more searchingly and continue to learn more about Spruce and his work. A recent symposium includes a list of the orchids he sent back to England.[168] He found most of them in the Upper Rio Negro and Orinoco river basins. Almost all the live plants died en route. Taxonomists worked from dried specimens.

Spruce collected 33 new species of orchid and one new variety. Because he was not working for a nursery firm, these were not rushed into "production" but did add immeasurably to scientific knowledge. Lindley described 11 of the new species in his great work. *Erythrodes sprucei*, Spruce number 6138, was finally described in 1978.

Spruce had very broad interests and did not just collect plants. He amplified the understanding of ethnobotany and was instrumental in sending the red-barked cinchona, *Cinchona succirubra*, to plantations in India. His work on the genus *Hevea*, source of rubber, has never been superseded. Spruce was adept at making maps, too, a most valuable skill.

The observations he made on native life and his drawings of the people turned out to be providential many years later. The murderous frenzy to leach rubber out of the forests ever more quickly led to the loss of entire tribes of natives. If it were not for Spruce's records, no one would know these tribes had ever existed.

Like many young people of the time, he was in poor health, though it is not clear precisely what was wrong with him. In most cases some form of tuberculosis underlay the mysterious symptoms. Cows were riddled with bovine *Mycobacterium tuberculosis*, and their milk contained the microbes. This illness did not hamper his early activities, and he lived to be quite old, but the last part of his life was seriously affected by his disease.

Much of the time he was in South America he was quite ill and on several occasions had to be nursed in unbelievably uncomfortable dwellings at the mercy of hostile and unsympathetic native women. They felt that the sick white stranger was a source of really bad luck. One woman did not hesitate to tell him to hurry up and die.

Spruce lived out the rest of his life in a small cottage a few miles from where he was born, patiently enduring a debilitating disease on an exiguous income. During his working years he had managed to save about £1,000, quite a lot of money in those days, but a clerk at the bank in South America to which he had entrusted the money embezzled it, leaving him almost destitute.

Wallace browbeat the British government into giving Spruce a minute pension, and a few years later had that augmented by another £50 a year. The justification was his work on cinchona for the empire. A handsome and reserved man, Spruce was never heard to complain about his poverty or ill health.

The Natural History Museum (British Museum, London) and the Royal Botanic Gardens at Kew are working jointly on the Spruce Project, restoring Spruce's collections and putting all his work in one permanent place, in light of its importance.

WILLIAM LOBB AND CHILOÉ

The island of Chiloé is the largest of a chain of islands off the coast of Chile. It is divided into two totally diverse sections by the central coastal range. The western section is open and largely uninhabited. It has temperate rainforest and beaches with dune habitat. The other half of the island is arable and heavily settled. The rainforest is made up of evergreen trees and shrubs, with myrtle, larch, and many others. These features make it a magnet for plant collectors.

Chiloé was isolated for much of Chile's history. It was cut off from the rest of the country by thick forest and hostile native peoples. Supplies came by ship at very rare intervals, rather like Alta California under the Spanish. The inhabitants survived by subsistence farming and fishing.

Darwin stopped there while on the *Beagle*. William Lobb found his way to the island in 1841, looking for the source of the monkey puzzle tree, *Araucaria araucana* (Figure 42). (See Chapter 13 for more information about William Lobb.) Lobb managed to find the *Araucaria* on ridges below the peaks in the southern Andes. Using his gun, he shot hundreds of cones from the trees, getting his native helpers to pick them up and help him pack them. They were happy to do this because the seeds were a source of food for them. More than 3,000 seeds were gathered in this manner.

The tree was an instant success in England when Veitch released the saplings. It turned out that Lobb was not the first collector to send back these seeds. Archibald Menzies had aced him, but he was the first one to send seed in quantity. Veitch cornered the lion's share of the market.[91]

Lobb's travels in South America took him right across the continent from the Atlantic to the Pacific coasts. He had been in Brazil and Argentina, had crossed the Andes, and had explored down the coast of Chile, all the time collecting plants with great horticultural potential. Lobb found *Escallonia*; *Begonia coccinea*, the ancestor of almost all future hybrid begonias; *Passiflora*; and *Tropaeolum*, as well as others listed in Table 3.

For a brief time, he rested back at home in Cornwall but was very anxious to go away again. This time he never returned.

SIR ROBERT SCHOMBURGK

The Horticultural Society drew on the talents and skill of scientists in related fields who happened to go to places where new plants could be found. Robert

Hermann Schomburgk (1804–1865) was a Ph.D. from Königsberg in East Prussia at a time when German science was becoming dominant. Nowhere else was the training so thorough and pragmatic. He and his brother both developed careers in England and were both knighted by Queen Victoria. Their careers bore a strong resemblance to Mueller's career in Australia.

The Royal Geographic Society sent Schomburgk to British Guiana where he laid down the fundamentals of the boundaries with Venezuela. Eventually he was appointed British Consul in Santo Domingo.

Schomburgk found many orchids in a journey to Siam (now Thailand), and there is a genus named for him. The plants resemble *Laelia*. He contributed numerous articles to the horticultural literature of his day.

CHAPTER 13

A PASSION FOR ORCHIDS

THE LURE OF EXOTIC ORCHIDS

Few other plants suggest tropical mystery and exoticism as much as the orchid. In some ways orchids epitomize the extraordinary change that was taking place in British horticulture. Important careers were founded on the orchid.

New terrestrial orchids had come from North America early. These presented very little difficulty to growers but were not showy or romantic. Many terrestrial orchids were found in England and northern Europe, but once again they were simple meadow plants.

One can view tropical, mainly epiphytic, orchids as a useful proxy for the forthcoming sweeping changes. Tropical orchids were hard to grow. The sexual characteristics were hidden and mysterious. Seed did not set easily. A person who could grow orchids was ready for almost any challenge. Commercial nurseries also expanded and grew as a result of the competition over orchids.

If one looks at the arc of the Veitch family's nursery as an example, it can be seen to follow the development and maturation of the orchid craze (Figure 43). Starting in about 1779, old John Veitch had a very respectable but conventional business in Devonshire, selling trees and shrubs all over the country. His youngest son James took over from him in the mid-1820s. By single-mindedly concentrating on exotic plants, principally orchids, collected by his own employees, James increased the business out of all recognition and took it into another level altogether.

There were other important nurseries, such that of Joachim Conrad Loddiges, a German (or possibly Dutchman) who moved to the village of Hackney, just north of London, and began his firm in 1771. He was succeeded by his son George and grandson Conrad. Loddiges quickly understood the significance of the nobility in building a business and had extraordinary stock available very early. Loddiges introduced the rhododendron. He sold orchids right at the beginning. The nursery was the acknowledged leader for more than 30 years.

Loddiges may have been the leader, but Veitch gave him a good run for his money. Before too long, the little provincial firm from Exeter invaded London, very close to Loddiges' own turf. By the late 1860s, Loddiges closed for good. Veitch went on for almost another 60 years.

Lee and Kennedy's Vineyard Nursery at Hammersmith was another major firm. They, too, had an extensive stock of rare and exotic plants and understood how to cater to the nobility.

John Lindley (1799–1865), a lower-middle-class boy, son of a nurseryman, rose to become professor of botany at London University by the age of 31 and secretary of the Horticultural Society because of his skill in the taxonomy of orchids. He wrote the monumental treatise *Genera and Species of Orchidaceous Plants*, issued over a period of 10 years from 1830 to 1840. The work is still relevant.[111]

Collectors in India soon began to send back orchids, but they were not the first to collect tropical orchids. Sir Charles Wager grew the first one in Britain in a hothouse in 1731. He revived a dried specimen sent to Peter Collinson.[41] This was a *Bletia* from the Bahamas, named for Monsieur Louis Blet, a pharmacist and amateur botanist in Algeciras.

Dr. William Houstoun sent orchids to Philip Miller from Jamaica. (Colonel James Oglethorpe of Georgia had hoped to make use of Houstoun's services, too.) The West Indies were the first reliable source of tropical orchids in England, before the floras of India and other hot countries were discovered.

The Dutch grew the first tropical orchid in Europe, a *Brassavola nodosa* (originally known as *Epidendrum nodosum*).[38] A picture of it appeared in Peter Hermann's *Paradisus Batavus* in 1698. Tropical orchids were not imported into England until 1731. Peter Collinson received *Bletia verecunda* from the Bahamas, though no one knows who sent it to him.

In 1782 *Epidendrum cochleatum* was the first tropical, epiphytic orchid to bloom at the Royal Botanic Gardens. A year later, *E. fragrans* also bloomed at Kew. The principal interest at Kew was botanical, not horticultural. Francis Masson sent dormant orchid tubers back to Kew from the Cape in 1787.

The first edition of *Hortus Kewensis* recorded 24 types of orchid, largely British and European. There were two exotic species: *Bletia tuberosa* and *Epidendrum retusum* (now *Rhynchostylis retusa*).[132]

In the third volume of the 1789 *Hortus Kewensis*, William Aiton noted that Kew had more than 40 species of orchids. The first tropical orchid to bloom in Kew's hothouse was *Phaius tankervilliae.*[132] Dr. John Fothergill had somehow obtained it from China and had given it to Kew.

When Aiton's son, William Townsend Aiton, reissued the *Hortus* in 1810, the list of orchids had grown to 48 genera and 115 species. More than half of the species were exotic varieties.[120]

Apart from the academic institutions, only a few enthusiasts paid much attention to orchids for a long time. William Cattley of Barnet, a village north of London, was one of them. He was commemorated in the genus *Cattleya.* The story about *Cattleya* is a classic one of chance favoring the prepared mind.

An English collector in Brazil, William Swainson, sent a large consignment of plants to Cattley in 1818. He had found most of them in the Organ Mountains near Rio de Janeiro. Swainson was interested in different plants and had used what he thought was an unimportant one as packing to protect the others.

Cattley hated to see any plant go to waste. He took care of the dried-out packing plants and nursed them back to life. His conscientiousness was rewarded. First some shoots appeared. A few months later they produced astonishing flowers.

Lindley was invited to see this plant. It impressed him deeply. He named the genus *Cattleya* to honor Cattley. The labellum was unusually prominent, and hence he chose the specific name *labiata*.

Another English enthusiast, James Bateman (1811–1897), botanist and orchid lover, used the adventurous George Ure Skinner (1804–1867) to collect orchids for him in Central and South America. Skinner found more than 100 species for Bateman. The latter called one of them *Cattleya skinneri*.

From 1837 to 1843, James Bateman brought out successive portions of his *The Orchidaceae of Mexico and Guatemala*.[169] These are exceptionally large and beautiful books. Only 125 copies were ever printed. Many years later Bateman produced *A Monograph of Odontoglossum*.[170]

The Horticultural Society started its flower shows in 1829. When *Cattleya labiata* was displayed at the Horticultural Society flower show, the public was collectively overwhelmed. There was a quantum leap in the public's attention to orchids. They became the rage.

Wealthy Frenchmen and Germans joined in this pastime, growing the same tropical epiphytic orchids from Asia, Australia, and the Americas. Many of the best-known collectors were German or Belgian, beside Scots and other Britons. One man stood out because he had previously lost his left arm and hand in an industrial accident. Benedict Roezel inspired great awe among the Amazonian tribesmen with his sinister hook.[41]

There was a rush to find more and more of the *Cattleyas*, but the original wild variety was not found again for 50 years. The places in which it had grown were all destroyed by indiscriminate logging for farm land. We bemoan the loss of wilderness in modern times, but it was also rife 150 years ago. The natural world seemed so vast that nothing could reduce its vitality or damage its ability to recreate itself.

The commercial collectors were responsible for a great deal of destruction themselves. There were two reasons. One was that they had to collect vast numbers of individual plants since reproduction back in Europe was still a very uncertain process. Many years later, Veitch's star collector, Ernest H. Wilson, dug up more than 10,000 *Lilium regale* bulbs in China. Although he packed them carefully in clay, most of them rotted before they reached their destination.[125]

The other reason was that once the collectors had what they needed, they did not want the competition to find the same plants. They uprooted everything they could, similar to clear-cutting. James Veitch wrote to William Hooker with the reverse problem. He hoped his collector William Lobb was not following in the footsteps of a Continental competitor, thereby finding plants that had already been discovered and sent home.

THE DUKE OF DEVONSHIRE

One man caused acceleration in this field though change was to be expected simply by the force of events. For much of his life, William George Spencer Cavendish, sixth Duke of Devonshire (1790–1858), was quite lukewarm about plants, simply going through the motions as befitted a man of his class. He never married, feeling slightly cut off from active social life by increasing deafness, and was known as the "bachelor duke." Then, in middle life, he fell prey to "orchidmania," becoming besotted with *Oncidium papilio* at a Horticultural Society flower show in 1833. Here at last was a gratifying pastime.

From then on nothing could stop him. The duke's head gardener, Joseph Paxton, shared his enthusiasm, designing and building the great hothouse at Chatsworth that started his phenomenal career (Figure 44). The duke had spotted Paxton at the Horticultural Society and asked him to work at Chatsworth.

Paxton suggested that the duke could accumulate many more varieties by sending someone directly to India. He recommended one of the estate's gardeners, a young man named John Gibson. This was the young man who was in such a funk when he arrived in Calcutta and whom Nathaniel Wallich soothed so successfully. Gibson vindicated this trust by finding more than 50 species of orchids in his travels. With these and all his other wealth of sources, the duke's collection became the largest in Britain.

ORCHIDS IN COMMERCE

The duke's passion indirectly spurred the nurseries to outdo themselves. His social prominence meant that anyone with a desire to shine in society had to emulate him. The middle classes also trotted close behind, trying to climb the social ladder. In response to these stimuli, certain nurseries undertook the risk and expense of sending their own collectors to the endemic sites instead of just waiting for chance to bring one or two new plants from time to time.

The reward was worth it. If a firm could capture the market for a desirable new orchid, it stood to make a lot of money, but if the gamble failed, the loss was acute. James Veitch participated in one syndicate in which everything was lost. This expedition was to the Pacific Northwest and Canada. Two of the Duke of Devonshire's other gardeners were killed during the search. The duke was horrified and never sent anyone out exploring again.

In spite of James Veitch's very cautious temperament, he was not deterred from continuing to send explorers at his own expense (Table 10). The possibility of making a huge fortune was just too exciting. His colleagues laughed at him behind his back but ceased laughing when he succeeded.

Table 10. Veitch's Travelers[1]

Traveler	Dates
William Lobb	1840–1857
Thomas Lobb	1843–1860
Richard Pearce	1859–1866
John Gould Veitch	1860–1870
David Bowman	1866
Henry Hutton	1866–1868
Carl Kramer	1867–1868
Gottlieb Zahn	1869–1870
George Downton	1870–1878
A. R. Endres	1871–1873
Gustave Wallis	1872–1874
Walter Davis	1873–1876
P. C. M. Veitch	1875–1878
Guillaume Kalbreyer	1876–1881
Christopher Mudd	1877
F. W. Burbidge	1877–1878
Charles Maries	1877–1879
Charles Curtis	1878–1884
Davis Burke	1881–1897
James H. Veitch	1891–1893
E. H. Wilson	1899–1905

1. Adapted from J. H. Veitch, 1906, *Hortus Veitchii*, privately printed.

In the search for more and more exotic orchids, the organized commercial hunt for plants entered a new phase. Individual nurseries behaved like states or large institutions. They paid for transport, porters, food, accommodation, supplies of all sorts, freight, and those other expenses that could not be anticipated but that often arose.

The owner of the nursery had to find a skilled person in robust health who was willing to face the most dreadful conditions for months on end, who knew what he was seeking, and who could be sensible and discreet. Hard-headed businessmen only paid these very superior men as much as they thought they could get away with and not a penny more.

Management had to trust the employee through months of silence and assume that all was going well. The employee similarly relied on the firm for all support. If misunderstanding developed, it festered badly. The correspondence between George Forrest and his employer A. K. Bulley early in the next century is a case in point.[171]

Forrest was laboring under terrible conditions in China with inadequate resources. He resented Bulley's chiding him over some minor miscalculation of his expenses while failing to send him the equipment he needed. Bulley had absolutely no idea just how miserable and uncomfortable Forrest really was. The same was true of other employers, such as James Veitch.

THE BROTHERS LOBB

James Veitch, John Veitch's youngest son and in charge of the business, first employed Thomas Lobb (1817–1894) in the nursery when the lad was 13 years old. He and his brother William (1809–1864) had grown up at Carclew in Cornwall, where their father was gamekeeper. The owner of Carclew, Sir Charles Lemon, an active botanist, saw that the boys were interested in natural history and encouraged them. They were also allowed to help take care of the plants in the hothouses.

William went to work as a gardener for the Williams family at Scorrier House near Falmouth, Cornwall. By 1840, William was 31 years old. In his free time he often met sea captains in the local pub and listened entranced to tales of faraway places. William had collected plants in the surrounding countryside and learned a lot about botany on his own. He longed to travel and see the beauties of the tropics. Thomas admired him greatly.

It may not be completely irrelevant that the boys grew up in Cornwall, the most temperate part of England. Plants that would not survive in the north can flourish in the mild climate. This might have given the Lobbs a more optimistic attitude toward growing exotic discoveries.

As James Veitch began to think about sending his own collectors, Thomas told him about William. This would be a very good choice. William was sensible and understood what was at stake. His knowledge of plants was adequate.

James dispatched him to London for a crash course and felt William would be able to pick much of the rest up himself. Ever cautious, James corresponded with Sir William Hooker at Kew, making sure he was doing things correctly. The directors of Kew had had the most experience with the care and feeding of plant collectors hitherto.

William collected very broadly, with a true horticulturist's eye for what would do well in a garden. Later Veitch sent Thomas to collect just orchids in the Malay Archipelago, the Philippines, and India. Thomas was said hardly ever to speak at all, and if he did, it was only about orchids. If Thomas said something was "very pretty," the men in the greenhouses knew it would be outstanding and gave it extra care.

Both Thomas and William knew how to cover their tracks. When Joseph Hooker met Thomas in India, he recognized that this was an upstanding young man but thought he was a real dullard who knew nothing at all about his business. He wrote to his father that the Veitches were wasting their money in employing him. How wrong could anyone be?

Thomas was making sure no one would find his sources. He feigned indifference and ignorance but, at the first opportunity, went his own way. The showy accoutrements of the great plant hunter were not for him. A few years later Joseph Hooker was quite indignant about the discrepancy between the government's and the commercial nurseries' funding for plant collection and the fact that firms like Veitch stood to make a lot of money from the discoveries. He realized belatedly how much Thomas had accomplished in his homespun fashion.

William's first efforts were very successful. Some of his plants still grow in English gardens. Many of them can only survive in hothouses and so have not been cultivated widely recently. Perhaps 10% were hardy.

By the end of William's final trip, in California, Veitch lost contact with him and became angry. All William's great acquisitions and skill at collecting suddenly seemed to have been forgotten. Veitch was horrified when William tried to bypass him and sell specimens directly to Hooker at Kew in 1858 without his knowledge or consent. The first Veitch knew about it was when Hooker wrote to him in some embarrassment. James sent his son to buy the specimens back from Hooker.

What had started out so strikingly and well ended badly for both brothers. William contracted an illness in America, possibly syphilis, and died alone and forgotten in San Francisco of "paralysis." (The tertiary stage of untreated syphilis was general paresis of the insane.) He was initially buried in the Lone Mountain Cemetery in San Francisco as an unknown stranger. This cemetery, one of the earliest to be elegantly landscaped in California, was created less than five years after the gold rush.

In 1940, Alice Eastwood, the remarkable woman who was in charge of the herbarium at the California Academy of Sciences, got the academy to move William's headstone to a crypt at Cypress Lawn Cemetery.

Thomas injured one of his legs irreparably in Indonesia. He managed to limp back to England and lived with the unhealed wound for a long time, but eventually had to have the leg amputated. It was done under anesthesia on the kitchen table at his sister's cottage in Cornwall.

James Veitch was succeeded by his son, also James. The son honored his father's contracts with the men in the field. The younger man was brilliant, but irascible and almost uncontrollable if anybody crossed him.

He, too, had a son. This young man was very promising but showed no signs of settling down to the grind of the family business. John Gould Veitch was handsome and very well-educated but needed time to get romantic and glamorous visions out of his system.

His father sent him to Japan and Southeast Asia, the first member of the family to explore for the firm. He took back excellent conifers that sold very well, but Spongberg says his most significant introduction was the "Boston" ivy, *Parthenocissus tricuspidata*.[62]

In 1869 James asked Thomas Lobb to come up to London for a meeting. John Gould Veitch was the obvious person to take over orchid collecting for the family firm, but he had tuberculosis and could not live much longer. He would die in 1870, leaving behind a widow and two-year-old son. James thought that, with John Gould clearly dying, he needed to look for someone else. Perhaps Thomas might be persuaded to collect orchids again. Faced with this request, Thomas probably felt very aggrieved at his own and his brother's neglect.

Harsh words were exchanged. Thomas took a late train back to the country. James Veitch died later the same night, maybe because his blood pressure shot up during the altercation. He was 54. James' other son Harry never referred to

OTHER BROTHERS IN PLANT COLLECTING

THE DRUMMONDS

Another set of brothers from Britain became plant collectors during much the same period but without the benefit of fixed employment. Thomas and James Drummond went in widely divergent directions. Thomas covered the Western Hemisphere, dying in Cuba in 1835. James went south, to western Australia (see Chapter 11). The genus *Drummondita* was named for James, but no genus of this name can be found today. The species names commemorate Thomas.

The first time Thomas Drummond visited North America he worked in Canada. On his next trip he collected on the eastern slopes of the Rocky Mountains and along the Columbia River.

Thomas later went to Texas, loosely attached to a trading expedition protected by the U.S. military. McKelvey points out that the dangers of the unexplored wilderness were so great that a prudent collector made sure he had some protection.[82]

Thomas survived an attack of cholera in Texas and traveled back east and eventually to Cuba. He died in Cuba, though it is not clear what happened to him. Presumably it was from disease, only too frequent at that period. The British consul sent his meager effects back to England but certainly would have commented if Thomas had died a violent death.

Sir William Hooker inserted a brief notice about Thomas' fate into one of Kew's journals. James, Thomas' brother, was so far away in western Australia that the first he knew about this loss was when the journal reached him about a year later. He wrote a poignant letter to Hooker. ". . . I learn by a newspaper that a Mr. Drummond died in one of the West India islands. . . I should be much obliged by your sending me an account of my brother which you allude to."[140]

Thomas Drummond introduced several valuable garden plants into Europe. *Phlox drummondii* and *Coreopsis drummondii*, now *C. basalis*, commemorate his name.

THE CUNNINGHAM BROTHERS

Allan and Richard Cunningham constitute a third set of brothers, after the Lobbs and the Drummonds, who both became plant collectors in remote places (see Chapter 11). They, too, were of Scottish descent, although they were born and brought up in London. Unlike the other sets of brothers, they had fairly dissimilar temperaments. Richard was diligent and sincere but rather irresponsible. He also had very little sense of direction. This led to his death in very unpleasant circumstances while exploring in Australia.

Allan was mature and exceedingly reliable. He died of tuberculosis, the scourge of even apparently healthy adults in the nineteenth century.

Thomas Lobb again nor had any contact with him. He considered him to have murdered his father.

Harry took charge of the firm after his father died. He sent more collectors to Asia and South America. Most of them were pretty conscientious, and some were brilliant, but the Lobb brothers had been pioneers, not knowing what to expect and constantly facing new difficulties without anyone or anything to help them in the wilderness.

In 1917 Harry shut the business down rather than sell it to someone who might sully his family's great reputation. He had no children and no obvious heirs. The relatives still in Exeter were not suited to his kind of commerce.

COLLECTING THE WILD PLANTS

Even tough, well-trained professionals had difficulties in collecting plants in faraway places. Philip Short assembled excerpts from collectors' diaries in the recently published *In Pursuit of Plants.*[86] To say that these excerpts make one's hair stand on end is an understatement. Death was not uncommon. Serious maiming such as Thomas Lobb's losing a leg also occurred quite frequently. The long-term consequences of malaria, dysentery, and other tropical diseases cannot be underestimated. In some case, the collector's mental equilibrium was disturbed, with very good reason. When an inexperienced novice like John Gibson was sent out, things were naturally far worse (see Chapter 8).

EFFECTS OF ORCHID COLLECTING

The results of nurseries collecting the plants for themselves were so rewarding that the practice continued for almost a century until World War I made great estates too expensive to maintain. With fewer people able to afford the rare and the exotic, nurseries refocused their efforts in the mainstream. Only the universities and syndicates continued exploration. Nurseries obtained new material from these trips.

Table 11 gives only a fraction of the enormous number of orchids imported into Britain during the nineteenth century. Even though they are not totally comprehensive, the sketchbooks of Thomas Day (1824–1888), a wealthy London wine merchant whose hobby was collecting, drawing, and painting orchids, show the rate at which imports were flowing.[172] Day's work includes 53 notebooks and more than 2,800 pictures, which the Royal Botanic Gardens at Kew have preserved since his death.

There was nothing sentimental about Day's passion for the plants. He bought and sold the orchids at auctions quite unromantically. Once he had finished with them, he sold many of them off to pay for the next installment. He patronized Stevens' auction rooms, but there were other orchid auctioneers.

Part of the impetus in orchid growing came from the very wealthy men who ran the Horticultural Society. They could afford the expensive hothouses needed

Table 11. Partial Timeline of Orchid Introductions in England, Principally in the Nineteenth Century[1]

Date	Plant	Collector	Place Found	Recipient
1698	*Epidendrum nodulus*		Curaçao	Fagel, Caspar
1731	*Bletia verecunda*		Bahamas	Collinson /Wager
1737	*Cypripedium pubescens*		N. America	Miller, Philip
1737	*C. spectabile*		Canada	Miller, Philip
1760	*Epidendrum rigidum*		West Indies	RBG Kew
1787	*Encyclia cochleata*			RBG Kew
1810	*Cattleya loddigesii*	Woodforde	Brazil, south	Liverpool Botanic
1818	*C. labiata*	Swainson, William	Brazil	Cattley, William
1821	*Cymbidium iridioides*	Wallich, Nathaniel	Nepal	
1826	*Cattleya intermedia*			Glasgow Botanical
1826	*Oncidium papilio*[2]			
1827	*Cattleya guttata*		Brazil, Rio de Janeiro	London
1830	*C. aurantiaca*	Skinner, George Ure	Guatemala	
1832	*Laelia albida*	Karwinski, Baron Wilhelm	Mexico, Oaxaca	
1832	*Oncidium crispum*	Brazil		
1835	*Odontoglossum bictoniense*	Skinner, George Ure	Guatemala	Bateman, John
1836	*Cattleya mossiae*	Skinner, George Ure	Venezuela	
1836	*C. skinneri*	Skinner, George Ure	Guatemala	
1837	*Cymbidium devonianum*	Gibson, John	India, NE	Duke of Devon-shire
1839	*Cattleya aclandiae*	Skinner, George Ure	Brazil, Bahia	
1839	*C. walkeriana*	Skinner, George Ure	Brazil, Bahia	
1840	*C. granulosa*	Skinner, George Ure	Brazil, Bahia	

(continued)

Table 11. Partial Timeline of Orchid Introductions in England, Principally in the Nineteenth Century[1] *(continued)*

Date	Plant	Collector	Place Found	Recipient
1840	*Cymbidium madidum*		Australia	
ca. 1845	*Aerides fieldingii*	Lobb, Thomas	Philippines	Veitch
ca. 1845	*Bulbophyllum lobbii*	Lobb, Thomas	Philippines	Veitch
ca. 1845	*Dendrobium albosanguineum*	Lobb, Thomas	Burma	Veitch
ca. 1845	*Paphiopedilum* spp.	Lobb, Thomas	Philippines	Veitch
ca. 1845	*Phalaenopsis ×intermedia*	Lobb, Thomas	Philippines? (natural hybrid)	Veitch
ca. 1845	*Vanda coerulea*	Lobb, Thomas	Assam (India)	Veitch
ca. 1848	*Cattleya dowiana*	Warscewicz, Jozef	Costa Rica	
1848/1849	*C. warscewiczii*	Warscewicz, Jozef	Colombia	
ca. 1849	*Dendrobium hookerianum*	Hooker, Joseph D.	Sikkim (India)	RBG Kew
1850	*Miltonia spectabilis*		Brazil, Organ Mountains	
1855	*M. regnellii*		Brazil, Minas Gerais	
1856	*Laelia jongheana*	Libon, Joseph	Brazil, mountains	
ca. 1860	*Cattleya dowiana* var. *aurea*	Wallis, Gustav	South America	Veitch
ca. 1860	*C. eldorado* var. *splendens*	Wallis, Gustav	South America	Veitch
ca. 1860	*Dracula wallisii*	Wallis, Gustav	South America	Veitch
ca. 1860	*Odontoglossum wallisii*	Wallis, Gustav	South America	Veitch
ca. 1860	*Oerstedella wallisii*	Wallis, Gustav	South America	Veitch
1863	*Cymbidium parishii*	Parish, Charles	Burma	
1866	*Odontoglossum crispum*	Bowman, David	Brazil	Veitch
1868	*Vanda insignis*[3]	Hutton, Henry	Indonesia	Veitch
1869?	*Cymbidium huttonii*	Hutton, Henry	Java	Veitch
1872	*Miltoniopsis vexillaria*		Colombia	
n.d.	*Cymbidium lowianum*	Low, Hugh	Borneo	
n.d.	*C. tigrinum*	Parish, Charles	Burma	
n.d.	*Dendrobium bellatulum*	Henry, Augustine	China, Yunnan	
n.d.	*D. brymerianum*		SE Asia, montane forest	
n.d.	*D. pendulum*	Parish, Charles	Burma	
n.d.	*Dimorphorchis lowii*	Low, Hugh	Borneo	
n.d.	*Paphiopedilum concolor*	Parish, Charles	Burma	
n.d.	*Paphiopedilum lowii*	Low, Hugh	Borneo	

n.d., no date known.

1. Compiled from Alice M. Coats, 1969, *The Plant Hunters,* New York: McGraw-Hill; Wilfrid Blunt, 1978, *In for a Penny: A Prospect of Kew Gardens,* London: Hamish Hamilton; Sue Shephard, 2003, *Seeds of Fortune: A Gardening Dynasty,* London: Bloomsbury Publishing; Merle A. Reinikka, 1972 (reissued 1995), *A History of the Orchid,* Portland, Oreg.: Timber Press.

2. Source of the Duke of Devonshire's "epiphany" in 1833.

3. Very rare.

to grow them and essentially competed with one another. Growing orchids was rather like playing polo or keeping a yacht. If you had to enquire how much it cost, it was clear you could not afford it, to paraphrase the wonderful maxim of J. P. Morgan.

In spite of the narrow elite involved, this internecine competition ultimately benefited science. It stimulated exploration and scientific study, and hastened many useful observations.

An example of this force is shown by the following story. The first professional nurseryman to open a business in California, in 1849, understood the effects of beneficent competition.[173] Colonel J. L. Warren lobbied the state legislature, nagging them to create state agricultural and horticultural boards. The boards offered prizes for specific projects.

Warren started the first California state fairs in which the public could compete for prizes in many categories. The fairs still take place each year and are still stimulating development. Horticulture is a major economic engine in California.

CARING FOR ORCHIDS

Caring for the orchids was a very hit-or-miss affair for many years. The nurserymen did somewhat better than amateurs. Thomas Lobb took a rare *Vanda* back to James Veitch, together with *Dendrobium*, *Coelogyne*, and *Aerides*. They all thrived because of Veitch's skill.

James Veitch was exceptionally patient and careful and understood the importance of controlling the plants' environment. Joseph Paxton stumbled onto the fact that cooling the plants and ventilating the hothouses often worked better than keeping them at uniformly tropical temperatures. Eventually hothouse experts realized they needed three separate rooms with gradients of temperature to succeed: cool, intermediate, and tropical.

Veitch harbored immense ambition and wanted to win prizes and recognition to further the growth of his business. As his collectors sent back incredible plants, he knew that he had made the right decision, that producing fabulous orchids was the way to go. Veitch created sensation after sensation with his blooms.

Part of his success lay in having enticed the leading nursery propagator in the region away from the rival firm of Lucombe and Pince in Exeter. John Dominy was Veitch's not so secret weapon. It was a sad day when Dominy left Veitch to work for a wealthy family in the neighborhood. Many years later Veitch was able to entice him back.

Dominy also created the first intentional hybrid orchid in the world for the Veitches, working slowly and patiently in the greenhouse. This was *Calanthe* ×*dominii*, the result of crossing *C. furcata* with *C. masuca*. It was named in his honor by John Lindley. Dominy taught John Seden his secrets, and Seden continued to produce hybrids after Dominy retired. Both men were awarded prizes by the Royal Horticultural Society, as it was later called, and eventually won the society's coveted Victoria Medal. A third man, John Heal, carried on their tradition as they grew older.

James Veitch's son, James Jr., was another valuable assistant. James Jr. was passionate about orchids. His father had ensured that he was very well trained in horticulture. It was while he was an apprentice in London at the Rollison Nursery that young James' taste for orchids was formed.

The young man worked in the family greenhouses side by side with Dominy. Before the orchids took up so much of their attention, they had groomed rhododendron, dahlias, fuchsia, and camellias to win prizes at the major flower shows.

ORCHID SCIENCE

Kew kept its eyes on science. Sir William Hooker reminded everyone that surface beauty was not the sole criterion for collecting an orchid. He was also anxious to offer orchids that ordinary people could afford. Loddiges sold him a special set of plants for this purpose.

Hooker and Darwin collaborated in important experiments on the pollination of orchids that Darwin was conducting at Down House. In his inimitable fashion Darwin sat and observed the unusual way in which *Catasetum* released its pollen in a catapult action. He described this phenomenon very carefully in a letter to Thomas Huxley, another leading scientist of the day. The latter is reported to have replied to Darwin, "Do you really think I can believe all that?" Unabashed, Darwin proceeded to write *The Various Contrivances by which British and Foreign Orchids are Fertilized by Insects* and published it in 1862.[174] It remains the standard text on the subject.

Another feature of the Kew collection was the Australian specimens. Captain Bligh sent some of them, as did Joseph Banks. Bligh contributed *Dendrobium linguiforme*, Banks *D. speciosum*. The Flinders expedition with Robert Brown on board found about 100 new species during their survey. Allan Cunningham and George Caley both collected many orchids in Australia.

Cunningham had previously collected in Brazil, where he found epiphytic orchids. Unfortunately, the garden staff did not take good care of the orchids obtained at such enormous cost. A few years later an expert gardener found them in a deplorable state. Many of them had died, probably unnecessarily.

Foreign scholars were welcome to visit the Royal Botanic Gardens and work on the collection of orchids. Ernst Pfitzer came from Heidelberg in 1880. Artists and photographers used them frequently. In a darker moment, the suffragettes saw the fragile orchids as a vulnerable target and smashed them up in 1912 as they struggled for women to be allowed to vote. One could call them "prototerrorists."

Kew tended to stand aloof from the frenzy of hybridization that began soon after the plants could be grown reliably. The private nurseries persevered with this. By 1859, *Cattleya* hybrids were being shown. There was a *Laelia* hybrid by 1864 and a *Paphiopedilum* (*Cypripedium*) by 1869. Lindley threw up his hands in mock horror when confronted with these displays and exclaimed, "Do you want to send all the taxonomists mad?"[41]

Another towering figure in mid-century orchid study was Heinrich Gustav Reichenbach (1823–1889), professor of botany and director of the botanical garden in Hamburg, Germany. He had cooperated pleasantly enough with Lindley and other English scholars, but somehow Robert Rolfe, a self-taught orchidologist appointed by Kew, got under his skin.

For the 20 years after Lindley died in 1865, Reichenbach was the undisputed leader in orchidology. When Thomas Day received an orchid without a name, he sent it to Reichenbach to be identified and named.[172] Reichenbach often used Day's drawings to illustrate his scientific reports.

Kew assumed Reichenbach would bequeath his monumental collection to them when he died. He had made some remarks to that effect. The Hookers were shocked when he left everything to the Imperial Hof Museum (now the Museum of Natural History) in Vienna. Some thought this was done to spite Rolfe. The herbarium had to remain sealed for 25 years "to avoid the destruction of the specimens," but it also prevented Rolfe from getting a good look at them.[41]

Robert Rolfe (1855–1921), curator of the orchid herbarium at Kew and editor of the *Orchid Review*, produced some very fine hybrids at Kew in the early 1900s, among them *Disa* ×*kewensis* and *Epidendrum kewense*. The *Kew Bulletin* for 1891 showed that the garden had 766 orchids. In 1896, 1,800 species covering 200 genera were noted in the garden's *Handlist of Orchids*. Rolfe held strong opinions and did not hesitate to express them.[41]

ORCHID GROWING IN THE UNITED STATES

In the United States, orchid lovers rapidly became as frenzied as the Europeans. The Massachusetts Horticultural Society held regular meetings. In June 1837, Marshall Wilder exhibited specimens of *Oncidium flexuosum*.

A Boston family, the Bootts, had one son, James, living in England. By 1838, James Boott had sent his brother John a fine collection of orchids. John may have had a few plants back in 1829. John pottered about in his small Bowdoin Square greenhouse after work in the evenings, tending his *Cattleyas*, *Dendrobiums*, and *Epidendrums*. When John died in 1845, he left his orchids to John Amory Lowell in Roxbury.

Ultimately, another wealthy Bostonian, Edward Rand, inherited the survivors of these plants. After showing them at meetings, he bequeathed the whole collection to Harvard. They ended up in the Cambridge Botanic Garden greenhouse. Rand had augmented the collection by importing more plants directly from Messrs. Hugh Low in Clapton, near London.

Orchid growing spread rapidly up and down the eastern seaboard: Isaac Buchanan and Thomas Hogg of New York, Cornelius van Voorst in Jersey City, Erastus Corning in Albany, and other successful men in Flushing, Hoboken, Newark, and Tarrytown all participated. The plants still primarily came from England.

Modern leadership in the orchid world had long been concentrated in Europe, with Lindley, Rolfe, Reichenbach, and later Rudolf Schlechter dominating the field of scholarship. By 1925 all these men had died, and Oakes Ames (1874–1950) stepped into this void as the leader.[41] He was the natural heir.

Ames was a descendant of Amyas Ames of Boston, a man who had made his fortune in ironmongery two centuries before. The great wealth Oakes Ames inherited allowed him to become a professor at Harvard and travel extensively, collecting orchids without worrying about money. He spent years in the Philippines and the rest of his life classifying the orchids and writing about them.

He also admired the ancient wisdom and plant lore of forest shamans. His lab at Harvard was filled with native artifacts from his travels. One of the students most influenced by this aspect of Ames' work was Richard Evans Schultes. Ames fostered students' enthusiasm and on occasion even paid them from his own pocket, carefully disguising it as a university stipend.

Ames had a competitive streak and wanted to be sure that Germany did not resume its role as sole arbiter in taxonomy. He was helped in his quest by Charles Lankester, an expatriate Englishman in Costa Rica who sent him valuable collections of Costa Rican orchids. Ames' grandson was the late George Plimpton, well-known writer and sports figure.

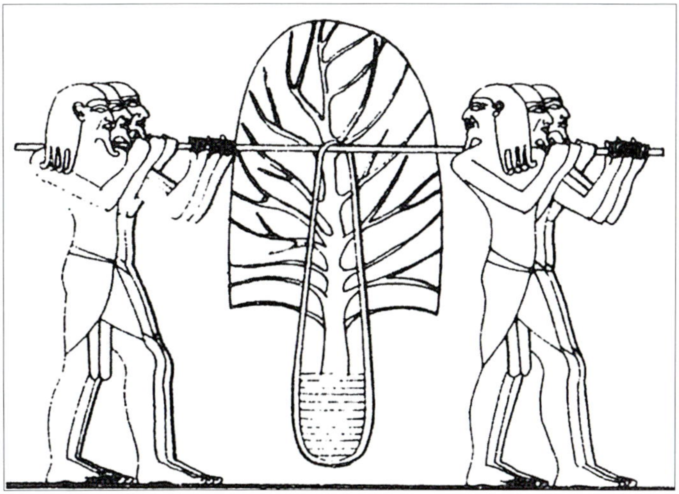

Figure 1. Line drawing from Queen Hatshepshut's tomb showing her plant collectors returning from the Land of Punt with frankincense trees. Reproduced by courtesy of Jules Janick, Purdue University.

Figure 2. *Helleborus niger,* the Christmas rose, introduced into Britain in Roman times. Reproduced by permission of Ian Grant.

Figure 3. Roman feast, with guests about to be suffocated by rose petals (*The Roses of Heliogabalus* by Sir Lawrence Alam-Tadema). Reproduced with permission: Bridgeman Art Institute.

Figure 4. Modern version of *Rosa gallica,* known to the Greeks and Romans and earlier civilizations. Reproduced by permission of Roger Phillips.

Figure 5. *Erysimum cheiri*, wallflowers, growing in the time-honored fashion, in the crevices of an old wall. Reproduced by permission of the Photo Library, New York.

Figure 6. *Agapanthus.* Reproduced by permission of Tresco Abbey Gardens.

Figure 7. Tulips. Reproduced by permission of Ian Britton.

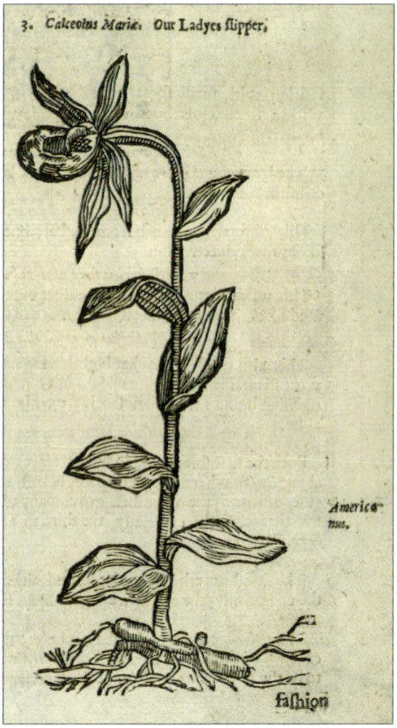

8

9

10

Figure 8. One of the first foreign orchids to be described in England, *Cypripedium* species. From John Parkinson's *Theatricum Botanicum*, 1640. Reproduced by permission of the Hunt Institute for Botanical Documentation.

Figure 9. Sir Hans Sloane. Reproduced by permission of the Hunt Institute for Botanical Documentation.

Figure 10. John Tradescant the Elder. Reproduced by permission of the Ashmolean Museum, Oxford.

Figure 11.
John Tradescant the Younger. Reproduced by permission of the Ashmolean Museum, Oxford.

Figure 12.
Sir Joseph Banks. Reproduced by permission of the Royal Botanic Gardens, Kew.

Figure 13. Sir William Hooker. Reproduced by permission of the Royal Botanic Gardens, Kew.

Figure 14. Sir Joseph D. Hooker. Reproduced by permission of the Royal Botanic Gardens, Kew.

15

16

17

Figure 15. Philip Miller. Reproduced by permission of the National Portrait Gallery, London.

Figure 16. John Bartram (posthumous rendering; no actual portrait of Bartram is known). Reproduced by permission of Bartram's Garden, Philadelphia.

Figure 17. Bernard de Jussieu. Reproduced by permission of the Hunt Institute for Botanical Documentation.

Figure 18. *Rosa laevigata*, the Cherokee rose. Reproduced by permission of Gerald F. Guala.

Figure 19. David Hosack. Reproduced by permission of the Hunt Institute for Botanical Documentation.

Figure 20. Asa Gray. Reproduced by permission of the Hunt Institute for Botanical Documentation.

Figure 21. Archibald Menzies. Reproduced by permission of the Hunt Institute for Botanical Documentation.

Figure 22.
David Douglas. Reproduced by permission of the Hunt Institute for Botanical Documentation.

Figure 23.
Thomas Nuttall. Reproduced by permission of the Hunt Institute for Botanical Documentation.

24

Figure 24.
Hedychium gardnerianum, Kahili ginger, collected by Nathaniel Wallich. Reproduced by permission of Special Collections, University of Hawaii.

Figure 25.
Meconopsis betonicifolia, blue poppy. Reproduced by permission of the Meconopsis Society.

Figure 26.
Rhododendron arboreum. Reproduced by permission of William Spohn and the American Rhododendron Society.

25

26

Figure 27. Dr. Emil V. Bretschneider. Reproduced with permission of T'oung pao, Krill NVD.

Figure 28. *Flowers in a Wooden Vase* by Jan Brueghel the Elder. Reproduced by permission of the Kunsthistorisches Museum, Vienna.

Figure 29.
Robert Fortune.
Reproduced by permission of the Royal Botanic Gardens, Kew.

Figure 30.
Francis Kingdon Ward.
Reproduced by permission of Oliver Tooley.

Figure 31. Dr. Augustine Henry. Reproduced by permission of the National Botanic Gardens, Glasnevin, Ireland.

Figure 32. *Hydrangea macrophylla* (Hortensia group), 'Generale Vicomtesse de Vibray.' Reproduced by permission of C. J. and D. M. van Gelderen, Esveld Nurseries, Boskoop, the Netherlands.

Figure 33. Captain William Dampier, by Thomas Murray, probably painted in 1698. Dampier is holding a copy of his first book. Reproduced by permission of the National Portrait Gallery, London.

Figure 34.
A specimen of *Willdampia formosa* (the desert pea) collected by William Dampier in 1699. Reproduced by permission of the Fielding-Druce Herbarium, Oxford University Herbaria.

Figure 35. A drawing of *Willdampia formosa* contemporary to Figure 34. Photograph by Alex George. Reproduced by permission.

Figure 36. A living specimen of *Willdampia formosa*. Photograph by Kevin Kenneally. Reproduced by permission of Alex George.

Figure 37.
William Colenso.
Reproduced by permission of the Alexander Turnbull Library, Wellington, New Zealand.

Figure 38.
Francis Masson. Photograph by Richard Todd-White. Reproduced by permission of the Linnean Society.

Figure 39.
Watercolor sketch of a *Stapelia* species by Francis Masson. Reproduced by permission of the Royal Botanic Gardens, Kew.

Figure 40.
Alexander von Humboldt. Reproduced by permission of the Hunt Institute for Botanical Documentation.

The blue orchid, Acacallis cyanea.

Figure 41.
Acacallis cyanea, blue orchid. Painting by Margaret Mee in Margaret Mee, 1988, *In Search of Flowers of the Amazon Forests.* Reproduced by permission of the Margaret Mee Trust at the Royal Botanic Gardens, Kew.

Figure 42. *Araucaria araucana*, the monkey puzzle tree. Reproduced by permission of Chris Booth.

FAMILY TREE

John Veitch.
Born 1752, at Jedburgh, Scotland.
Died 1839.
Rented land at Lower Budlake, near Killerton, near Exeter, in 1808.
Acquired additional land in 1810.
Moved to Mount Radford in 1832.
Succeeded in 1837 by

James Veitch.
Born 1792.
Died 1863.
Nurseryman at Mount Radford.
James Veitch and James Veitch junior, as James Veitch & Son, of Exeter, acquired the business and rented the land of Messrs. Knight & Perry, Chelsea, in 1853.
James Veitch remained at Mount Radford, Topsham Road, Exeter.
James Veitch junior moved to Chelsea in 1853.

James Veitch junior.
Born 1815.
Died 1869.
Nurseryman at Chelsea, Coombe Wood, &c., as James Veitch & Son.
Ceased all interest in the Exeter firm in 1864.

Robert Toswill Veitch.
Born 1823.
Died 1885.
Nurseryman at New North Road and High Street, Exeter, in 1864, on the death of James Veitch in 1863, as Robert Veitch & Son.

Peter C. M. Veitch.
Nurseryman at Exeter as Robert Veitch & Son.

John Gould Veitch.*
Born 1839.
Died 1870.

Harry James Veitch.*

Arthur Veitch.*
Born 1844.
Died 1880.

*Nurserymen at Chelsea, Coombe Wood and Langley, as James Veitch & Sons.

James H. Veitch. **John Gould Veitch.**
Nurserymen at Chelsea, Coombe Wood, Langley and Feltham, as James Veitch & Sons, Ltd.

Figure 43. Veitch family tree. From the *Hortus Veitchii*, 1906.

Figure 44. Joseph Paxton. Reproduced by permission of the Trustees of Chatsworth Estates.

Figure 45. Frank N. Meyer (right) converses with renowned plant explorer Dr. David Fairchild. Reproduced by permission of Frank N. Meyer Collection, Special Collections, National Agricultural Library.

CHAPTER 14

EXOTIC PLANTS ARRIVE IN NORTH AMERICA

THE CONTINUALLY expanding flow of ornamental plants from Asia began to replace American dominance in European gardens. When the discoveries in Central and South America, Australia, New Zealand, and South Africa are factored in, the mid-nineteenth century seems to be the turning point. Gardeners and horticulturists seethed with excitement. Magazines and journals abounded.

Landscape design reflected the new abundance. What we now think of as rather horrible ideas, such as carpet bedding, were really a reveling in plenty, the declaration of the possible. Designers and gardeners had not previously been able to assemble thousands of brilliantly colored and varied annual plants in the way they could now.

NEW PLANTS IN THE UNITED STATES

During this period, ornamental plants were also starting to travel across the Atlantic more rapidly. The roses moving between France and the southern United States had paved the way. Geraniums, irises, and many other foreign plants were offered for sale in North America almost as soon as they arrived in England or Europe.

The interval between exotic plants being reproduced commercially in England or France and reaching North America was getting shorter and shorter. Plants would soon go directly to the United States from their foreign sources, but during the heyday of great plant collectors that did not happen very often.

It is not clear why Americans did not explore energetically for plants in Asia or the Southern Hemisphere during the Victorian era. There was some activity. Bretschneider mentions a few American missionaries in China who were interested in plants. The government sponsored large-scale voyages of exploration in which botany was part of the agenda, but nothing was on the scale of the British.

One explanation may be that North America is so vast that scientists had their hands full just finding out about their own territory. A more likely one is the organization of the nursery and florists' trades and existence of a market. The wealthy owners of very large estates in the British Isles and Europe had an insatiable appetite for floral novelties. Wealthy and aristocratic customers dominated the nurseries' business to such an extent that it paid a firm to seek ever more exotics.

Businessmen knew that they could sell whatever they found. This predetermined market was not such a strong feature of American society at the time.

Direct importation of new plants into the United States started in the 1860s but began increasing in the 1890s. By the early twentieth century, this movement was very significant. Much of the impetus came from David Fairchild and Charles Sprague Sargent.

In addition, two other American collectors, Frank Meyer and Joseph Rock, made major contributions. Both were foreign born.

DAVID FAIRCHILD

David Fairchild (1869–1954) was like a force of nature—brilliant, dedicated, effective, and able to get his ideas adopted (Figure 45). He started the Office of Foreign Seed and Plant Introduction, Bureau of Plant Industry, in the U.S. Department of Agriculture. The new office was devoted to expanding and improving American agriculture by a worldwide search for food and other useful plants that offered improved yields, resisted disease, or exhibited some other valuable quality. Fairchild himself had visited the East many times, always finding useful economic plants.[175]

Fairchild was born into a family of scholars and married Alexander Graham Bell's daughter. It was his idea to establish an experimental site for plant introduction in Florida, reasoning that imports from the tropics would perform better there than in other parts of the country.

His dealings with scientists and rugged plant collectors as well as distinguished public officials all show a man whose personality was pleasant and well integrated. This quality made him beloved and allowed him to accomplish much more than if he had been prickly and difficult. This informal and relaxed quality comes through in his books.

Inevitably, ornamental plants came in with the useful ones. Fairchild found many himself, and other explorers sent specimens to him in droves. The bulletins he published from 1900 on contain huge numbers of plant introductions.[176] Mixed in with the berries, pomes, stone fruits, nuts, citrus, vegetables, grains, fiber sources, and fodder grasses were flowers of all sorts and many ornamental trees and shrubs.

Fairchild shared the seeds and bulbs with other botanical research centers beside his own. The collectors he trained often participated in exploring or botanical expeditions sent by the universities, or at least helped in their activities. Frank Meyer was unhappy that he had to do this on one occasion because it took him away from his own work.[177]

FRANK MEYER

Frank Meyer (1875–1918) initially moved to the United States from the Netherlands in 1900. His actual name was Frans Meijer, but he later anglicized it. He had undergone rigorous training in gardening and botany under Hugo de Vries in Amsterdam. De Vries was very important in the history of biology because of his part in rediscovering the work of Gregor Mendel and his own theory of mutation.

Fairchild came across Meyer in the Missouri Botanical Garden and realized that the rather strange young man could be useful for plant collection in China. Meyer's principal task was to find economic, chiefly food, plants, but in the course of doing that he took ornamental plants to America that had not been known there before, even though they had previously been grown in Britain. He spent three years in the East on three occasions and died in China during a fourth term.

Meyer found a fine catalpa, *Catalpa bungei*, on his first trip in 1905. That year he also sent *Pinus bungeana*, *Juniperus chinensis* 'Columnaris,' and *Aesculus chinensis*. Fairchild tried to keep in close touch with him, for Meyer, though physically strong and rugged, was unexpectedly sensitive and moody. Many of his letters and notes indicate a constant level of depression that he only kept at bay by intense work.

Fairchild also tried to shield Meyer from the bureaucratic nonsense that afflicted anyone on the federal payroll. Illiterate laborers had not signed receipts for the 20 cents he paid them, and the federal auditors refused to reimburse Meyer for costs totaling $1.40.[177]

In his second year, Meyer found *Spinacia oleracea*, a large-leafed spinach that saved the spinach canning industry after the American spinach was decimated by fungi. Fairchild considered that this one introduction alone covered the entire cost of Meyer's journeys. For people who enjoy cooking, the Meyer lemon, *Citrus meyerii*, commemorates his name permanently.

Meyer sent prodigious material, carefully packing it himself and supervising its passage at every step. He found a new type of flowering cherry, *Prunus pseudocerasus*, which blossomed very early, 10 days before other species. (This had been found much earlier in China and was grown in John Reeves' garden in Canton before going to England in the 1830s.) Zoysia grass, *Zoysia japonica*, was another remarkable discovery that benefited millions of homeowners.

No one knows exactly what happened to Meyer on a steamer going down the Yangtze River in 1918, but his dead body was found in the water one morning. It was a terrible loss. He had introduced more than 2,500 species and would have gone on to do even more.

JOSEPH ROCK

Joseph Rock (1884–1962) was born in Vienna but worked as professor of Chinese and of botany at the University of Hawaii. He was self taught in both disciplines. David Fairchild met him in 1920 and once again recognized a man who could carry out his agenda in Asia. The principal reason Rock went to Southeast Asia in 1922 was to find the seeds of the chaulmoogra tree. Chaulmoogra oil was the only medicine that had any effect on leprosy at that time.

Rock made four journeys to China, the last one ending in 1949. After that he returned to Hawaii and spent the remainder of his life teaching and writing. Like so many of the truly amazing men who had explored in China, from Père David to Frank K. Ward, Rock was a polymath. Besides botany, he left classic texts on ethnography, zoology, and ornithology.

Not all of his discoveries turned out to be new species. There are two reasons for this. Many stretches of territory had been thoroughly combed by his predecessors. He sent the seed of 500 species of rhododendron to the United States, but most of them were repeats.[178] The other reason was that the unexplored places he reached tended to be barren.

Two spruces and a tree peony redeemed all this for him. *Paeonia suffruticosa*, or Rock's variety, supposedly came from a lamasery in Cho-ni, on the Tibetan border. Later this plant was to be called *P. rockii*. The monks had cultivated it for a long time. Farrer had found a wild form but had been unable to introduce it. Rock was successful. The story is that Rock stayed at the lamasery in the winter of 1925–1926 and that after it burned down in 1928, destroying the courtyard and plants, Rock was able to send the monks clones of their peonies when they rebuilt.

Recent scholarship has cast doubt on this delightful story. William McLewin in *Peony rockii and Gansu Mudan* carefully dissects all the statements and shows that nothing happened the way it has been handed down to us.[179] The dates were improbable, and it was not likely that Rock saw a peony in full flower in the depths of winter. The plant Rock actually found was not a species *Paeonia rockii* but a classical hybrid form cultivated in China for centuries.

Picea asperata and *P. likiangensis* var. *purpurea* came from the Richthofen Mountains (now the Qilian Mountains) in that same area. Both turned out to be hardy and useful in northern sections of the United States.

Rock prepared the definitive flora of the Hawaiian Islands' native plants while the great influx of foreign plants continued. The climate permitted tender tropical plants from all over the world to flourish. Visitors can be forgiven if they think that orchids, ginger, and bromeliads are native to Hawaii.

CHARLES SPRAGUE SARGENT

Charles Sprague Sargent (1841–1927), a cousin of John Singer Sargent, the portrait painter, was blessed by having great wealth and high social status in addition to talent and skill. That meant he could arrange things the way he wanted once he became director of the Arnold Arboretum.

The Arboretum resulted from the merger of a restricted bequest of $100,000 by James Arnold and a piece of land in Jamaica Plain given by Benjamin Bussey in 1863.

After complex negotiations between the City of Boston and Harvard University, the Arboretum was authorized in 1872, and Sargent was appointed to be the first director. It opened in 1881. Frederick Law Olmsted Sr. laid out the grounds and slotted it into his "Emerald Necklace," the elegant system of parks he created surrounding Boston. The Arboretum is dedicated to the study of woody plants and also supports taxonomists and other basic scientists.

In spite of the seemingly large endowment, the new institution only allowed its director a stipend of $3,000 per annum, insufficient to support someone without a private income. Sargent worked extremely hard at his new tasks, ignoring many inconsequential social duties. He was only interested in trees and their value to society. His legacy is an unprecedented collection of plants and a famous library, built around the books he donated and later bequeathed himself.

Sargent organized scientific expeditions to the East and traveled to Japan himself on more than one occasion. It was a sign of his open mind and skill at organization that he recognized how effective Ernest H. Wilson would be as his successor.

One of Sargent's disciples was Beatrix Jones Farrand (1872–1959), a very influential landscape architect in her day. She knew that, as a woman, she needed impeccable credentials to develop her career. She was able to apprentice herself to Sargent because of her social connections in old New York society. With Sargent's name behind her, wealthy patrons were reassured about her ability to carry out their desires.

ESTABLISHING DATES THAT PLANTS ENTERED THE UNITED STATES

Denise Wiles Adams has performed an inestimable service for the historian trying to establish when particular plants were imported into the United States.[70] This is a very difficult task. The most obvious source is nursery catalogs, but as Mrs. Adams points out, there are many fallacies in accepting any entry in an old catalog at face value. Despite these caveats, catalogs are still the most fruitful place to look.

In the earlier years, the plants came from Europe or from the Middle East indirectly through Europe. A few Asian plants had made an appearance in the eighteenth century.

Nurserymen and private individuals brought in new plants from many sources in England, Europe, and other countries, but it was done haphazardly, according to fashion and the whim of the public. Advertisements in newspapers and stories in magazines reflect these events. This gives the historian another place to search.

Records in California indicate how nurseryman after nurseryman obtained plants from other parts of the United States or directly from Australia, across the Pacific Ocean (see Chapter 15).[173] It was said to be faster to go from Australia to California than from New York to California before steamships and railroads. Ships were chartered for the purpose and the cargo auctioned at the wharf. On the basis of these and other records, the late Harry M. Butterfield could say with some certainty when exotic plants were introduced into California.

It was different in England and Europe. Centralized horticultural institutions sent collectors to the East with fairly specific instructions. Large firms of nurserymen, such as Veitch or Bulley, also supported expeditions. This meant that their discoveries were identified, named, and cataloged and the results published systematically. Even the plants collected in privately sponsored expeditions were processed in the same way. *Hortus Kewensis*, started by William Aiton and continued by his son, is still in existence, modified into a cumulative index, *Index Kewensis.* There are other European botanical indexes.

The same was not true for the United States until the end of the nineteenth century and the beginning of the twentieth. David Fairchild laid the foundations for official American collecting, part of his work at the U.S. Department of Agriculture, Bureau of Plant Industry. Although his task was to find valuable economic plants, he and his collectors found many ornamental plants as well. They were all recorded in the Bulletins of the Bureau of Plant Industry, issued annually from 1901 until 1953.

Charles Sprague Sargent visited Japan himself a number of times in the last quarter of the nineteenth century and also sent Ernest H. Wilson, under the auspices of Harvard University. Sargent's *Plantae Wilsonianae*, a compendium of the thousands of plants ostensibly introduced by Wilson, is an authentic source of data about plants coming into the United States directly. Moreover, several competent taxonomists worked at the Arboretum. The results of their labors were published in various journals.

While the great ferment in exotic plants was stirring, the leading botanists at Harvard, Philadelphia, and other centers were working to capacity during the mid-nineteenth century naming and categorizing the huge flora of this country. At various times state horticultural societies helped to spread information about imported plants. Using the new official names, Dana Hovey's *Magazine of Horticulture* in Massachusetts listed new plants, but it only lasted about 30 years before closing in 1868.[180]

There was no consistent source of scientific information neatly packaged for the future investigator. Looking for these data involves following many avenues. There were literally dozens of ephemeral magazines that appeared for a year or two and then ceased. Liberty Hyde Bailey (1858–1954) included biographies and information about nurserymen and horticultural journals of the past in his *Standard Cyclopedia of Horticulture*, first published in 1914.[71] Very few other men have had such a productive career over such a very long life as Bailey. As professor and chairman of Agricultural Science at Cornell University, he changed the way in which botany and horticulture were taught and applied.

The American Horticultural Society issued its first national journal in 1922, under the title *National Horticultural Magazine.* This publication, which was later renamed *The American Gardener*, is intended for the public and is very accessible, but it does not publish taxonomic studies.[181]

The U.S. National Arboretum, administered by the U.S. Department of Agriculture, was founded in 1927. It publishes data about new plant introductions, whether from abroad or bred in its own grounds. Each year the arboretum releases about five or six new plants into circulation. There is less and less about foreign introductions now.

The Germplasm Resources Information Network database (GRIN), supported by the Agricultural Research Services Department of the U.S. Department of Agriculture, is encyclopedic, as is TROPICOS from the Missouri Botanical

Garden. Both of these contain reliable taxonomic information, but neither lists the date when plants first entered the United States. The principal date recorded both in GRIN and TROPICOS is when the plant was first described in the botanical literature.

In spite of all the obstacles to establishing accurate dates, it can be seen that the importation of exotic Asiatic and Southern plants into the United States was very active in the middle years of the nineteenth century. Appendix 4 contains an adaptation of lists compiled by Denise Adams, sorted by date of introduction.

WEST COAST OF THE UNITED STATES

An informal source of imports was seen on the West Coast of the United States. Wealthy men commissioned sea captains of their acquaintance to bring specific trees or shrubs from the Orient.

In California's Marin County, Albert Dibblee asked his friend Admiral Shufeldt to buy trees in China, Japan, and Korea for him. The admiral had "opened" Korea and had been the first American minister to that country under President Ulysses S. Grant.

Nurserymen in San Francisco established contact with Australian colleagues very soon after the gold rush. By 1865, *Araucaria heterophylla* from Norfolk Island, the Norfolk Island pine, was a standard article of commerce in San Francisco. No garden that pretended to high social status was without one. James Walsh, an enterprising San Francisco nurseryman, commissioned boatloads of Australian plants and auctioned them at the wharf.

William Walker, a lawyer from Alabama who started one of the earliest nurseries in San Francisco right after the gold rush, left a handwritten version of his 1858 catalog showing the source of many of his offerings. Fortunately, this valuable document was saved by a colleague, Charles Abraham, and is now safely in the California Historical Society archives.

Walker bought many plants from the Guilfoyle Nursery in Melbourne (see Chapter 11). Nurserymen all over the United States developed ties with foreign sources. It was easier in the eastern states, but as communications improved, others followed suit. Competitors could work together when the industry as a whole would benefit. Societies and organizations instigated by the nursery trade characterized this watershed period.

Part Three

GLOBAL TRANSCENDENCE

CHAPTER 15

BOTANICAL AND HORTICULTURAL INSTITUTIONS AND PUBLICATIONS

THE WORD botany is derived from the Greek βοτάνη, meaning "grass" or "pasture." In time its meaning grew to incorporate the systematic collecting and labeling of wild plants and attendant science. Horticulture comes from the Latin *hortus*, meaning "a garden."

Emphasizing these differences is not mere pedantry. Most of the time the words are used somewhat interchangeably without any problem, but it is necessary to remember the distinction when it comes to plant collecting and exploration.

In the very earliest years scientists wanted to see the plants in distant places. Collecting them even without the seed or roots helped them to pursue their studies. They were deeply concerned with plants that had medicinal properties. These were carefully sown and observed, but physicians and botanists were not primarily concerned with the appearance of their gardens or anything else so frivolous.

Ordinary people, whether wealthy or not, did not share this austere attitude. The sheer beauty of the new finds did begin to affect people. Clusius, for example, brought back numerous flowers just for their appearance. The delicate charm of crown imperial, narcissus, hyacinth, and the other small bulbs led to their rapid spread across Europe. The tulip was an extreme case of this. At that epoch one could say that botany and horticulture were still merged because everything was so new.

By contrast, Pierre Belon took nothing back to France after his travels in the sixteenth century. He contented himself with describing what he had seen and trying to fit everything into his system. Belon was a botanist, perhaps the first to explore for new plants intentionally, according to Alice Coats.[1]

Succeeding plant explorers emphasized one or another aspect of these sciences. Even when someone like Philibert Commerson took thousands of specimens back to France, the purpose was to analyze them and study them, not to put them in a garden. Even if he had wanted to use them in a garden, the inefficiency of the

Jardin des Plantes precluded anything of the sort. The huge bundles of samples were just stuck in a basement and allowed to molder. Commerson himself did not select anything special or pay much attention to his specimens after he delivered them to headquarters.

Some explorers were more attuned to garden needs. Tradescant (the elder) looked for plant novelties to improve the ducal and royal gardens he tended. His son sent back suitable plants for parks and gardens from Virginia. The ultimate in horticultural exploration was Harry Veitch's sending Ernest H. Wilson to China in 1899 to bring back the seeds of one particular tree and that tree alone.

As a new discipline emerged, it had to overcome its radical aspect and prove itself, but eventually it would be coopted into the academy. By the seventeenth century, botany as a separate discipline had become sufficiently respectable to warrant a chair at Oxford. The inclusion of botany as a discipline in universities accelerated the spread of knowledge. An institution like a university unites like-minded people who can exchange ideas and offer support or criticism of one another's work. With rare exceptions, the original scholars were men.

Even though a university department in the sixteenth and seventeenth centuries was not what we recognize today, a framework now existed to absorb the new specimens of plants coming from so many places. Resemblances and differences were fitted into various schemes. When attempts to shoehorn everything into a known or fixed scheme finally failed, the schemes were adjusted to account for new findings.

NOMENCLATURE

Classification was at an impasse. Vernacular names were too inconsistent, repetitive, and meaningless. The scholars immediately turned to Latin for ease of communication. Careful observation of plant growth and at least some idea of the effects of plant extracts on human physiology codified previously haphazard folklore. Still, without a unified conceptual basis for their studies, even the best efforts of the scholars were not consistent with modern scientific thought. Until Carolus Linnaeus (1707–1778) published his *Species Plantarum* in 1753, they had no rhyme or reason to choose one name over another.[57] In spite of that, their work represented a marked improvement over what had gone before.

Linnaeus cut the Gordian knot. He was a Swede, educated in Holland. His father was a poor Protestant pastor. In his late teens, Linnaeus made a famous journey to northern Lapland on foot, by himself, to collect plants. Hermann Boerhaave, one of the leading physicians in Leiden, recognized the young Swede's gifts and introduced him to other scientists. He also found him a job so he could survive in Holland. Once Linnaeus put forward the sexual system, scholars could place a plant into a particular class on the basis of the number of stamens. With that, advance naming became less of a random business.

The concept of a scientifically arranged garden for observation and teaching spread across Europe from Italy, and within a relatively short time more botanical gardens were founded.

A wealthy Dutch banker, George Clifford, had a remarkable garden. Linnaeus was employed to classify its contents and prepare a catalog. When he returned to Sweden, he was given a chair in botany at the university in Uppsala. His wife would have preferred him to practice medicine and make more money. She is said to have made his life a bit miserable.

Linnaeus' publication of *Philosophia Botanica* in 1751 and *Species Plantarum* in 1753 turned out to be only the beginning.[182] His was an "artificial" system based on the plants' sexual organs and did not account for other variables in plant anatomy. Counting stamens was too simple to stand as the last word. Linnaeus' permanent contribution was giving every organism a binomial label: genus and species.

Linnaeus built on the work of his predecessors, Andrea Cesalpino (1519–1603), who laid the philosophical foundations of scientific classification, and Joachim Jung (1587–1657), who studied plant morphology and created definitions that assisted all subsequent botanists. Gaspard Bauhin (1560–1624) had been the first to glimpse the possibilities of a binomial system. Tournefort took that idea forward but did not believe that the root, stalk, or leaf would contribute to adequate classification.

The true hallmarks of science are reproducibility and consistency. Linnaeus' early education in logic allowed him to conceive of a taxonomic framework that would be internationally acceptable and permit botanists to reach the same conclusions when dealing with the same plant. This would foster valid comparison and take the science forward.

When he went to England, Linnaeus visited the botanical garden in Oxford. He spoke no English but could communicate with his colleagues in Latin. Of the approximately 2,000 specimens in the garden, more than 1,400 were exotics, imported from abroad.

Linnaeus and his son Carl handled many of the Chinese plants coming into Europe. They eventually named and described 160 of them. Bretschneider laments that the father often omitted the names of the donors and was muddled about geography. He was not clear about where China was and frequently mixed it up with India. He named the double-flowered China rose *Rosa indica* because he received it from India. He did not know its true origin.[118] Carl tried his best but only outlived his father by five years, dying in 1783.

The elder Linnaeus first came into contact with exotic Chinese plants while studying in Holland. He saw them in Clifford's garden, at de Hartecamp. In the two years he was there, Linnaeus grew 10 plants from Chinese seed and described them in 1737.

Sweden was not yet ready to support major scientific work at that period. Linnaeus was an important native son but an exception. A few other Swedes took botany forward. Perhaps the best known after Linnaeus was Carl Thunberg (1743–1828). He had a colorful career all over the globe, collecting plants and creating trouble for other people because of his personality. Finally, Thunberg settled down to become professor of botany at Uppsala.

SOCIETIES

LINNEAN SOCIETY

Linnaeus' widow inherited the family's scientific papers, both those of her husband and son. Sir Joseph Banks would have liked to buy them, but her price was too high. He referred her to a wealthy physician-botanist, J. E. Smith (1759–1828). Dr. Smith bought them in 1784, using connections at court to forestall the king of Sweden's insisting they remain at Uppsala. Fru Linnaeus needed money and knew the university would not meet her price. This episode has a very modern ring to it.

Smith founded the Linnean Society in 1788, as an adjunct to the Royal Society. It remains a leading botanical institution both in England and internationally. In 1814, George III knighted Smith and he became Sir James. He analyzed Linnaeus' herbarium and published extensive commentary on these specimens. He did not make a very good bargain as much of the collection was in poor condition and had not been well maintained.

HORTICULTURAL SOCIETY (LATER ROYAL HORTICULTURAL SOCIETY)

A few congenial friends clubbed together and founded the Horticultural Society in London in 1804.[65] This was neither an academic society nor did it derive from a royal charter. Its antecedents lay in the Royal Society, that association of like-minded men who explored the natural sciences. Joseph Banks, James Dickson, William Aiton the younger, John Wedgwood, Charles Greville, William Forsyth, and Richard Salisbury met at Hatchard's bookshop in Picadilly to plan the organization. Dickson, Aiton, and Forsyth were Scottish nurserymen and gardeners.*

Newly discovered plants were eventually named for all of the founders except Wedgwood and Aiton. Aiton's father had been honored with *Aitonia capensis. Banksia, Dicksonia, Grevillea,* and *Forsythia* are well known. Salisbury introduced *Ginkgo biloba,* which was brought from South Africa in 1777. For a time it was called *Salisburia.*

Aristocrats and members of the upper middle class who shared a passion for plants needed a headquarters for their activities. Ultimately there were 1,500 "fellows" (i.e., members). A very few of these were women.

It was even possible for working-class people to rub shoulders with the aristocracy at the flower shows. If a carpenter grew better delphiniums, the dukes were right there to encourage him and perhaps be a little envious. There was a superficial *bonhomie* based on some level of respect, but the wise working-class man knew where the boundaries were and avoided taking liberties.

*In 1904, the society hung a centennial plaque at Hatchard's bookshop.

Joseph Sabine (1770—1837) was the secretary initially. John Lindley (1799–1865) began as a gardener but studied botany assiduously until he transformed himself into a powerful scientific figure. He was Sabine's assistant at first, then took over as secretary for many years. He finally became professor of botany at the new University of London.

The organization later became so successful that Prince Albert arranged for it to be chartered as the Royal Horticultural Society in 1861, the name by which it is still known. He had toured the society's gardens at Chiswick and was deeply impressed.

The Horticultural Society served as a very efficient distributor of new introductions. Once the plants had been proven to grow in British soil and survive in its climate, the society propagated them vigorously. For example, in 1825, they gave away 350,000 seed packets, 90,000 plants, and 50,000 parcels of cuttings from their testing grounds at Chiswick.[65]

Sabine and Lindley were very hardheaded in one regard, though Sabine was extravagant and caused the society to go into debt. They both knew that they could not afford to antagonize the professional nurserymen. Nothing was given away that could be bought in nurseries. They concentrated on rare exotics that had not yet become generally available. Mueller in Melbourne was not so astute (see Chapter 11).

GARDENS FOR TEACHING

Botanical institutions began primarily as physic gardens, considered necessary for instructing medical students. Gardens organized for teaching purposes derived from the monasteries' physic gardens.

The development of a teaching garden became possible because there was a body of knowledge to pass onto students. Plants were laid out in careful rows, attempting to show family resemblances. Gradually, physic gardens became botanical gardens with a broader goal than just teaching doctors. They were clearinghouses and centers of learning for botany on its own terms. Horticulture was not important at first. The gardens were not arranged to be decorative. The constant addition of newly discovered plants from much of the globe meant that there was not enough room to grow everything in even the biggest garden, always assuming that certain plants would grow in an alien environment. One solution was to dry the specimens and collect them methodically in cabinets. This was the beginning of the herbarium, an essential tool for studying the anatomy of a plant and giving it a name.

EARLY BOTANICAL GARDENS IN EUROPE

Italy

The first teaching gardens in western Europe were in Pisa and Padua.[163] There is a dispute about the dates when they were founded. The garden at Padua's medical school was authorized in 1545 by the Senate in Venice. Great physicians such as William Harvey, who discovered the circulation of the blood, were trained at Padua. The Orto Botanico at Pisa claims to have been founded in 1544 under the auspices of Cosimo I de' Medici.

This may well be the case. The leading botanist in Pisa was Luca Ghini (1490–1556). He was one of the earliest teachers to realize that the available wild plants from one region could not supply all the material needed for a good pharmacopoeia. He is said to have come up with the idea of planting as many specimens as possible in a garden.

Ghini is also credited with the invention of the herbarium sheet for the same reason. This allowed botanists to look at plants and their organs across the seasons. William Turner studied with Ghini for a time.

Florence and Bologna followed. It is not surprising that this movement began in Italy. Scholars fleeing the destruction of Constantinople by the Ottoman Turks in 1453 carried their precious Greek manuscripts with them to Venice and other Italian cities. Renaissance ideals spread to all branches of learning.

A fine garden exists in Palermo in Sicily, built with Bourbon money in 1795. One of its treasures is a grove of extraordinary *Ficus* trees. These trees are now more than two centuries old. Before long, every province of Italy had a botanical garden, including Genoa and Bari.

France

In France, Henri IV founded the first botanical garden at the University of Montpellier in 1593. The royal garden in Paris, Jardin des Plantes (initially Jardin Royal des Plantes Medicinales), was adapted for the purpose in 1635. Louis XIII's physician, Guy de la Brosse, kept pressing the government to let him create a medicinal or botanical garden, but it took him seven or eight years until the authorities agreed.

Britain

In England, a series of important botanical institutions developed. At Oxford a botanical garden began in 1621, though it did not function until 1632. The garden was planted on land leased from Magdalen College, and it is still in the same place. Inigo Jones designed the handsome gate that leads into it from the High Street. Jacob Bobart, a German scholar, was appointed the director.

Robert Morison, a Scotsman from Aberdeen who was the first botanical professor in England, had been trained in Paris at the Jardin des Plantes. His mentor was Vespasien Robin, son of the royal gardener Jean Robin, colleague of the Tradescants. In a class-ridden and rigid society, it says something about the quality of the Robin family that students from a different social level would subject themselves to their guidance and teaching.

At that time Scotland itself was not in the forefront of botanical studies even though its citizens made such large contributions to the science in remote places. Another Scottish physician returning from training in Paris, Robert Sibbald, was disgusted by the absence of a teaching garden in his home town of Edinburgh. In 1670, a year when many significant botanical findings emerged, Sibbald and his colleague Andrew Balfour bought a small plot of land and planted a thousand specimens of medicinal plants. This was the origin of the Royal Botanic Garden at Edinburgh.

The institution has had a most distinguished history. Noted plant collectors and horticulturists such as Archibald Menzies and George Forrest were trained there. These are but two of the best known.

The Society of Apothecaries of London created their garden, the Chelsea Physic Garden, in 1673 next to the river Thames at Chelsea Reach. All traces of that original property have been obliterated by development. Between Sir Hans Sloane's money and Philip Miller's skill, this garden excelled. Once they were gone, the Chelsea Physic Garden declined and Kew began its ascent.

When George III was a boy, he spent a lot of time at Kew House under the tutelage of the Marquess of Bute. His parents, Frederick, Prince of Wales, and Princess Augusta, had bought Sir Henry Capel's modest house next to the river Thames. By 1736, it was established as a royal palace with a noted garden.

Bute guided the Princess' gardening endeavors even after her husband died and George became king. Bute understood plants and had very practical ways of making the garden work.

As a minister of the crown, his ideas on the role of the colonies and colonial self rule were tragically much less effective. Bute's political advice to his master King George III was not helpful.

Philip Miller suggested that Bute appoint William Aiton, another Scot, to be superintendent of Kew Gardens in 1759. Aiton and later his son, also William, prepared the first catalog of the plants in the garden, *Hortus Kewensis*, the forerunner of *Index Kewensis*. When the intellectual and financial heft of Joseph Banks was added to Kew's governance, it roared forward.

Royal Botanic Gardens at Kew

Kew ceased to lead the horticultural and botanical world after Banks' death in 1820. The king was insane. The queen had died in 1818. No one was left to care for the gardens. Control was transferred to successive government departments, ending with the Commissioners of Woods and Forest. Pettifogging bureaucracy stifled the place.

With Kew dormant, the Horticultural Society grew rapidly (see "Societies," this chapter), but eventually the members of the society realized that Kew served very important functions and fought to have it restored and properly run. England had no official botanical garden in its capital city. This put the English at a disadvantage when comparing themselves with large continental cites.

Robert Brown (1773–1858), the botanist who went to Australia with Captain

Robert Flinders in 1802, took charge of Banks' library at the end of the latter's life. Brown retired to the library and laboratory to classify and organize the great volume of material that was flowing into England.

Brown is remembered for having noticed the ceaseless movement of minute particles in cells, "Brownian motion," and other important scientific observations.[183] Banks left him his personal herbarium, the books in the library, his house in London, and his collection of drawings.

All this was Brown's as long as he lived. He became the keeper of the herbarium and made a profound contribution to botany. On his death the property reverted to the British Museum. It now forms the museum's botanical department.

Under Banks' guidance, Kew had financed eight significant plant explorations in as many countries. The explorers had sent back a total of 7,000 new species in Banks' lifetime. After Banks died, the foundation passed from the royal family to the nation under a special act of Parliament in 1821. The result was that no one paid much attention to the property.

The royal family used it as a kitchen garden. It ceased to have its former scientific functions without the driving energy of Banks. The complex was about to be totally dismantled in 1840 when John Lindley recommended William Hooker, professor of botany at the University of Glasgow, to take charge and revitalize the place. The Horticultural Society had sponsored a commission on Kew, to restore it to greatness.

The Horticultural Society's Garden at Chiswick

In 1821 the Horticultural Society leased land from the Duke of Devonshire for an experimental garden in Chiswick, just west of London. Ever gracious, the duke reduced the rent from £400 per annum to £300 during a rough patch.

There were periods of poor financial management when the men in charge expanded too rapidly and took on too much expense. By the 1850s, things were so bad that the library and many of the finest plants in the gardens were sold off. Even the herbarium was sold. It was really all for nought, as the total amount they raised was trivial in comparison with their needs. Fortunately, some of the wealthier members came to their rescue.

Other British Gardens

In 1837, when Queen Victoria ascended to the throne, there were 22 botanical gardens, but only 10 were really working. In the British Isles, excellent gardens in Edinburgh, Dublin, and Belfast fluctuated inversely in their importance as Kew rose and fell. Provincial cities managed to create and maintain botanical gardens because of enthusiastic citizens. Liverpool is a case in point (see "Bees Ltd. and A. K. Bulley," Chapter 9).

The Royal College of Physicians, founded by Thomas Linacre in Henry VIII's time, did not have its own garden. It was beneath the college's dignity. Medical practice at that epoch was divided into three distinct tiers: physicians at the top, surgeons at the bottom, and apothecaries more or less acting as pharmacists. Apothecaries compounded the prescriptions written by physicians. Surgeons were still based in the barbershops and not considered professional. They were not supposed to prescribe at all.

Ireland has an old teaching garden, founded in 1790 at Glasnevin near Dublin by the efforts of Lord James Oriel. It was intended to encourage economic botany, but it also had a *Hortus Linnaeaensis* to demonstrate the Linnaean system. This garden later emphasized horticulture as well, issuing diplomas for students every year since the 1840s. In the modern period it is a showplace for Irish botany and horticulture.

Trinity College, Dublin, had a fine teaching garden, which was moved to three different sites before finally succumbing to modern development. Nothing remains of it except for the iron and granite walls and some holm oaks.

In 1816, a group of professional gardeners in Dublin got together and founded a horticultural society. It prospered for a time, but as the gardeners were not socially prominent, it faded. They wrote new rules in 1830 that favored the aristocracy. This newly polished organization became the Royal Horticultural Society of Ireland the year after Queen Victoria ascended the throne.

The Netherlands

The Dutch lured Clusius away from his employment at Frankfurt managing the Landgrave of Hesse's botanical garden. He was 67 when he moved lock, stock, and barrel to Leiden in 1593, becoming *Horti Praefectus*. There he created the botanical garden for a new medical school.

William of Orange had been very grateful to the citizens of Leiden for their support in a major battle against the Spanish and offered them the choice of two rewards. One was a partial commutation of taxes for one year; the other was the foundation of a university. They very wisely chose the university.

The leaders of the newly founded university knew that a botanical garden was necessary to keep up with the times. The faculty of medicine was indifferent to this responsibility, and so the *curatores* approached Clusius.

At first he refused to go. He was old, he had a permanent injury from a fall, and he was tired of moving about. In further negotiations he requested, and was granted, the assistance of a practical gardener to do the heavy physical labor.

Clusius' work was continued by a series of professors of whom perhaps the best known was Boerhaave. Linnaeus spent two years in Holland learning medicine and botany.

German-Speaking Countries

The Hapsburg emperors started to build the palace of Schönbrunn in Vienna in 1700. Adrian van Steckhoven from the Netherlands designed its splendid gardens.

The Belvedere Palace alpine garden was started at the end of the eighteenth century. Its purpose was to display all the native plants in Austria.

In Berlin, the botanical garden of Berlin-Dahlem goes back to 1815. At first the garden was in Berlin itself, based on previous smaller efforts by the great botanist Carl Willdenow (1765–1812). Alexander von Humboldt was one of his most distinguished pupils. They created the discipline of botanical geography. It was at the Berlin-Dahlem botanical garden that Roberto Burle Marx had his epiphany about the native flowers of his homeland, Brazil, early in the twentieth century.[184] Not until he got to Germany did he realize just how amazing the Brazilian native flora was. The Germans were using Brazilian plants to great effect but not the Brazilians. At the time, Brazilian landscape design depended mainly on plants of European descent.

Munich's botanical garden was founded in 1812. There are about 60 more botanical gardens in Germany itself, but the ones mentioned previously are the oldest.

Scandinavia

The influence of Linnaeus was of course reflected in Uppsala. A very old garden in which he worked was replaced in 1787 by the University Botanical Garden. King Gustavus III donated part of his royal park to create adequate space for this expansion. Very much later, in 1916, another garden was founded in Göteborg. The Copenhagen Botanical Garden dates to 1871.

Russia

St. Petersburg

Peter the Great instructed his physicians to lay out an herb garden in St. Petersburg in 1714. This was the second garden of its kind in Russia. Moscow's garden had been created a few years before. Both contained native and foreign plants and were intended for medicinal purposes. The produce was dried and sold in pharmacies all over town.

In about 50 years, the garden at St. Petersburg had about 300 species of plants. At the turn of the nineteenth century, it had advanced to the point at which it no longer simply supplied botanicals to apothecaries but was becoming a true botanical garden. The name was changed to the Imperial Botanical Gardens.

Russia currently has about a 150 botanical gardens. They have been amalgamated into a network. The principal herbarium is in St. Petersburg.

Moscow

The nucleus of the garden in Moscow was Count Razumovky's estate at Gorenki. Alexei Grigorievich Razumovsky (1709–1771) rose from being a lowly Ukrainian shepherd boy to a courtier and lover of Tsarina Elisabeth because of his exquisite voice as a child. There are even indications that he married the tsarina in secret.

Razumovsky continued to be involved in music, and his name lives today because he commissioned six string quartets from Beethoven. He collected plants on such a sumptuous scale that eventually the estate became a public institution. The present university botanical garden in Moscow is not at the former Razumovsky estate.

BOTANICAL GARDENS IN NORTH AMERICA

THE UNITED STATES

There were some botanical gardens in early America, besides the Bartrams' property. Dr. David Hosack's Elgin Botanical Garden in New York was important for teaching medical students at King's College, the forerunner of Columbia University. Rockefeller Center stands on the land where the Elgin garden used to be.

By about 1860 St. Louis, Missouri, ceased to be a frontier town and rapidly developed into a metropolis. For a long time it had been a military outpost and the point of departure for emigrants' wagon trains heading west.

A wealthy St. Louis businessman, Henry Shaw (1800–1889), wanted to found a botanical garden as a civic gesture. He had made his fortune very young and retired from active commerce at the age of 39.

Shaw was born in England. In 1851 he went to see the Great Exhibition in London. While in England he saw the Duke of Devonshire's estate at Chatsworth. He was so struck by its grandeur and beauty that he decided to create a park on the same scale in St. Louis.

Once he had arranged the disposition of the land, he wrote to Sir William Hooker at Kew for advice on how to proceed. Hooker told him that one of the premier botanists in the whole country, Dr. George Engelmann (1809–1884), was living and working quite close to him.

Shaw and Engelmann collaborated on the garden but not without some tension and difficulties. Engelmann was a physician and scientist; Shaw was more of a showman. A basic disagreement arose right at the beginning. Shaw's land was a long way outside the city limits, and Engelmann did not want to have to travel so far. In spite of these rough patches, the institution survived to become a major scientific establishment of the present day.

Harvard College created a botanical garden in Cambridge, but did not arrange for any funds to maintain it (see Chapter 6). James Arnold had previously left a very large sum of money to the university for agricultural or horticultural purposes, but it could not be used for the original garden, which was run by Asa Gray, professor of botany. This sum increased to over $100,000 by 1872, yet none of this money was available to Gray.

The Arnold Arboretum came about because a benefactor named Bussey bequeathed 125 acres of land to the university in Jamaica Plain toward the end of the nineteenth century.

The City of Boston undertook to build roads and other infrastructure for the new arboretum, but the authorities needed a wealthy man to take charge as there still was insufficient money to pay a director adequately. Charles Sprague Sargent became the first director. He was a good choice.

CANADA

A national botanical garden was not established in Canada until 1941, while the country was a dominion of Britain. The Royal Botanic Gardens at Hamilton, Ontario, has been the home of many distinguished scientists.

BOTANICAL GARDENS IN THE SOUTHERN HEMISPHERE

South Africa

The Dutch East India Company established a working garden in Cape Town in 1657 to grow fruit and vegetables for the sailors returning from the East.[185] One of the company's employees, a carpenter named Leendert Cornelissen, offered to grow food on a freelance basis rather than take wages any more. In the documents that have survived, it was stipulated that he should "order a wife from the Fatherland" and remain at the Cape for 15 years. Leendert did not last too long as an independent farmer. By 1661, his drinking and brawling were so egregious that he was relieved of his position.

The first site turned out to be a mistake. Its climate was inclement, so the garden was pulled up and rebuilt a few miles away, in the lee of the mountain.

At a later date, when plant explorers were traveling through South Africa, the old company garden was useful as a base. After that declined, three collectors formed a botanical garden at the Cape in 1848. It was poorly maintained and had become a general park by the time the city of Cape Town assumed its control in 1891. The park did not have any specimens of the extraordinary flora of South Africa.

In 1903, a brilliant and energetic botanist, Harold Welch Pearson (1870–1916), left Kew to become professor of botany at Cape Town. Less than 10 years later he had persuaded an influential group of citizens that the country needed a first-class botanical garden and that Kirstenbosch was the place to build it. Pearson noticed that some plants in South Africa were already endangered by then and might become extinct if no action were taken.

They created a permanent site at Kirstenbosch in 1913, incorporating the late Cecil Rhodes' house and garden, which had been left to the nation. The prime minister of South Africa at the time, Jan Smuts, was a rarity among politicians, a successful general who had been trained as a botanist.

Australia

The botanical gardens in Melbourne and Sydney were the first botanical gardens in Australia. From these two small beginnings, the number of botanical gardens in Australia has grown to 80.

Melbourne

Sir Ferdinand Mueller held sway in the botanical garden in Melbourne until his arrogance led to a citizens' revolt (see Chapter 11). The antagonism of the commercial nurseries was an important factor in his downfall, and it is possible that they fomented trouble.

Mueller was very public spirited. He considered it his duty to share fine seeds and plants with the public at no charge, and of course he was correct. The garden was supported by public funds. Many small towns and villages had avenues of handsome trees in their parks and streets because of his generosity, but the commercial nurserymen were alarmed. His behavior undercut their livelihood.

Sydney

A botanical garden was established in Sydney in 1816.[146] Allan Cunningham was the director for a short time; he was followed by other important explorers and collectors. Charles Moore, who was appointed in 1848, remained in position for nearly 50 years, giving the garden continuity and a firm basis.

New Zealand

The town of Dunedin in Otago, named nostalgically for Edinburgh, appears to have been the first one in New Zealand to create a botanical garden. It took form in 1863 after a very long gestation period. Although the incredible riches of New Zealand's flora were beginning to emerge, the early garden did not deal with Colenso or the other explorers (see Chapter 11).

Sir Joseph Hooker, who took over as director of Kew after his father died, made good use of this resource, asking the later directors to collect plants for him.

South America

Brazil

The Portuguese royal family fled from Lisbon before Napoleon could imprison them. They wanted to escape the fate of the Spanish king. In 1808, after the family settled in Brazil, Don João VI, the Prince Regent, created the Real Horto, now the Jardim Botânico of Rio de Janeiro. Tropical trees are its principal feature, but many valuable economic introductions came through the garden, including tea.

Argentina

The botanical garden in Buenos Aires opened in 1892. Unlike other cities, Buenos Aires has multiple similar gardens in addition to its main one. The first director undertook to plant several other parks and gardens in the same fashion.

HORTICULTURAL PUBLICATIONS

Publications served the institutions and ultimately took on the characteristics of an institution in themselves. The vast accumulation of new information needed to be recorded systematically and the results disseminated. Individuals corresponded voluminously with their colleagues. They refined their ideas and worked out problems in this manner. Many letters have been preserved that shed considerable light on the thought processes of men like Sir William Hooker and Charles Darwin.

Scientists developed specific and highly technical means of communication between themselves, heavily indebted to Latin but in a jargon form of the vernacular. Scientific language reached its apogee in the twentieth century, effectively shutting out the general public. The very word "scientist" was coined in the nineteenth century after several other, more awkward terms were discarded.

The terminology took some time to mature. In the early period, science was not yet complex enough for it to be restricted to the academy. The public assumed it would participate in scientific experimentation. Intelligent Londoners could spend an afternoon at the Royal Society watching the experiments. These were quite comprehensible, both intellectually and linguistically. Benjamin Franklin had real box office appeal in his day.

The time-honored way of codifying new information and establishing the author's claims was to write a book. Individuals labored mightily to produce a book that incorporated the information they had accumulated. This was slow, and, unfortunately, almost as soon as the book appeared, its contents were superseded by new information.

The Royal Society, at first only an association of dedicated scientific amateurs, published its transactions from the beginning. Gradually the notion of an intermittent publication, a cross between a newspaper with its topical news and the static book, emerged. It was called a "magazine," based on the name of a storehouse or shop with many items—the French *magasin*. The first magazines were political and literary, but science quickly took up the idea.

ENGLAND

Curtis's Botanical Magazine

An enterprising employee at the Chelsea Physic Garden, William Curtis (1746–1799), Demonstrator in Botany, had the idea of starting a monthly magazine devoted to the new plants rapidly coming into England. The publication was very successful and still exists today.[186]

Curtis gauged his market very clearly. He was soon able to give up his day job. Starting in 1787, *The Botanical Magazine; or Flower-Garden Displayed* capitalized on the public's fascination with the exotic new finds. In its turn, the magazine vastly stimulated further interest.

Each issue featured three or four new plants in detail. A hand-colored picture was accompanied by a brief description of the plant, its characteristics, its name, and any other pertinent information. These articles were indispensable to garden enthusiasts who were trying to grow the new plants in various parts of the country.

Peter Collinson had written in his time that as soon as he had sent his associates their share of Bartram's discoveries, he received plaintive letters asking desperately for help, for example, "My gardener is an ignorant fellow. Pray tell me what he is to do with this." A publication such as *The Botanical Magazine* would have offered a solution.

In that first period Sydenham Teast Edwards drew all but 75 of the 1,720 plates that were published. Edwards died in 1819.

The magazine has been through six different incarnations in the past two centuries and had new names for a while, but it resumed its earlier name, *Curtis's Botanical Magazine*, in 1995. Handsome colored pictures of a few featured plants and a useful, concise text are still its hallmark.

Index Kewensis: Royal Botanic Gardens at Kew

William Aiton the elder (1731–1793) produced the first catalog of Kew's plants in *Hortus Kewensis* in 1789. The garden contained more than 5,500 plants. Aiton was the director of Kew from 1759 until his death. His son, the younger William Aiton, took over the post and remained in it until 1845. He died in 1849.

The *Index Kewensis* is the outgrowth of this beginning. This was perhaps the first, modern comprehensive listing for plant names and presages the modern online taxonomic indexes the International Plant Names Index (IPNI) and TROPICOS. In fact, IPNI builds from the historic nomenclature of Aiton's *Index Kewensis*, and incorporates the Gray Card Index of the Harvard University Herbaria as well as the Australian Plant Names Index from the Australian National Herbarium. What started as Aiton's listing of plant names in the descriptive literature, which included the the fascinating plants brought back worldwide by Kew botanists, has transformed the plant world with the now truly global nature of the online indexes of IPNI, TROPICOS, and others.

In addition to the invaluable resource of *Index Kewensis*, Kew Gardens also publishes many other valuable monographs and journals.

Horticultural Society

The Horticultural Society had a series of publications that came and went. In 1807, very soon after the society started, one of its principal activities was to publish a journal, *Transactions of the Horticultural Society of London*, with descriptions of the myriad new plants coming into Britain from all over the world. This was very important. The journal became the gold standard. It established the name and characteristics of each new plant unequivocally.

Lindley did most of the identification and naming, supplying accurate taxonomic information. He studied all the plants that arrived, not just the orchids. Very few men have had such concentrated influence. He could see all the minute detail in the flowers even though he had lost the use of one eye in childhood.[111]

Transactions ceased publication in 1848, no doubt because of the other magazines that began to appear.[65] John Lindley and Joseph Paxton started *The Gardeners' Chronicle* in 1841. The *Journal of the Horticultural Society of London* began in 1846. In 1975 the journal's name was changed to *The Garden.*

Joseph Paxton edited *Paxton's Magazine of Botany* and *The Horticultural Register* in the 1830s and 1840s. *Paxton's* was considered to be authoritative. For example, nurserymen far away in California relied on it to identify new plants and to find out what to do with the ones they had. William Walker in San Francisco looked up new plants that came from Australia and noted they were "not in Paxton."

Paxton was almost wholly occupied with engineering and landscaping commissions for great families such as the Rothschilds and yet kept his magazines up to date. Every word had to be written by hand. It was a grueling routine.

The society had 1,500 fellows (members) at the time. These magazines would go from hand to hand and be very widely disseminated, beyond the membership alone.

J. C. Loudon

John Claudius Loudon (1783–1843) was yet another Scottish gardener with immense talent and energy. In his case these qualities manifested themselves in journalistic skill. He brought out *The Encyclopaedia of Gardening* in 1822 and revised it frequently.[187] *The Encyclopaedia of Plants* came out in 1829.[188] He also founded *The Gardener's Magazine.* His publications became the arbiters of horticultural taste. If Loudon did not like you, things could be very grim indeed. Years could go by before he might change his mind.

He had moved to London as a very young man and noticed the dreary gloom of London's squares. They were planted only with native British conifers. Loudon suggested using the newly acquired plane trees, sycamore trees, and almond trees instead. He believed that these would all tolerate London's polluted air.

In a very short time he received landscaping commissions and was able to put his ideas into action. He was also in demand as an enlightened farmer, adopting modern approaches at a time when agriculture was stuck in the mud. Excellent as he was, it was the encyclopedia that made his name and has kept it fresh. Loudon's wife Jane also published valuable books on horticulture and had a considerable reputation of her own.

Because of his stubborn and righteous attitude, Loudon very frequently got into the most unedifying rows with other garden writers, and I am sure the gardening public reveled in all the backbiting and sharp invective. It was better than going to the theater.

Dozens of small gardening periodicals started and abruptly ended within the decades 1820 to 1840. Editors moved from one to the other at the whims of the owners and backers. The great horticultural historian Ray Desmond made a detailed study of them.[189]

UNITED STATES

It is hardly surprising that horticulture moved forward most quickly in the states that had been established the longest. They had reached levels of prosperity and civic complexity that made this possible.

Massachusetts

Charles H. Hovey started a journal for the Massachusetts Horticultural Society in 1838. After his death in 1868, the journal no longer appeared. This was the first significant horticultural periodical in the country. Many smaller journals started, but none persisted. L. H. Bailey listed most of them in his extraordinary work *Standard Cyclopedia of Horticulture*, which first appeared in 1914.[71] Unlike other encyclopedists, Bailey listed nurserymen and journals, not just plants.

Hovey was an important figure in nineteenth-century American horticulture. Another Hovey, Dana, went to California with the gold rush and started a nursery. It is not clear whether they were related. Charles' brother was Phineas Hovey.

For a very long time there was no standard national American horticultural publication covering the entire country. In 1904 the magazine *Horticulture* was started as a weekly trade journal for florists and gardeners on estates and in greenhouses. The Massachusetts Horticultural Society acquired the magazine in 1923, to use as a journal for its members. The magazine did not take on its present incarnation until 1978. The relationship between the Massachusetts Horticultural Society and the magazine *Horticulture* later dissolved, but *Horticulture* is still published. It is a lively magazine for the general gardening public.

The American Horticultural Society

The American Horticultural Society was founded in 1922. The same year, a group called the National Horticultural Society was also formed. The National Horticultural Society's initial publication was called the *National Horticultural Magazine*, and the first issue was published in August 1922.

In 1926 the two organizations merged as the American Horticultural Society and the publication continued as the *National Horticultural Magazine* until 1959, at which time it was renamed the *American Horticultural Magazine*. In 1972 the magazine was renamed again as *American Horticulturist*. Then, in 1996, the magazine became *The American Gardener*, which is how it is known today. *The*

American Gardener is the official membership publication of the American Horticultural Society, and is targeted to experienced gardeners, Master Gardeners, and professional horticulturists or landscape designers.[190]

Pennsylvania Horticultural Society

The Pennsylvania Horticultural Society published intermittent reports, starting in 1830.[191] The society was founded in 1827. A yearbook began in 1926. David Landreth, a noted Philadelphia nurseryman, was active with the society and was authorized to subscribe to Loudon's *Gardener's Magazine*. In 1829, the society held its first flower show. The records indicate that two Chinese azaleas were on display, as well as a Chinese tree peony. Other imports were seen there very early.

United States Federal Government

The federal government, through the U.S. Department of Agriculture, maintains the U.S. National Arboretum and publishes its work in plant introduction. From the beginning of the twentieth century, the Department of Agriculture has published annual lists of new plants entering the country. These are predominantly useful, economic plants, but mixed with them are large numbers of ornamental plants that arrive at the same time. The originator of this series was David Fairchild, Director of the Office of Foreign Plant Introduction, Bureau of Plant Industry, and a noted plant explorer himself.

Taxonomic reports on new plants nowadays are published in the scientific literature. These periodicals do not have the mass circulation of *Horticulture* and *The American Gardener*, the older English journals such as *Curtis's*, or the early periodicals of the Royal Horticultural Society, and their emphasis is not the same as that of the earlier work. Scientific descriptive botany will always remain necessary and relevant but it is not the dominant discipline anymore. Molecular biology has unseated it. The new methods offer definitive answers to many questions of plant relationships which were previously subject to uncertainty.

CHAPTER 16

COMING FULL CIRCLE

The Gardener

THE POWERFUL combination of great institutions, great publications, and individual courage and skill has resulted in the present ease and convenience with which gardeners can carry out any ideas they have.

In the present age, there is no limit to what we can grow, no matter where it originates (see Appendix 5). The variety of choices is bewildering, and yet this abundance imposes an unexpected uniformity and homogeneity on domestic and urban landscapes. Postage stamp–sized front gardens in Auckland, New Zealand, look very much like those in Columbus, Ohio; Streatham in London; or Toronto.

Nurseries developed and expanded into large businesses in Britain, Europe, and North America with the great stimulus of the imports. Some even took on the qualities of institutions, sending out their own explorers into Asia and South America. Their collectors had different instructions from those of the scientific establishments, but many made important contributions nevertheless.

Ernest Wilson and Frank Kingdon Ward are striking examples of these collectors. Both specialist and general nurseries benefited, changing out of all recognition.

Sometimes things got out of control. Wilson once dug up 10,000 lily bulbs in China. Sadly, almost all of them rotted in transit and ended up useless. In the 1950s cut cattleya orchids were all the rage as corsages. As noted in a previous chapter, the existing hybrids could not be grown fast enough to meet the enormous demand. Even with tissue culture and meristem breeding, it still took seven years to bring an orchid cultivar into flower.[192] Florists turned to importing species cattleyas by the thousands from Central and South America. This was done with such intensity that the sources were depleted. Episodes like that contributed to the need for the CITES (Convention on International Trade in Endangered Species of Wild Fauna and Flora) treaty.

Not everything that was imported was benign. The very success of some imported plants has led to a backlash. Some of them are invasive and have overwhelmed native plants' habitats. One thinks of kudzu and shudders. Native plant societies flourish wherever invaders threaten the preexisting flora. Members of these societies hope to restore the ecosystems that existed before the appearance of the invaders. The ecosystems support many other forms of life, not just plants.

Horticulture, together with many other industries, evolved in response to economic forces. Applying the techniques of mass production has always led to increasing numbers of people being able to obtain attractive goods that had previously not been available to them and thus reducing large social differences.

The garment industry is a striking example. For centuries, only the rich had nice clothes or more than one dress or shirt. All clothes had to be laboriously made by hand and cost far more than the lower classes could afford. With the invention of the sewing machine and the herding of seamstresses into factories, the price of clothing fell sharply. By the 1920s a poor servant girl or shop assistant could buy a pretty and fashionable dress that blurred the distinction between her and her employers. This apparently frivolous change had significant social and political consequences.

When applied to the nursery industry, mass production led to the garden center. The extraordinary success of Claude Hope in breeding impatiens in many sizes and colors is one of the foundations of this phenomenon. Many general nurseries have grown into the huge garden centers that can accommodate the suburban homeowner who reluctantly has to pay attention to the garden or the driven fanatic who pursues special needs all his or her own. The centers are supplied by enormous centralized wholesale growers and propagators.

Plants have become commodities. It is not necessary to go to a nursery to buy what one wants, but if one does, the price is far lower than previously. Plastic containers with six little impatiens plants ready to go can be bought at supermarkets, pharmacies, hardware stores, and other unexpected outlets, just as if they were a newspaper or chocolate bar.

A good analogy is that of the grocery business. The supermarket grew from the amalgamation of small individual shops selling a single commodity such as milk, vegetables, or meat. The extraordinary choices in the stock of a supermarket reflect the major role played by imported foods.

It is not farfetched to argue that these trends have had an impact on society, allowing it to be more democratic than at any time in the past. No longer is owning a beautiful garden the prerogative of the wealthy. The lower costs of bulk growing are passed on to the public as lower prices, putting even quite handsome plants in the hands of anyone who wants them.

Some very thoughtful people mourn the fact that plants have become mere commodities. Something is always lost in the transition, but the wide availability of healthy attractive plants and the pleasure this gives to large numbers of people largely counteract the negative aspects of these changes. One thing that bothers the dedicated gardener is that choice is limited by profitability. If only a few people buy a particular plant, the nursery ceases to carry it the following year.

This is a real problem but one that cannot easily be solved. It takes energy, knowledge, and persistence to bypass the obvious choices and use different, perhaps native, plants instead. As soon as a homeowner or gardener becomes a little ambitious, the tendency is to choose more rather than fewer exotic plants, but the nurseries are ready for that, too. Nurseries respond quickly to changes in demand.

A new specialty nursery can result when a hobbyist turns an amateur venture into a business.

The "race to the bottom" is not over yet. Garden centers are concerned about the growth of "megastores" or "big boxes," which undercut their prices significantly. Scientific advances in the propagation of hybrids and the efficiency of distribution are still evolving and leading to even lower costs.

In addition, some very large corporations have turned to the breeding of exotic flowers as a way of diversifying their product lines. The Taiwan Sugar Company in Taiwan is a formidable competitor in the orchid business. The profits from the sugar division can be used to support the orchid business while it grows and develops. It is hard for even a large specialty plant nursery to compete on these terms.

As a result of these trends, the public's attitude toward orchids has changed. Once an orchid symbolized luxury and wealth. An orchid plant was bought after much consideration or given as a gift to express serious respect. The recipient tended the plant carefully for as long as possible. Today a person can buy an orchid on impulse in the supermarket together with the groceries, do very little to maintain it, and, when it dies, simply throw it out with the garbage.

Nothing ever stands still. Plant businesses have to adapt to survive. The results of such seemingly modest activities as carrying a few new plants back to England or France hundreds of years ago have coalesced and led to immense changes in science, commerce, and even political or social behavior.

APPENDIX 1

TAXONOMIC LIST*

A

Abelia R. Br.
Abelia chinensis R. Br.
Abelia floribunda (M. Martens & Galeotti) Decne.
Abelia rupestris Lindl.
Abelia schumannii (Graebn.) Rehder
Abelia uniflora R. Br.
Abies balsamea (L.) Mill.
Abies bracteata (D. Don) Poit.
Abies concolor (Gordon & Glend.) Lindl. ex Hildebr.
Abies concolor subsp. *lowiana* (Gordon) E. Murray
Abies grandis (Douglas ex D. Don) Lindl.
Abies jezoensis Siebold & Zucc.
Abies kaempferi Lindl.
Abies magnifica A. Murray
Abies mariesii Mast.
Abies procera Rehder
Abies spectabilis (D. Don) Spach
Abies venusta (Douglas ex Hook.) K. Koch
Acacallis cyanea Lindl.
Acacia caleyi A. Cunn ex Benth.
Acacia pendula A. Cunn. ex G. Don
Acacia spectabilis A. Cunn. ex Benth.
Acalypha conferta Roxb.
Acanthus mollis L.
Acanthus spinosus L.
Acer L.
Acer negundo L.
Acer palmatum Thunb.
Acer platanoides L.
Acer pseudoplatanus L.
Acer rubrum L.
Acer saccharum Marshall
Achillea ptarmica L.
Aconitum autumnale Lindl.
Aconitum chinense Siebold ex Paxton
Aconitum napellus L.
Acorus calamus L.
Acorus gramineus Aiton
Actinidia chinensis Planch.
Actinidia kolomikta Maxim.
Adamia versicolor Fortune
Adenanthera pavonina L.
Aerides Lour.
Aerides fieldingii Jenn.
Aerides odoratum Lour.
Aesculus californica Nutt.
Aesculus chinensis Bunge
Aesculus hippocastanum L.
Aesculus parviflora Walter
Aesculus pavia L.
Agapanthus africanus Hoffmanns.
Agathis australis (D. Don) Loudon
Agave L.
Agave virginica L.
Ageratum L.
Ageratum houstonianum Mill.
Aglaia odorata Lour.
Aglaonema Schott
Agrimonia eupatoria L.
Agrostis maxima Roxb.
Ailanthus Desf.
Ailanthus altissima (Mill.) Swingle
Ailanthus glandulosa Desf.

*This listing provides the nomenclatural authorities for accepted taxa discussed in the text and tables. Taxa provided with authorities in Appendices 3, 4, and 5 are listed here only if they also appear elsewhere in the text.

Ajuga reptans L.
Akebia quinata Decne.
Alcea rosea L.
Aletris chinensis Lam.
Allium sativum L.
Aloysia triphylla Royle
Alpinia calcarata Roscoe
Alstroemeria L.
Althaea rosea (L.) Cav.
Alyssum L.
Amaranthus L.
Amaranthus caudatus L.
Amaranthus cruentus L.
Amaranthus oleraceus L.
Amaranthus tricolor L.
Amaranthus tristis L.
Amaryllis aurea L'Hér.
Amaryllis radiata L'Hér.
Amelanchier Medik.
Amelanchier laevis Wiegand
Amelanchier oblongifolia M. Roem.
Amelanchier ovalis Medik.
Amelanchier racemosa Lindl.
Amherstia nobilis Wall.
Ammannia baccifera L.
Amomum zingiber L.
Amorpha fruticosa L.
Amygdalus pumila L.
Amyris punctata Roxb.
Anaphalis margaritacea (L.) Benth. & Hook. f.
Anchusa officinalis L.
Andromeda chinensis Lodd.
Androsace lanuginosa Wall.
Anemone L.
Anemone coronaria L.
Anemone hupehensis Lemoine
Anemone japonica Siebold & Zucc.
Anemone nemorosa L.
Anemone rivularis Buch.-Ham. ex DC.
Anemone vitifolia Buch.-Ham. ex DC.
Angelica archangelica L.
Anigozanthos Labill.
Annona hexapetala L. f.
Anredera cordifolia (Ten.) Steenis
Anthemis nobilis L.
Anthriscus cerefolium Hoffm.
Anthurium Schott
Antirrhinum L.
Antirrhinum majus L.
Apios americana Medik.
Apios fortunei Maxim.
Aquilegia L.
Aquilegia canadensis L.
Aquilegia fragrans Benth.
Aquilegia vulgaris L.
Arachis hypogaea L.
Aralia papyrifera Hook.
Araucaria Juss.
Araucaria araucana (Molina) K. Koch
Araucaria bidwillii Hook.
Araucaria heterophylla (Salisb.) Franco
Araucaria imbricata Pav.
Arbutus andrachne L.
Arctostaphylos uva-ursi (L.) Spreng.
Aristolochia macrophylla Lam.
Aristolochia sinarum Lindl.
Armeria latifolia Willd.
Armeria maritima Willd.
Artemisia abrotanum L.
Artemisia capillifolia Lam.
Artemisia indica Willd.
Artemisia minima L.
Artemisia sinensis (Pamp.) Ling & Y. R. Ling
Arum divaricatum L.
Arundina chinensis Blume
Asclepias syriaca L.
Asclepias tuberosa L.
Asclepias variegata L.
Asimina triloba (L.) Dunal
Aster L.
Aster chinensis L.
Aster novae-angliae L.
Aster turbinatus S. Moore
Astilbe Buch.-Ham. ex D. Don
Astragalus chinensis L. f.
Astragalus sinicus L.
Astranthus cochinchinensis Lour.
Atriplex hortensis L.
Aubrietia DC.
Aucuba japonica Thunb.
Aurinia saxatilis (L.) Desv.
Averrhoa carambola L.
Azalea crispiflora Hook.
Azalea indica L.

Azalea narcissiflora Fortune ex Planch.
Azalea obtusa Lindl.
Azalea ovata Lindl.
Azalea ramentacea Lindl.
Azalea sinensis Lodd.
Azalea squamata Lindl.
Azalea vittata Dum. Cours.

B

Balduina Nutt.
Bambusa Schreb.
Bambusa nana Roxb.
Banksia L. f
Banksia caleyi R. Br.
Baptisia australis (L.) R. Br.
Barthea chinensis Hook. f.
Bauhinia corymbosa Roxb. ex DC.
Beesia Balf. f. & W. W. Sm.
Begonia L.
Begonia boliviensis A. DC.
Begonia coccinea Hook.
Begonia davisii Veitch ex Hook. f.
Begonia discolor R. Br.
Begonia pearcei Hook.
Begonia veitchii Hook. f.
Bejaria Mutis ex L.
Bellis perennis L.
Berberidopsis corallina Hook. f.
Berberis L.
Berberis aggregata C. K. Schneid.
Berberis aquifolium Pursh
Berberis candidula C. K. Schneid.
Berberis consanguinea Fortune
Berberis darwinii Hook.
Berberis fortunei Lindl.
Berberis ilicifolia L. f.
Berberis thunbergii DC.
Berberis trifurca Lindl. & Paxton
Berberis verruculosa Hemsl. & E. H. Wilson
Berberis vulgaris f. *atropurpurea* Regel
Berberis wilsonii Hemsl.
Bergenia ciliata (Haw.) Sternb.
Bergenia ligulata Engl.
Beta L.
Betula L.
Betula pendula Roth
Bidens chinensis Willd.
Bignonia grandiflora Thunb.
Blackwellia padiflora Lindl.
Bletia Ruiz & Pav.
Bletia hyacinthina R. Br.
Bletia tuberosa (L.) Ames
Bletia verecunda (Salisb.) R. Br.
Bocconia cordata Willd.
Borago L.
Borago officinalis L.
Boronia molloyae J. Drumm.
Brassavola nodosa (L.) Lindl.
Brassica L.
Brassica chinensis L.
Brassica oleracea L.
Brassica violacea L.
Broussonetia papyrifera Vent.
Bryonia tenella Roxb.
Bryoria fremontii (Tuck.) Brodo & D. Hawksw.
Buddleja L.
Buddleja alternifolia Maxim.
Buddleja auriculata Benth.
Buddleja colvillei Hook. f. & Thomson
Buddleja davidii Franch.
Buddleja globosa Hope
Buddleja lindleyana Fortune
Buddleja salviifolia Lam.
Bulbophyllum bicolor Lindl.
Bulbophyllum lobbii Lindl.
Buxus L.
Buxus fortunei Carrière
Buxus sempervirens L.

C

Cactus chinensis Roxb.
Calanthe furcata Bateman ex Lindl.
Calanthe masuca (D. Don) Lindl.
Calanthe ×*dominii* Lindl.
Calathea G. Mey.
Calendula officinalis L.
Callicarpa purpurea Juss.
Callicarpa reevesii Wall. ex Walp.
Callistephus chinensis Nees
Caloscordum exsertum Herb.
Calycanthus floridus L.
Calycanthus praecox L.
Calystegia pubescens var. *flore pleno* Lindl.
Camellia cuspidata (Kochs) C. H. Wright ex H. J. Veitch

Camellia japonica L.
Camellia oleifera C. Abel
Camellia reticulata Lindl.
Camellia saluenensis Stapf ex Bean
Camellia sasanqua Thunb.
Camellia sinensis (L.) Kuntze
Campanula carpatica Jacq.
Campanula grandiflora Jacq.
Campanula medium L.
Campanula nobilis Lindl.
Campanula pyramidalis L.
Campsis chinensis (Lam.) Voss
Campsis fortunei Seem.
Campsis grandiflora K. Schum.
Campsis radicans (L.) Seem. ex Bureau
Camunium chinense Roxb.
Canarium album Raeusch.
Canna indica L.
Cannabis gigantea Crevost
Capparis acuminata Willd.
Caragana chamlagu Lam.
Cardiospermum halicacabum L.
Carpenteria Torr.
Carpesium abrotanoides L.
Carthamus tinctorius L.
Carum carvi L.
Cassia L.
Cassia pumila Lam.
Castanea Mill.
Castanea dentata Borkh.
Castanea pumila (L.) Mill.
Catalpa Scop.
Catalpa bignonioides Walter
Catalpa bungei C. A. Mey.
Catasetum Rich. ex Kunth
Cattleya Lindl.
Cattleya aclandiae Lindl.
Cattleya aurantiaca (Bateman) P. N. Don
Cattleya dowiana Bateman
Cattleya dowiana var. *aurea* (Linden) B. S. Williams & T. Moore
Cattleya eldorado var. *splendens* Linden ex B. S. Williams
Cattleya granulosa Lindl.
Cattleya guttata Lindl.
Cattleya intermedia Graham
Cattleya labiata Lindl.
Cattleya loddigesii Lindl.
Cattleya mossiae C. Parker ex Hook.
Cattleya skinneri Bateman
Cattleya walkeriana Gardner
Cattleya warscewiczii Rchb. f.
Ceanothus lobbianus Hook.
Ceanothus thyrsiflorus Eschsch.
Ceanothus veitchianus Hook.
Cedrela sinensis Juss.
Cedrus deodara (Roxb.) G. Don
Cedrus libani A. Rich.
Celastrus scandens L.
Celosia L.
Celosia argentea L.
Celosia argentea var. *cristata* (L.) Kuntze
Celosia coccinea L.
Centaurea L.
Centaurea cineraria L.
Centaurea cyanus L.
Centranthus ruber DC.
Cephalotaxus fortunei Hook.
Cercis L.
Cercis canadensis L.
Cercis canadensis var. *mexicana* (Rose) M. Hopkins
Cercis chinensis Bunge
Chaenomeles Lindl.
Chamaecyparis pisifera (Siebold & Zucc.) Endl.
Chamaerops excelsa Thunb.
Chapmannia Torr. & A. Gray
Chaptalia Vent.
Chelidonium majus L.
Chenopodium atriplicis L. f.
Chenopodium purpurascens Jacq.
Chimonanthus fragrans Lindl.
Chionanthus retusus Lindl. & Paxton
Chionanthus virginicus L.
Chirita sinensis Lindl.
Chloranthus inconspicuus Sw.
Chloranthus monostachyus R. Br.
Choripetalum obovatum Benth.
Chrysanthemum L.
Chrysanthemum indicum L.
Chrysanthemum morifolium Ramat.
Chrysanthemum sinense Sabine
Cibotium barometz (L.) J. Sm.
Cichorium intybus L.
Cichorium endivia L.

Cinchona L.
Cinchona ledgeriana (Howard) Bern. Moens ex Trimen
Cinchona succirubra Pav. ex Klotzsch
Cineraria L.
Cinnamomum camphora (L.) J. Presl
Cistus L.
Cistus salviifolius L.
Citrus L.
Citrus acida Pers.
Citrus aurantium L.
Citrus decumana L.
Citrus inermis Roxb.
Citrus japonica Thunb.
Citrus meyerii Yu. Tanaka
Citrus nobilis Lour.
Citrus sinensis Osbeck
Cladrastis kentukea (Dum. Cours.) Rudd
Clarkia Pursh
Clarkia unguiculata Lindl.
Clematis L.
Clematis flammula L.
Clematis fortunei T. Moore
Clematis lanuginosa Lindl.
Clematis montana Buch.-Ham. ex DC.
Clematis terniflora DC.
Clematis virginiana L.
Clematis ×*jackmanii* T. Moore
Clerodendron canescens Wall.
Clerodendron cyrtophyllum Turcz.
Clerodendron foetidum Bunge
Clerodendron fortunatum L.
Clerodendron fortunei Hemsl.
Clerodendron fragrans (Vent.) Willd.
Clethra alnifolia L.
Cleyera fortunei Hook. f.
Clianthus puniceus (G. Don) Banks & Sol. ex Lindl.
Clivia Lindl.
Cnicus benedictus L.
Cobaea scandens Cav.
Cocculus DC.
Codiaeum Rumph. ex A. Juss.
Codonopsis Wall.
Coelogyne Lindl.
Colocasia esculenta (L.) Schott
Colutea arborescens L.
Combretum chinense Roxb. ex G. Don
Commelina virginica L.
Consolida ajacis (L.) Schur
Convallaria japonica L. f.
Convallaria majalis L.
Convolvulus bilobatus Roxb.
Convolvulus chinensis Ker Gawl.
Convolvulus reptans L.
Convolvulus tricolor L.
Conyza Less.
Conyza chinensis L.
Conyza hirsuta L.
Conyza patula Dryand.
Cookia punctata Sonn.
Corchorus L.
Corchorus capsularis L.
Corchorus japonicus Thunb.
Coreopsis L.
Coreopsis basalis (A. Dietr.) S. F. Blake
Coreopsis drummondii Torr. & A. Gray
Coreopsis lanceolata L.
Cornus L.
Cornus florida L.
Cornus nuttallii Audubon
Cornus stolonifera Michx.
Cortaderia selloana (Schult. & Schult. f.) Asch. & Graebn.
Corydalis bungeana Turcz.
Corydalis longiflora (Willd.) Pers.
Corydalis sheareri S. Moore
Corylopsis pauciflora Siebold & Zucc.
Corylopsis spicata Siebold & Zucc.
Cosmos pinnatus Jacq. ex Steud.
Cotinus coggygria Scop.
Cotoneaster Medik.
Cotoneaster frigidus Wall. ex Lindl.
Cotoneaster microphyllus Wall. ex Lindl.
Cotoneaster rotundifolius Wall. ex Lindl.
Crataegus L.
Crataegus glabra Thunb.
Crataegus indica L.
Crataegus laevigata DC.
Crataegus phaenopyrum (L. f.) Medik.
Crinum bulbispermum (Burm. f.) Milne-Redh. & Schweick.
Crinum sinicum Roxb. ex Herb.
Crocus L.
Crocus flavus Weston
Crocus sativus L.

Crocus vernus (L.) Hill
Crotalaria elliptica Roxb.
Croton L.
Croton aromaticus L.
Croton lachnocarpus Benth.
Croton sebiferum L.
Cryptomeria japonica (Thunb. ex L. f.) D. Don
Cucumis L.
Cucumis acutangulus L.
Cucurbita pepo L.
Cunninghamia sinensis R. Br.
Cupressus L.
Cupressus funebris Endl.
Cupressus sempervirens L.
Cuscuta chinensis Lam.
Cycas revoluta Thunb.
Cyclamen L.
Cydonia oblonga Mill.
Cydonia sinensis Thouin
Cymbidium devonianum Paxton
Cymbidium huttonii Hook. f.
Cymbidium iridioides D. Don
Cymbidium lowianum Rchb. f.
Cymbidium madidum Lindl.
Cymbidium parishii Rchb. f.
Cymbidium sinense Willd.
Cymbidium tesselatum Sw.
Cymbidium tigrinum Parish ex Hook. f.
Cynara scolymus L.
Cynoglossum grandiflorum Benth.
Cyperus pangorei Rottb.
Cyphomandra betacea (Cav.) Sendtn.
Cypripedium L.
Cypripedium calceolus L.
Cypripedium pubescens Willd.
Cypripedium spectabile Salisb.
Cyrtandra fortunei C. B. Clarke
Cyrtomium fortunei J. Sm.

D

Dahlia Cav.
Dalechampia parvifolia Lam.
Daphne fortunei Lindl.
Daphne genkwa Siebold & Zucc.
Daphne mezereum L.
Daphne odora Thunb.
Daphne sinensis Lam.
Daphniphyllum calycinum Benth.
Datura ferox L.
Daucus carota L. subsp. *sativa* (Hoffm.) Archangeli
Davidia Baill.
Davidia involucrata Baill.
Delphinium L.
Delphinium cardinale Hook.
Delphinium elatum L.
Delphinium grandiflorum L.
Dendrobium Sw.
Dendrobium albosanguineum Lindl. & Paxton
Dendrobium auriferum Lindl.
Dendrobium bellatulum Rolfe
Dendrobium brymerianum Rchb. f.
Dendrobium catenatum Lindl.
Dendrobium hookerianum Lindl.
Dendrobium linguiforme Sm.
Dendrobium pendulum Roxb.
Dendrobium speciosum Sm.
Desmanthus natans Willd.
Deutzia Thunb.
Deutzia crenata Siebold & Zucc.
Deutzia gracilis Siebold & Zucc.
Dianthus barbatus L.
Dianthus caryophyllus L.
Dianthus chinensis L.
Dianthus japonicus Thunb.
Dianthus plumarius L.
Dicentra Bernh.
Dicentra spectabilis (L.) Lem.
Dicksonia L'Hér.
Dictamnus albus L.
Dieffenbachia Schott
Dielytra spectabilis G. Don
Dienia congesta Lindl.
Digitalis purpurea L.
Dimocarpus litchi Lour.
Dimocarpus longan Lour.
Dimorphorchis lowii Rolfe
Dioscorea batatas Decne.
Diospyros kaki Thunb.
Diospyros virginiana L.
Disa ×kewensis Hort. ex W. Wats.
Dodecatheon meadia L.
Dolichos obcordatus Roxb.
Dolichos phaseoloides Roxb.
Dolichos sinensis L.
Dracaena Vand. ex L.

Dracaena ferrea L.
Dracula wallisii (Rchb. f.) Luer
Drummondita Harv.
Dryandra cordata Thunb.

E

Echinacea Moench
Echinacea purpurea (L.) Moench
Echium vulgare L.
Edgeworthia chrysantha Lindl.
Edgeworthia gardneri (Wall.) Meisn.
Elaeagnus latifolia L.
Elaeagnus umbellata Thunb.
Elaeaodendron fortunei Turcz.
Elaeocarpus decipiens Hemsl.
Elliottia Muhl. ex Elliott
Encyclia cochleata (L.) Dressler
Enkianthus reticulatus Lindl.
Epacris purpurascens Sims
Epidendrum cochleatum L.
Epidendrum ensifolium L.
Epidendrum fragrans Sw.
Epidendrum nodosum L.
Epidendrum retusum L.
Epidendrum rigidum Jacq.
Epipremnum Schott
Eriobotrya japonica Lindl.
Erycibe obtusifolia Benth.
Erysimum cheiri (L.) Crantz
Erythrodes sprucei Garay
Escallonia Mutis ex L. f.
Eschscholzia californica Cham.
Eucalyptus globulus Labill.
Eugenia michelii Lam.
Euonymus L.
Euonymus americanus L.
Euonymus gracillimus Hemsl.
Euonymus radicans Siebold ex Miq.
Eupatorium fortunei Turcz.
Eupatorium reevesii Wall.
Euryale ferox Salisb.
Exacum L.
Exochorda racemosa (Lindl.) Rehder

F

Fagara avicennae Lam.
Fagara nitida Roxb.
Fagopyrum esculentum Moench
Fagraea fragrans Roxb.
Fagus sylvatica L.
Farfugium grande Lindl.
Ficus L.
Ficus benghalensis L.
Ficus hirta Vahl
Ficus nitida Thunb.
Ficus pertusa L. f.
Ficus pumila L.
Ficus religiosa L.
Ficus stipulata Thunb.
Filipendula vestita (Wall.) Maxim.
Filipendula vulgaris Moench
Flacourtia chinensis Clos
Flemingia prostrata Roxb. f. ex Roxb.
Fontanesia fortunei Carrière
Forsythia Vahl
Forsythia fortunei Lindl.
Forsythia viridissima Lindl.
Fortunaea chinensis Lindl.
Fothergilla gardenii L.
Fragaria L.
Fragaria indica Andrews
Franklinia alatamaha W. Bartram ex Marshall
Fraxinus L.
Fraxinus americana L.
Fraxinus chinensis Roxb.
Fraxinus retusa Champ. ex Benth.
Fremontodendron californicum (Torr.) Coville
Fritillaria imperialis L.
Fritillaria verticillata Willd.
Fuchsia L.
Fuchsia colensoi Hook. f.
Fuchsia procumbens R. Cunn. ex A. Cunn.
Fumaria caudata Lam.
Fumaria spectabilis L.
Funkia fortunei Baker

G

Gaertnera obtusifolia Roxb.
Galanthus nivalis L.
Galega officinalis L.
Galium verum L.
Gardenia J. Ellis
Gardenia augusta Merr.
Gardenia florida L.
Gardenia florida var. *fortuniana* Lindl.
Gardenia micrantha Thunb.

Gardenia radicans Thunb.
Gardenia spinosa Thunb.
Gazania Gaertn.
Gelsemium sempervirens (L.) J. St.-Hil.
Gentiana farreri Balf. f.
Gentiana fortunei Hook.
Geranium lambertii Sweet
Geranium nepalense Sweet
Geranium wallichianum D. Don ex Sweet
Geranium sibiricum L.
Gerbera L.
Geum rivale L.
Geum urbanum L.
Ginkgo biloba L.
Gladiolus L.
Gladiolus purpureo-auratus Hook. f.
Glechoma hederacea L.
Gleditsia L.
Gleditsia horrida Salisb.
Gleditsia macrantha Hort.
Gleditsia sinensis Lam.
Gleditsia triacanthus Mill.
Glochidion fortunei Hance
Glossaspis antennifera Rchb. f.
Glottidium Desv.
Gloxinia L'Hér.
Glycine chinensis Sims
Glycine sinensis Lindl.
Glyptostrobus heterophyllus Endl.
Gomphrena globosa L.
Gossypium L.
Gossypium hirsutum L.
Grevillea R. Br. ex Knight
Grevillea acanthifolia A. Cunn.
Grevillea caleyi R. Br.
Grewia nitida Juss.
Grislea tomentosa Roxb.
Gymnocladus chinensis Baill.
Gypsophila elegans M. Bieb.
Gypsophila paniculata L.

H

Halesia fortunei Hemsl.
Hartwegia Nees
Hastingia coccinea Sm.
Hebe Comm. ex Juss.
Hedera L.
Hedera helix L.
Hedychium gardnerianum Roscoe
Hedyotis paniculata Lam.
Hedysarum lagopodioides G. Forst.
Hedysarum tomentosum Thunb.
Helianthus L.
Helianthus annuus L.
Helianthus tuberosus L.
Helichrysum bracteatum (Vent.) Haw.
Helicteres spicata Colebr. & G. Don
Heliotropium arborescens L.
Helleborus niger L.
Hemerocallis L.
Hemerocallis cordata Thunb.
Hemerocallis fulva L.
Hemerocallis lilioasphodelus L.
Hemerocallis plantaginea Lam.
Hesperaloe Engelm.
Hesperis matronalis L.
Heuchera L.
Hevea Aubl.
Hibiscus L.
Hibiscus chinensis Mast.
Hibiscus manihot L.
Hibiscus moscheutos L.
Hibiscus mutabilis L.
Hibiscus syriacus L.
Hibiscus syriacus var. *chinensis* Lindl.
Hippeastrum Herb.
Hordeum L.
Hosta lancifolia Engl.
Hosta plantaginea Aschers.
Hosta sieboldiana Engl.
Hosta ventricosa (Salisb.) Stearn
Hovea celsii Bonpl.
Hovenia dulcis Thunb.
Hoya carnosa R. Br.
Humea Sm.
Hyacinthoides italica (L.) Roth.
Hyacinthus L.
Hyacinthus orientalis L.
Hydrangea L.
Hydrangea chinensis Maxim.
Hydrangea hortensia Siebold
Hydrangea hortensis Sm.
Hydrangea macrophylla (Thunb.) Ser.
Hydrangea paniculata Siebold
Hydrangea paniculata var. *grandiflora* Siebold
Hydrangea quercifolia W. Bartram
Hypericum chinense Osbeck

Hypericum erectum Thunb.
Hypericum hookerianum Wight & Arn.
Hypericum monogynum L.
Hypericum uralum Buch.-Ham. ex D. Don
Hyssopus lophanthus L.
Hyssopus officinalis L.

I

Iberis sempervirens L.
Ilex cornuta Lindl. & Paxton
Ilex leptacantha Lindl. & Paxton
Ilex microcarpa Lindl. & Paxton
Ilex opaca Sol.
Ilex reevesiana Fortune
Ilex terago Anom.
Impatiens L.
Impatiens balsamina L.
Impatiens glandulifera Royle
Impatiens walleriana Hook. f.
Incarvillea Juss.
Incarvillea sinensis Lam.
Indigofera decora Lindl.
Ionidium heterophyllum Vent.
Ipomoea L.
Ipomoea batatas (L.) Lam.
Ipomoea pileata Roxb.
Ipomoea purpurea (L.) Roth
Ipomoea quamoclit L.
Iris L.
Iris biglumis Vahl
Iris chinensis Curtis
Iris crocea Jacq.
Iris ensata Thunb.
Iris fimbriata Vent.
Iris germanica L.
Iris latifolia (Mill.) Voss
Iris tectorum Maxim.
Iris xiphium L.
Isatis indigotica Fortune
Isatis tinctoria L.
Ischaemum aristatum L.
Ixia chinensis L.
Ixora coccinea L.
Ixora stricta Roxb.

J

Jacaranda ovalifolia R. Br.
Jacksonia scoparia R. Br.
Jasminum L.
Jasminum affine Royle ex Lindl.
Jasminum hirsutum (L.) Willd.
Jasminum nudiflorum Lindl.
Jasminum officinale L.
Jasminum paniculatum Roxb.
Jasminum pubescens (Retz.) Willd.
Juglans nigra L.
Juglans regia L.
Juncus effusus L.
Juniperus aquatica Roxb.
Juniperus californica Carrière
Juniperus cernua Roxb.
Juniperus chinensis L.
Juniperus communis L.
Juniperus dimorpha Roxb.
Juniperus sphaerica Lindl.
Juniperus virginiana L.
Justicia picta L.

K

Kalanchoe Adans.
Kalanchoe spathulata DC.
Kalmia latifolia L.
Kerria DC.
Kerria japonica (L.) DC.
Kerria japonica var. *pleniflora* Witte
Koelreuteria Laxm.
Koelreuteria paniculata Laxm.

L

Lablab purpureus (L.) Sweet
Laburnum anagyroides Medik.
Laelia Lindl.
Laelia albida Bateman ex Lindl.
Laelia cinnabarina Bateman ex Lindl.
Laelia jongheana Rchb. f.
Lagerstroemia L.
Lagerstroemia chinensis L.
Lagerstroemia indica L.
Lamium chinense Benth.
Lantana camara L.
Larix decidua Mill.
Larix leptolepis (Siebold & Zucc.) Gordon
Latania chinensis Jacq.
Lathyrus latifolius L.
Lathyrus odoratus L.
Laurus aggregata Sims
Laurus dulcis Roxb.
Laurus glauca Thunb.

Lespedeza ciliata Benth.
Leucanthemum ×superbum (Bergmans ex J. W. Ingram) D. H. Kent
Leucojum vernum L.
Lewisia Pursh
Lewisia rediviva Pursh
Liatris spicata Willd.
Ligustrum L.
Ligustrum coriaceum Hort. ex Decne.
Ligustrum lucidum W. T. Aiton
Ligustrum ovalifolium Hassk.
Ligustrum sinense Lour.
Ligustrum vulgare L.
Lilium L.
Lilium auratum Lindl.
Lilium brownii F. E. Brown ex Miellez
Lilium canadense L.
Lilium candidum L.
Lilium chalcedonicum L.
Lilium concolor Salisb.
Lilium formosanum Wallace
Lilium fortunei Lindl.
Lilium japonicum Thunb.
Lilium lancifolium Thunb.
Lilium martagon L.
Lilium pumilum Delile in Redouté
Lilium regale E. H. Wilson
Lilium sinicum Lindl. & Paxton
Lilium speciosum Thunb.
Lilium superbum L.
Lilium tigrinum Ker Gawl.
Lilium wallichianum Schult. f.
Lilium wallichianum var. *neilgherrense* (Wight) H. Hara
Lilium ×testaceum (Lindl.) Turrill
Limodorum tankervilleae Aiton
Limonia trifoliata L.
Linum perenne L.
Linum stelleroides Planch.
Liparis nervosa (Thunb.) Lindl.
Liquidambar L.
Liquidambar styraciflua L.
Liriodendron tulipifera L.
Lobelia L.
Lobelia cardinalis L.
Lobelia erinus L.
Lobelia radicans Thunb.
Lobularia maritima (L.) Desv.
Lonicera aureo-reticulata Hort. ex T. Moore
Lonicera flava Sims
Lonicera flexuosa Thunb.
Lonicera fragrantissima Lindl. & Paxton
Lonicera japonica Thunb. ex Murray
Lonicera maackii (Rupr.) Maxim.
Lonicera periclymenum L.
Lonicera sempervirens L.
Lonicera standishii Jacques
Lonicera tatarica L.
Lupinus polyphyllus Lindl.
Lychnis chalcedonica L.
Lychnis coronaria (L.) Desr.
Lychnis coronata Thunb.
Lychnis flos-cuculi L.
Lychnis grandiflora Jacq.
Lycium barbarum L.
Lycium chinense Mill.
Lycium foetidum L. f.
Lycium halimifolium Mill.
Lycopodium aristatum Humb. & Bonpl. ex Willd.
Lycopodium cernuum L.
Lycoris Herb.
Lycoris radiata Herb.
Lycoris straminea Lindl.
Lysimachia candida Lindl.
Lysimachia klattiana Hance
Lysionotus pauciflorus Maxim.

M

Maclura pomifera (Raf.) C. K. Schneid.
Macranthera Nutt. ex Benth.
Magnolia L.
Magnolia acuminata L.
Magnolia conspicua Salisb.
Magnolia discolor Vent.
Magnolia fuscata Andrews
Magnolia grandiflora L.
Magnolia kobus DC.
Magnolia obovata Thunb.
Magnolia pumila Andrews
Magnolia purpurea Curtis
Magnolia stellata (Siebold & Zucc.) Maxim.
Magnolia tomentosa Thunb.
Magnolia tripetala L.
Magnolia virginiana L.
Magnolia ×soulangeana Hort.

Mahonia aquifolium (Pursh) Nutt.
Malus Mill.
Malus ioensis (Alph. Wood) Britton f. *plena* Rehder
Malus sargentii Rehder
Malva chinensis Mill.
Malva verticillata L.
Mandragora officinarum L.
Mangifera indica L.
Manisuris granularis (L.) Sw.
Maranta L.
Massonia Thunb. ex L. f.
Mastacanthus sinensis Endl. ex Walp.
Matricaria indica Mill.
Matthiola incana var. *annua* (Sweet) Voss
Mayaca Aubl.
Mazus rugosus Lour.
Meconopsis Vig.
Meconopsis betonicifolia Franch.
Meconopsis napaulensis DC.
Medicago denticulata Willd.
Melastoma malabathricum L.
Melia azedarach L.
Melissa officinalis L.
Menispermum hexagynum Roxb.
Mentha perilloides Lam.
Menziesia ferruginea Sm.
Mesembryanthemum L.
Mesembryanthemum crystallinum L.
Mespilus japonica Thunb.
Microdesmis caseariifolia Planch.
Miltonia regnellii Rchb. f.
Miltonia spectabilis Lindl.
Miltoniopsis vexillaria (Rchb. f.) God.-Leb.
Mimosa fruticosa Roxb.
Mirabilis jalapa L.
Mogorium multiflorum Lam.
Momordica balsamina L.
Monarda didyma L.
Monochasma savatieri Franch. ex Maxim.
Moraea chinensis Thunb.
Morus L.
Morus alba L.
Morus atropurpurea Roxb.
Morus nigra L.
Morus rubra L.
Murraya exotica L.
Musa coccinea Andrews
Musa textilis Née
Mussaenda L.
Myosotis L.
Myrica nagi Thunb.
Myrica pensylvanica Mirb.
Myrtus L.
Myrtus tomentosa Vahl

N

Nandina domestica Thunb.
Narcissus L.
Nauclea orientalis (L.) L.
Nelumbium speciosum Willd.
Nepenthes distillatoria L.
Nepeta L.
Nepeta cataria L.
Nephelium litchi Cambess.
Nephthytis Schott
Nerium oleander L.
Nicotiana L.
Nicotiana fruticosa L.
Nicotiana rustica L.
Nidularium Lem.
Nigella hispanica L.
Nomocharis Franch.
Nothofagus Blume
Nothofagus menziesii (Hook. f.) Oerst.
Nyctanthes arbor-tristis L.
Nymphaea pygmaea Aiton

O

Ocimum basilicum L.
Odontoglossum bictoniense (Bateman) Lindl.
Odontoglossum crispum Lindl.
Odontoglossum wallisii Linden & Rchb. f.
Oenothera macrocarpa Pursh
Oerstedella wallisii (Rchb. f.) Hágsater
Olea fragrans Thunb.
Olearia Moench
Oncidium crispum Lodd.
Oncidium flexuosum (Kunth) Lindl.
Oncidium papilio Lindl.
Ophioglossum vulgatum L.
Opuntia humifusa Raf.
Ormosia pachycarpa Champ. ex Benth.
Ornithogalum umbellatum L.
Oryza L.

Osmanthus aquifolium Siebold ex Siebold & Zucc.
Osmanthus fragrans Lour.
Oxydendrum arboreum (L.) DC.

P

Paederia foetida L.
Paeonia L.
Paeonia moutan Sims
Paeonia rockii (S. G. Haw & Lauener) T. Hong & J. J. Li
Paeonia suffruticosa Andrews
Palmae Juss.
Panax L.
Pandorea pandorana (Andrews) Steenis
Panicum lineare L.
Papaver orientale L.
Papaver somniferum L.
Paphiopedilum Pfitzer
Paphiopedilum concolor Pfitzer
Paphiopedilum lowii (Lindl.) Pfitzer
Parthenocissus quinquefolia (L.) Planch.
Parthenocissus tricuspidata (Siebold & Zucc.) Planch.
Passiflora L.
Passiflora caerulea L.
Paulownia imperialis Siebold & Zucc.
Pelargonium L'Hér. ex Aiton
Pelargonium inquinans (L.) L'Hér. ex Aiton
Pelargonium ×hortorum L. H. Bailey
Pelargonium zonale L'Hér.
Penstemon barbatus (Cav.) Roth
Pergularia odoratissima (Lour.) Roxb. ex Sm.
Pericallis D. Don
Perilla ocymoides L.
Periploca graeca L.
Petunia Juss.
Petunia axillaris (Lam.) Britton, Stern & Poggenb.
Petunia integrifolia (Hook.) Schinz & Thell.
Petunia nyctaginiflora Juss.
Petunia violacea Lindl.
Phaius tankervilliae (Banks ex L'Hér.) Blume
Phalaenopsis Blume
Phalaenopsis ×intermedia Lindl.
Phaseolus L.
Phaseolus minimus Roxb.
Phaseolus radiatus L.
Philadelphus L.
Philadelphus coronarius L.
Phillyrea paniculata Roxb.
Philodendron Schott
Philydrum lanuginosum Banks ex Gaertn.
Phlox L.
Phlox drummondii Hook.
Phlox subulata L.
Pholidota chinensis Lindl.
Phormium J. R. Forst. & G. Forst.
Phormium colensoi Hook. f.
Phormium cookianum Le Jol.
Photinia fortuneana Maxim.
Photinia glabra (Thunb.) Maxim.
Photinia prunifolia Lindl.
Physalis alkekengi L.
Physostegia virginiana Benth.
Picea A. Dietr.
Picea abies (L.) H. Karst.
Picea asperata Mast.
Picea glauca (Moench) Voss
Picea likiangensis var. *purpurea* (Mast.) Dallim. & A. B. Jacks.
Picea pungens Engelm.
Picea smithiana (Wall.) Boiss.
Pieris D. Don
Pieris floribunda (Pursh) Benth. & Hook. f.
Pimpinella anisum L.
Pinckneya Michx.
Pinus attenuata Lemmon
Pinus bungeana Zucc.
Pinus coulteri Lamb. ex D. Don
Pinus gerardiana Wall. ex D. Don
Pinus lambertiana Douglas
Pinus lanceolata Lamb.
Pinus monticola Douglas ex D. Don
Pinus nepalensis Chambray
Pinus nigra Aiton
Pinus palustris Mill.
Pinus ponderosa Douglas ex Lawson & C. Lawson
Pinus radiata D. Don
Pinus sabiniana Douglas ex D. Don
Pinus sinensis D. Don
Pinus strobus L.
Pinus sylvestris L.

Pinus wallichiana A. B. Jacks.
Piper nigrum L.
Pistacia terebinthus L.
Pittosporum Banks ex Gaertn.
Pittosporum colensoi Hook. f.
Pittosporum glabratum Lindl.
Pittosporum tobira (Thunb.) W. T. Aiton
Platanthera galeandra Rchb. f.
Platanthera mandarinorum Rchb. f.
Platanus acerifolia (Aiton) Willd.
Platanus hispanica Münchh.
Platanus occidentalis L.
Platanus orientalis L.
Platycladus orientalis (L.) Franco
Platycodon grandiflorus A. DC.
Plectranthus L'Hér.
Plectranthus amethystoides Benth.
Plumbago larpentae Lindl.
Plumbago scandens L.
Plumeria rubra L.
Poa cylindrica Roxb.
Poinsettia Graham
Polemonium caeruleum L.
Polyanthes tuberosa L.
Polycoelium chinense A. DC.
Polygonum chinense L.
Polygonum hydropiper L.
Polygonum tinctorium Aiton
Polypodium dichotomum Thunb.
Polypodium fortunei Kunze ex Mett.
Polypteris Nutt.
Polystichum munitum (Kaulf.) C. Presl
Populus L.
Populus alba L.
Populus deltoides Bartram ex Marshall
Populus nigra L.
Populus orientalis (L.) Franco
Portulaca L.
Portulaca grandiflora Hook.
Potentilla L.
Potentilla reptans L.
Premna microphylla Turcz.
Primula L.
Primula auricula L.
Primula denticulata Sm.
Primula eburnea Balf. f. & R. E. Cooper
Primula elizabethiae Ludlow ex W. W. Sm.
Primula japonica A. Gray
Primula sinensis Lour.
Primula veris L.
Prunus L.
Prunus cerasus L.
Prunus dulcis (Mill.) D. A. Webb
Prunus glandulosa Thunb.
Prunus japonica Thunb.
Prunus mexicana S. Watson
Prunus persica (L.) Batsch
Prunus pseudocerasus Lindl.
Prunus salicina Lindl.
Prunus sargentii Rehder
Prunus serrulata Lindl.
Prunus sinensis Pers.
Prunus spinosa L.
Prunus triloba Lindl.
Pseudolarix amabilis (J. Nelson) Rehder
Pseudotsuga menziesii (Mirb.) Franco
Psychotria reevesii Wall.
Pteris serrulata Forssk.
Pterostigma grandiflorum Benth.
Pueraria DC.
Punica granatum L.
Pyracantha M. Roem.
Pyrus L.
Pyrus cathayensis Hemsl.
Pyrus chinensis Spreng.
Pyrus spectabilis Aiton

Q

Quercus L.
Quercus alba L.
Quercus bambusifolia Fortune
Quercus chinensis Abel
Quercus inversa Lindl. & Paxton
Quercus rubra L.
Quercus sclerophylla Lindl. & Paxton

R

Raphanus chinensis Mill.
Raphanus sativus L.
Rauvolfia L.
Rauvolfia serpentina (L.) Benth. ex Kurz
Reevesia thyrsoidea Lindl.
Reseda odorata L.
Rhamnus L.
Rhamnus theezans L.
Rhaphiolepis indica (L.) Lindl.

Rhapis flabelliformis L'Hér. ex Aiton
Rheum L.
Rheum compactum L.
Rheum palmatum L.
Rheum undulatum Pall.
Rhododendron L.
Rhododendron anthopogon D. Don
Rhododendron arboreum Sm.
Rhododendron calendulaceum (Michx.) Torr.
Rhododendron campanulatum D. Don
Rhododendron campylocarpum Hook. f.
Rhododendron catawbiense Michx.
Rhododendron cinnabarinum Hook. f.
Rhododendron falconeri Hook. f.
Rhododendron ferrugineum L.
Rhododendron fortunei Lindl.
Rhododendron griffithianum Wight
Rhododendron hirsutum L.
Rhododendron indicum (L.) Sweet
Rhododendron kaempferi Planch.
Rhododendron nuttallii Booth ex Nutt.
Rhododendron occidentale (Torr. & A. Gray) A. Gray
Rhododendron ponticum L.
Rhododendron thomsonii Hook. f.
Rhododendron viscosum (L.) Torr.
Rhus L.
Rhus chinensis Mill.
Rhus javanica L.
Rhus microlepis Turcz.
Rhus semialata Murray
Rhus succedanea L.
Rhus typhina L.
Rhynchospermum jasminoides Lindl.
Rhynchostylis retusa (Sw.) Blume
Ribes L.
Ribes fasciculatum Siebold & Zucc.
Ribes rubrum L.
Ribes sanguineum Pursh
Ribes speciosum Pursh
Ribes uva-crispa L.
Ricinus communis L.
Robinia L.
Robinia chamlagu L'Hér.
Robinia hispida L.
Robinia pseudoacacia L.
Rosa L.
Rosa acicularis Lindl.
Rosa alba L.
Rosa anemoniflora Fortune
Rosa banksiae W. T. Aiton
Rosa blanda Aiton
Rosa bracteata J. C. Wendl.
Rosa brunonii Lindl.
Rosa canina L.
Rosa carolina L.
Rosa centifolia L.
Rosa chinensis Jacq.
Rosa damescena Mill.
Rosa gallica L.
Rosa gigantea Collett ex Crep.
Rosa glanduligera Kult.
Rosa indica L.
Rosa laevigata Michx.
Rosa macrophylla Lindl.
Rosa microphylla Desf.
Rosa moschata Herrm.
Rosa multiflora Thunb.
Rosa odorata var. *pseudoindica* (Lindl.) Rehder
Rosa palustris Marshall
Rosa rugosa Thunb.
Rosa semperflorens Curtis
Rosa sericea Lindl.
Rosa setigera Michx.
Rosa sinica L.
Rosa ternata Poir.
Rosa triphylla Roxb.
Rosa wichuraiana Crép.
Rosa ×fortuniana Lindl. & Paxton
Rubus L.
Rubus chamaemorus L.
Rubus fruticosus L.
Rubus idaeus L.
Rubus parvifolius L.
Rudbeckia L.
Rudbeckia laciniata L.
Ruellia indigotica Fortune
Rumea chinensis Wall. ex Steud.
Rumex sanguineus L.

S

Sabatia stellaris Pursh
Sabia japonica Maxim.
Saccharum sinense Roxb.

Sagittaria latifolia Willd.
Sagittaria obtusifolia L.
Saintpaulia H. Wendl.
Salisburia adiantifolia Sm.
Salix L.
Salix babylonica L.
Salix caprea L.
Salpiglossis sinuata Ruiz & Pav.
Salsola L.
Salvia L.
Salvia fortunei Benth.
Salvia plebeia R. Br.
Salvia sclarea L.
Salvia splendens Ker Gawl.
Sambucus canadensis L.
Sanguinaria canadensis L.
Sanguisorba officinalis L.
Sansevieria carnea Andrews
Sapium sebiferum (L.) Roxb.
Saponaria officinalis L.
Sarracenia flava L.
Sarracenia purpurea L.
Sassafras albidum (Nutt.) Nees
Satureja calamintha (L.) Scheele
Saururopsis chinensis Turcz.
Saxifraga fortunei Hook.
Saxifraga sarmentosa L. f.
Scabiosa atropurpurea L.
Schefflera J. R. Forst. & G. Forst.
Schizanthus pinnatus Ruiz & Pav.
Sciadopitys verticillata Siebold & Zucc.
Scilla esculenta Ker Gawl.
Scilla peruviana L.
Scirpus tuberosus Roxb.
Scolopia chinensis Clos
Scutellaria rivularis Wall.
Scytalia longan Roxb.
Sedum L.
Sempervivum tectorum L.
Senecio divaricatus L.
Senecio pseudochina L.
Senna marilandica (L.) Link
Sequoia sempervirens (D. Don) Endl.
Sequoiadendron giganteum (Lindl.) J. Buchholz
Serissa foetida (L. f.) Poir. ex Lam.
Serratula cinerea Roxb.
Sesamum orientale L.
Shortia galacifolia Torr. & A. Gray
Sida abutilon L.
Sida chinensis Retz.
Siegesbeckia orientalis L.
Silene fortunei Vis.
Sinapis brassicata L.
Sinapis chinensis L.
Sinapis juncea L.
Sinningia Nees
Sinningia speciosa (Lodd.) Hiern
Skimmia japonica Thunb.
Skimmia reevesiana Fortune
Smilax aspera L.
Smilax bona-nox L.
Smilax china L.
Smyrnium olusatrum L.
Soja hispida Moench
Solanum betaceum Cav.
Solanum chinense Dunal
Solanum decemdentatum Roxb.
Solanum melongena L.
Solenostemon Thonn.
Solenostemon scutellarioides (L.) Codd
Sophora japonica L.
Sorbus aucuparia L.
Spathiphyllum Schott
Spathoglottis fortunei Lindl.
Spinacia oleracea L.
Spiraea L.
Spiraea bella Sims
Spiraea callosa Thunb.
Spiraea japonica L. f.
Spiraea palmata Thunb.
Spiraea prunifolia Siebold & Zucc.
Spiraea pubescens Turcz.
Spiraea ×*vanhouttei* (Briot) Carrière
Spiranthes stylites Lindl.
Sprekelia formosissima Herb.
Stachys aspera var. *glabrata* Benth.
Stachys officinalis Franch.
Statice fortunei Lindl.
Sterculia balanghas L.
Sterculia platanifolia L. f.
Stewartia ovata (Cav.) Weath.
Stewartia pseudocamellia Maxim.
Stillingia sebifera Michx.
Strychnos paniculata Champ. ex Benth.
Styrax fortunei Hance

Symphoricarpos albus (L.) S. F. Blake
Symphoricarpos orbiculatus Moench
Symplocos congesta Benth.
Syngonium Schott
Syringa L.
Syringa chinensis Willd.
Syringa emodi Wall. ex G. Don
Syringa oblata Lindl.
Syringa villosa Vahl
Syringa vulgaris L.
Syringa ×persica L.

T

Taenitis chinensis Desv.
Tagetes L.
Tagetes erecta L.
Tagetes patula L.
Tanacetum parthenium Sch. Bip.
Taxodium distichum (L.) Rich.
Taxus L.
Taxus cuspidata Siebold & Zucc.
Taxus macrophylla Thunb.
Taxus nucifera L.
Terminalia vernix Lam.
Teucrium chamaedrys L.
Teucrium fortunei Benth.
Teucrium nepetifolium Benth.
Thalictrum fortunei S. Moore
Thea bohea L.
Thea sinensis L.
Thea viridis L.
Thuja L.
Thuja occidentalis L.
Thuja orientalis L.
Thuja plicata Donn
Thujopsis dolabrata (Thunb. ex L. f.) Siebold & Zucc.
Thujopsis standishii Gordon
Thunbergia alata Bojer ex Sims
Tilia L.
Tilia americana L.
Tilia platyphyllos Scop.
Torenia concolor Lindl.
Torreya grandis Fortune ex Lindl.
Tradescantia virginiana L.
Trapa bicornis L. f.
Trichosanthes anguina L.
Tricyrtis hirta Hook.
Trifolium stellatum L.
Triteleia uniflora Lindl.
Triticum L.
Tropaeolum L.
Tropaeolum majus L.
Tropaeolum minus L.
Tropaeolum peregrinum L.
Tsuga (Endl.) Carrière
Tsuga canadensis (L.) Carrière
Tulipa L.

U

Ulmus americana L.
Ulmus virgata Wall. ex Planch.
Urena lobata L.
Urtica nivea L.
Uvaria odorata Lam.
Uvularia chinensis Ker Gawl.

V

Vaccinium L.
Vaccinium angustifolium Aiton
Vaccinium myrtillus L.
Vaccinium oxycoccus L.
Vanda R. Br.
Vanda coerulea Griff. ex Lindl.
Vanda insignis Blume
Vanda multiflora Lindl.
Vangueria spinosa Roxb.
Verbena L.
Verbena platensis Spreng.
Vernonia fortunei Sch. Bip.
Veronica L.
Viburnum L.
Viburnum dilatatum Thunb.
Viburnum hanceanum Maxim.
Viburnum macrocephalum Fortune
Viburnum opulus var. *roseum* L.
Viburnum plicatum Thunb.
Viburnum sempervirens K. Koch
Vinca L.
Vinca major L.
Vinca minor L.
Viola L.
Viola odorata L.
Viola tricolor L.

Vitex chinensis Mill.
Vitex incisa Lam.
Vitex negundo L.
Vitex ovata Thunb.
Vitis L.
Volkameria kaempferi Jacq.

W

Weigela Thunb.
Weigela florida A. DC.
Weigela rosea Lindl.
Wellingtonia Meisn.
Welwitschia Hook. f.
Wikstroemia nutans Champ. ex Benth.
Wisteria frutescens (L.) Poir.
Wisteria sinensis Sweet

X

Xanthium chinense Mill.
Xanthium orientale L.

Y

Youngia dentata DC.
Yucca L.
Yucca aloifolia L.
Yucca filamentosa L.
Yucca gloriosa L.

Z

Zantedeschia aethiopica (L.) Spreng.
Zanthoxylum avicennae (Lam.) DC.
Zanthoxylum trifoliatum (Sw.) Krug & Urb.
Zea mays L.
Zea vulgaris Mill.
Zingiber Mill.
Zinnia L.
Zinnia elegans Jacq.
Ziziphus albens Roxb.
Ziziphus nitida Roxb.
Ziziphus sinensis Lam.
Ziziphus vulgaris Lam.
Zoysia japonica Steud.
Zygocactus K. Schum.

APPENDIX 2

NORTH AMERICAN GARDEN FLORA

DATABASE

I have chosen, perhaps arbitrarily, to examine the North American garden flora for the past 150 years to see when the Asian and Southern Hemisphere plants took the lead. It has to be remembered that this is against a background of earlier imports.

The figures in the following table are in descending order of frequency of origin.

Table A1. Origins of Garden Plants Grown in the United States at Present[1]

Asia (including China)	1,878 species	(China alone: 663 species)
Europe	1,222 species	
North America (including Canada)	1,122 species	(Canada alone: 7 species)
Africa (including South Africa)	770 species	(South Africa alone: 453 species)
South America	631 species	(Brazil alone: 215 species)
Mexico	348 species	
Australia	298 species	
New Zealand	190 species	

1. Calculated from Christopher Brickell and Judith D. Zuk, eds., 1997, *The American Horticultural Society A–Z Encyclopedia of Garden Plants,* New York: DK Publishing, Inc.

RESULTS

A cursory glance at any garden shows that most of the flora is based on exotic imports. The data validate this observation. Asia and the Southern Hemisphere have supplied more species for our American gardens than North America itself, 5,343 compared with 1,122. The surprise is the number of plants that came from Europe. The European imports were either indirectly from the colonial period or more recent, with the preponderance being more recent.

Mexico (together with Central America) and South America combined are the source of 979 species, almost as many as natives. Two hundred of these come from Brazil alone.

The Antipodes (Australia and New Zealand with their associated islands) add up to 488 species, a not insignificant number.

Source of Data for Table A1

Compiling the data from primary sources presented a very difficult task. I prepared brief databases from plant explorers' monographs and other texts where possible. It was not possible to be sufficiently comprehensive in some cases, and in others (such as Bretschneider's and Hemsley's monumental works) the data were overwhelming. The idea of trying to turn the Index Kewensis or some of the U.S. Department of Agriculture's primary lists into a database was also too daunting.

At that point, I decided to use a standard horticultural encyclopedia for the provenance of current garden plants. There are several excellent volumes, but in the end I chose *The American Horticultural Society A–Z Encyclopedia of Garden Plants*, 1997 edition.[193]

Not only is this encyclopedia handsomely produced and illustrated, but it lists the original source of every species described. This information was not available in some of the other encyclopedias, although all of them have reliable botanical and horticultural material, written and edited by experts.

The American Horticultural Society A–Z Encyclopedia of Garden Plants has about 15,000 entries, approximately the same as other encyclopedias. Its goal is to educate the public about currently grown plants. The editors have carefully pruned out any reference to plants that are not in the modern repertoire. For this reason, the work is a compromise between a complete and comprehensive listing of all species ever brought to the United States from abroad and the necessarily briefer lists in other sources. The editors of the encyclopedia present a valid selection based on their knowledge of plants actually used by gardeners and available in nurseries.

By confining my attention to species plants and not including hybrids or cultivars, I felt that the database I prepared from the encyclopedia was reasonably reliable and representative. It reflects plant species grown in all parts of the United States where they will survive.

In a few cases, it turned out that the original plant sent back by the collector was a hybrid, but if there was any doubt, I omitted the entry. A very small number of species plants had no comment about origin, not even to say there was some uncertainty. I omitted these plants, too.

I included all the principal synonyms in the original entries, thus making sure nothing was left out unintentionally. In the end, the database incorporated about 6,750 entries or slightly less than half of all the entries in the encyclopedia. The remaining plants are naturally or artificially engendered hybrids and cultivars.

Modern gardeners do not confine themselves to species plants. Such plants are probably in a minority, but species are a more dependable way of recording the imports. In the real world, new and improved cultivars attract far more attention and account for far more nursery sales. They are tougher, perform better, and are quite often more colorful and attractive. They can also be patented, a new concept.

NURSERIES AND HORTICULTURE IN THE UNITED STATES

An anecdotal aperçu is a point of departure, extremely valuable for taking an argument forward. Statistics establish the validity of the idea.

The U.S. Department of Agriculture began to collect data on the nursery, horticulture, and floriculture sectors of American industry in 1890.[194] The Bureau of the Census also studied what were called "horticultural specialty crops." These ventures reflected the energy and initiative of Norman Jay Colman, first U.S. Secretary of Agriculture, from 1885 to 1889.

Colman had been president of the American Association of Nurserymen twice in the early 1880s. The American Association of Nurserymen had been established by an Act of Congress in 1875.

One of his most important achievements was supporting the passage of the Hatch Act, clearing the way for government agricultural experiment stations.

The Society of American Florists was also a well-organized group at that time. The dean of American horticulturists, Liberty Hyde Bailey at Cornell, considered this society to be the strongest agricultural organization in the country in 1893.

From March 1890 to October 1891, the president of the American Association of Nurserymen, J. Howard Hale, was attached to the census office specifically to tabulate the number of greenhouses and nurseries counted in the preceding census.

Over the next 20 years, further efforts to count greenhouses, horticultural crops, and nurseries were started under Bailey's supervision, but few were published. In that period, powerful changes were occurring in agriculture. Horticulture began to lose its leadership. Large, mechanized farm equipment changed the way crops were grown and harvested.

In the nursery world, the key changes were from a switch from fruit tree stock to ornamental woody plants, a switch from vegetable crops to flowers in many greenhouses, and the increased production of bulbs, flower seeds, and vegetable seeds.

In 1953, M. Truman Fossum, himself an executive with the Society of American Florists before joining the Department of Agriculture, prepared a compendium of statistics between 1890 and 1950 showing these changes.[194] After 1890, the next census that included floriculture was in 1929, followed by 1949, 1959, 1970, and 1980. Since then, there have been analyses based on statistical samples in intermediate years. The Census of Horticultural Specialties has become decennial. The Census and its results are analyzed at frequent intervals.[195] A possible source of confusion is the existence of two sets of data, one from the Department of Agriculture, the other from the Bureau of the Census. In more recent years, the Economic Research Service of the Department of Agriculture has issued several reports on nursery products, with very useful tabulations.

Fossum found that nurseries concentrating almost solely on ornamental plants increased dramatically between 1910 and 1930. Definitions of what constitutes a horticultural specialty crop were to bedevil comparisons from one epoch to another, but the trend toward an expansion of commercial ornamental horticulture was unmistakable.

It could be measured in numbers of businesses, numbers of people employed by them, or the value of the goods they sold. For example, the profit per square foot of greenhouse space increased from 40 cents in 1889 to $1.00 in 1949. Even if this only reflects inflation, the difference is in a positive direction. When measured as a rate, there was one commercial greenhouse per 12,500 persons in the population in 1890 but one per 8,000 in 1950.

The industry contributed a noticeable amount to the gross domestic product, and this ensured its being included in the studies. By 1939, ornamental horticulture constituted almost 2% of the farm aggregate. It hovered around this amount for the next 10 years, dropping very slightly during World War II. By 1950, the aggregate contribution of ornamental horticulture was 0.5% of the nation's retail volume.

After 1930, an additional trend emerged, growing flowers outdoors on a very large scale, not just in greenhouses. Floriculture expanded greatly in the warmer sections of the country, such as Florida and California.

At present, ornamental horticulture is a $3,000,000,000 industry. This figure includes some minor anomalies such as the indoor culture of mushrooms but otherwise reflects mainly woody and herbaceous products of the nursery trade.

It is worth taking a little time to review what products the Department of Agriculture census was counting and how they were tabulated. The initial census in 1890 was at a very high level, dealing with structures and entities, not specifying any particular plants. The only exception was rose bushes. Evidently, they were already a significant section of the trade.

FLORICULTURE

As the census was repeated, a principal category was floriculture, distinct from nursery products. This is still a key category today. The main structure of this category was clear by 1949. The $1,000,000,000 earned by the floral industry in 1949 was split: floriculture 65% and nursery products 35%.

Within floriculture, the distinction is between "finished" and "unfinished" stock. The reason for discussing this in some detail is to show which plants were chosen as guides for the nurserymen in completing the questionnaires. The species and cultivars named were sold in the highest volume. Many others could have been chosen but were sold in a descending order of frequency. This observation will be examined further later.

FINISHED STOCK

Under "finished" stock, there were five subcategories: cut flowers, potted flowering plants, green or foliage plants, cultivated cut florists' greens, and "other" (various minor specialties). "Finished stock" meant that the plants were ready for purchase and enjoyment without further action by the retailer or customer. (In the enumeration below, the country of origin noted is the one that has supplied the varieties or cultivars currently grown in the United States, not necessarily the sole place in which the species is endemic.)

Cut flowers included roses (China), carnations (France), chrysanthemums (Japan), gladiolus (South Africa), orchids (tropics and South America), and gardenias (Africa and Asia).

Potted flowering plants included hydrangeas (Japan), poinsettias (Mexico), philodendron (South America), lilies (Asia), African violets (East Africa), calceolaria (Mexico and South America), and cyclamen (southern Europe and Near East).

Green or foliage plants included cacti and succulents (Mexico and American Southwest), philodendron (South America), ferns (worldwide), palms (tropics), and ivy (North America).

Cultivated, cut florists' greens include various asparagus ferns (South Africa), adiantum (worldwide), holly (worldwide), acacia (Australia), and privet (Japan).

The "other" category contained such novelties as ornamental banana (tropics) and dwarf citrus (semitropics).

UNFINISHED STOCK

This category is also divided, into noncommercial consumption and commercial consumption. These two principal categories include those immature plants prepared by vegetative processes and those grown from seed. The customer purchases these seedlings in quantity, by the dozen, the half dozen, or by the flat.

Noncommercial consumption means that the customer purchases the plant in an immature state and continues the growing process until the plants mature. This category includes mainly bedding plants.

Commercial consumption indicates that the wholesaler starts the process and another business entity continues it before selling the end product to the public.

Noncommercial Consumption

Vegetative stock includes begonia (South America), geranium (*Pelargonium*, South Africa), heliotrope (South America), lantana (Africa and South America), canna (tropical and subtropical South America), and vinca (Africa).

Seedlings include petunia (South America), marigold (Africa), and salvia (North and South America).

Commercial Consumption

Vegetatively reproduced stock includes poinsettias (Mexico), chrysanthemums (Japan), azalea (Asia), carnation (France), hydrangea (Japan), and geranium (*Pelargonium*, South Africa).

Seedlings include antirrhinum (Iberian Peninsula), stock (Mediterranean), calceolaria (Mexico and South America), primula (Asia), cineraria (Atlantic islands), and cyclamen (southern Europe and Near East).

Other unfinished stock includes dormant rose bushes (China), azaleas for forcing (Asia), rhododendrons for forcing (Asia), delphinium clumps (Asia, Europe, and North America), astilbe clumps (Asia and North America), and various flowering trees and shrubs for forcing.

Obviously, plants derived from foreign species, modified in many different ways, supplied the most popular commercial stock for the nursery trade.

ORNAMENTAL HORTICULTURE

The preceding notes show that the line between a florist and nurseryman was often blurred. In order to create the finished or unfinished stock, something resembling a nursery was required. It might be a commercial greenhouse alone or a combination of greenhouse and outdoors plats in which seedlings could be propagated. The distinction was really artificial.

Under the heading "ornamental horticulture," the census listed nine categories for woody and herbaceous plants:

1. Deciduous shade trees, such as oak, elm, maple, linden, ash, and beech
2. Deciduous shrubs, such as azalea, privet, forsythia, viburnum, and barberry
3. Evergreens, such as azalea, rhododendron, camellia, juniper, spruce, and yew
4. Roses
5. Ornamental vines, such as honeysuckle, wisteria, ivy, clematis, and pachysandra
6. Deciduous fruit and nut trees
7. Subtropical fruit trees, such as citrus
8. Other woody nursery plants
9. Herbaceous nursery plants, such as perennials, chrysanthemums, delphiniums, sedum, and asparagus fern

CENSUS 1998

Plants of foreign descent clearly dominated in the nursery business, whether for cut flowers, decorative baskets, or garden plants. Some figures from the 1998 census of horticultural specialties underscore this point even more clearly.

A major category is "bedding plants," principally annual flowers replaced every season. The census itemized the top-selling 21 varieties of annuals in detail and then lumped all the rest together in "other." I have chosen to count the volume of plants sold and not the total price because prices shift and vary frequently. The numbers that follow are all retail. Wholesale figures are also available.

BEDDING PLANTS

The data were listed in two ways: by volume of actual potted plants sold and by the flat. The plants themselves overlapped, but they had different places in the respective lists. For example, geraniums (*Pelargonium*) are more likely to be purchased in a pot at a more mature phase than planted from a flat.

Annuals

The distinction between an annual and a perennial often depends on the climatic zone. Geraniums do not die at the end of the summer in the milder sections of California, nor do impatiens. Fuchsia is also not an annual plant, yet these are all considered to be bedding plants.

Twenty-one varieties of plant accounted for about three quarters of the retail sales. "Other" was the aggregate of the remaining varieties sold in much smaller amounts.

The following is a list of plants sold in flats (in thousands):

Impatiens	19,687
Petunia	13,702
Viola (pansy)	8,650
Begonia	6,319
Tagetes (marigold)	6,132
Geranium (all types) (*Pelargonium*)	3,897
Salvia	2,145
Alyssum	1,843
Antirrhinum	1,478
Portulaca	1,304
Ageratum	1,300
Centaurea (dusty miller)	1,259
Coleus	1,020
Celosia	942
Verbena	897
Vinca	554
Zinnia	554
Gazania	293
Nicotiana	246
Fuchsia	150
Gerbera	114

The total is 72,486,000 flats. By contrast the total of "other" is 26,853,000 flats.

The following are plants sold in pots (in thousands):

Geranium (all types) (*Pelargonium*)	127,996
Impatiens (all types)	48,347
Viola (pansy)	35,216

Begonia	18,156
Petunia	14,474
Vinca	8,036
Tagetes (marigold)	6,377
Coleus	4,439
Gerbera	4,388
Salvia	3,715
Antirrhinum	2,829
Fuchsia	1,878
Centaurea (dusty miller)	1,800
Zinnia	1,772
Verbena	1,732
Celosia	1,407
Portulaca	1,299
Alyssum	1,233
Ageratum	1,222
Gazania	854
Nicotiana	469

The total is 287,639,000 pots. By contrast the total of "other" is 96,800,000 pots.

The original sources for the top 21 plants in the previous lists are as follows:

Ageratum	Mexico (cultivars of *A. houstonianum*)
Alyssum	Europe, North Africa, Asia
Begonia	Tropical and subtropical, mainly Brazil and Central America
Celosia	Equatorial tropics
Solenostemon (coleus)	Tropical Africa, Asia
Plectranthus (coleus)	Tropics
Dusty miller	Italy (*Centaurea cineraria* chiefly)
Fuchsia	Central and South America, New Zealand
Gazania	Africa
Geranium (*Pelargonium*)	South Africa
Gerbera	Africa, Indonesia
Impatiens	Africa, Southern Hemisphere
Tagetes (marigold)	New Mexico to Argentina, Africa
Nicotiana	South America (varieties most frequently used in the U.S.)
Viola (pansy)	Worldwide
Petunia	South America
Portulaca	South America (*P. grandiflora*)
Salvia	Worldwide
Antirrhinum	Iberian Peninsula
Verbena	Worldwide
Zinnia	Mexico (*Z. elegans*); southeastern United States, Mexico (*Z. haageana*)

Herbaceous Perennials

There are similar findings in the category of herbaceous perennial plants.

The following are plants sold in flats (in thousands):

Chrysanthemum	782
Dianthus	253
Hosta	170
Phlox	153
Delphinium	83
Hemerocallis	82
Ornamental grasses	79
Rudbeckia	72
Coreopsis	66
Heuchera	64
Sedum	51
Salvia (perennial)	49
Echinacea	44
Aquilegia	40
Astilbe	37
Veronica	31
Dicentra	29
Ferns (hardy garden)	18
Clematis (nonclimbing)	12
Iris	n/a
Paeonia	n/a

The total of listed plants sold in flats is 2,115,000. The total of other plants sold in flats is 2,994,000.

The dispersion here is much wider than in the bedding plants. The omission of iris and peony does not affect this result. The top 21 plants account for 42% of the total, while the "other" category accounts for 58%.

The following are plants sold in pots (in thousands):

Chrysanthemum	62,049
Hosta	8,230
Ornamental grasses	7,908
Hemerocallis	7,531
Ferns (hardy garden)	3,585
Dianthus	3,482
Coreopsis	3,342
Sedum	2,839
Rudbeckia	2,794
Phlox	2,712

Astilbe	2,212
Echinacea	2,143
Salvia (perennial)	1,987
Delphinium	1,636
Heuchera	1,427
Aquilegia	1,294
Iris	1,199
Veronica	975
Dicentra	944
Paeonia	825
Clematis (nonclimbing)	337

The total of listed plants sold in pots is 1,194,510. The total of other plants sold in pots is 1,020,520.

Original sources of the top 21 perennials are as follows:

Aquilegia	Northern Hemisphere
Astilbe	Asia
Chrysanthemum	Asia, Mediterranean, Arctic
Clematis	Worldwide
Coreopsis	North and Central America
Delphinium	Asia, North America
Dianthus	Europe, Asia
Dicentra	Asia, North America
Echinacea	North America
Ferns	Worldwide
Ornamental grasses	Worldwide
Hemerocallis	Asia
Heuchera	North America
Hosta	Asia
Iris	Northern Hemisphere
Paeonia	Asia, Iberian Peninsula
Phlox	North America
Rudbeckia	North America
Salvia	Europe, Africa, Mexico, America
Sedum	Europe, Asia
Veronica	Europe

POTTED FLOWERING PLANTS

The following are potted flowering plants (in thousands):

Poinsettia	64,661
Chrysanthemum	25,745
Saintpaulia (African violet)	20,254
Rosa	12,509
Azalea (*Rhododendron*)	10,606

Lilium (Easter lily)	10,005
Kalanchoe	7,360
Tulipa	5,960
Cyclamen	5,738
Primula	5,339
Orchid (other)	4,728
Hyacinthus	4,138
Hibiscus	3,970
Begonia	3,191
Narcissus (daffodil)	2,912
Gerbera	2,827
Lilium (other)	2,810
Orchid (*Phalaenopsis*)	2,408
Zygocactus	2,386
Orchid (*Dendrobium*)	2,022
Hydrangea	1,797
Gardenia	1,570
Gloxinia	1,361
Exacum	990
Cineraria	680
Pelargonium (regal)	533
Anthurium	484
Alstroemeria	405
Bromeliad (*Nidularium*)	80

The total of listed potted flowering plants sold is 25,861,700. The total of other potted flowering plants sold is 32,566,000. (I excluded two categories, "other spring bulbs" and "combination planters," because their contents could not be clearly defined for the purposes of this study.)

Original sources of potted flowering plants are as follows:

Saintpaulia (African violet)	Africa
Alstroemeria	South America
Anthurium	Tropics, South America
Azalea (*Rhododendron*)	Asia, eastern North America
Begonia	South and Central America
Bromeliad (*Nidularium*)	South and Central America
Chrysanthemum	Asia, Arctic, Mediterranean
Cineraria (*Pericallis*)	Atlantic islands, Europe, North Africa
Exacum (Persian violet)	Temperate Asia
Gardenia	Africa, Asia
Gerbera	South Africa
Gloxinia (*Sinningia*)	Central and South America
Hibiscus	Warm temperate zones
Hyacinthus	West and Central Asia
Hydrangea	East Asia, eastern America
Kalanchoe	Africa (Madagascar and Namibia)

Lilium (Easter lily)	Japan
Lilium (other)	Asia
Orchid (*Dendrobium*)	Tropics
Orchid (*Phalaenopsis*)	West Pacific
Orchid (other)	Tropics
Poinsettia (*Pinckneya*)	Southeastern United States
Primula	Himalayas
Pelargonium (regal)	South Africa
Rosa	China, Europe
Tulipa	West Asia, southeastern Europe
Zygocactus	Mexico, southwestern United States

FOLIAGE PLANTS SOLD IN POTS

The following are foliage plants sold in pots (in thousands):

Hedera	19,236
Dieffenbachia	13,347
Cacti and succulents	11,216
Epipremnum	10,901
Dracaena	10,857
Palmae	9,973
Spathiphyllum	9,260
Philodendron	7,592
Nidularium (bromeliad)	6,712
Croton	6,195
Ficus	5,718
Ferns	5,690
Schefflera	4,354
Syngonium (*Nephthytis*)	3,719
Maranta (*Calathea*)	3,116
Aglaonema	2,832
Anthurium	608

The total of the listed foliage plants sold in pots is 131,326,000. The total of other foliage plants sold is 61,110,000. (I have omitted combination planters.)

The original sources of the plants in the previous list are as follows:

Aglaonema	Tropical Asia
Anthurium	Tropics, South America
Nidularium (bromeliad)	Central and South America
Cacti and succulents	Mexico, Central and South America, southwestern United States
Croton (*Codiaeum*)	Malaysia, eastern Pacific
Dieffenbachia	Brazil
Epipremnum	Southeast Asia, western Pacific
Fern	Worldwide

Ficus	Tropics and subtropics
Hedera	Africa, Europe, Asia
Maranta (*Calathea*)	Central and South America
Palmae	Tropics
Philodendron	Mexico, Central and South America
Schefflera	Southeast Asia, Central and South America
Spathiphyllum	Tropical Asia, tropical North and South America
Syngonium (*Nephthytis*)	Central and South America

NURSERY PLANTS

The follow are deciduous shade trees sold (in thousands):

Oak (*Quercus*)	2,007
Maple, red (*Acer*)	1,679
Maple, Japanese (*Acer*)	1,272
Ash (*Fraxinus*)	1,228
Birch (*Betula*)	1,079
Maple, other (*Acer*)	1,000
Maple, Norway (*Acer*)	918
Willow (*Salix*)	764
Honey locust (*Gleditsia*)	671
Sweet gum (*Liquidambar*)	618
Maple, sugar (*Acer*)	588
Linden (*Tilia*)	571
Poplar (*Populus*)	559

The total of listed deciduous shade trees sold is 12,954,000. The total of other deciduous shade trees sold is 8,574,000.

The top 13 varieties were 50% more of the volume sold than all the other varieties together.

The original sources of deciduous shade trees are as follows:

Ash (*Fraxinus*)	Europe, Asia, North America
Birch (*Betula*)	North America, Asia, Europe
Honey locust (*Gleditsia*)	Asia
Linden (*Tilia*)	Europe, Asia, North America
Maple, Japanese (*Acer*)	Asia
Maple, Norway (*Acer*)	Europe
Maple, other (*Acer*)	North America, Asia
Maple, red (*Acer*)	North America
Maple, sugar (*Acer*)	North America
Oak (*Quercus*)	Europe, North America
Poplar (*Populus*)	Europe
Sweet gum (*Liquidambar*)	Asia, North America
Willow (*Salix*)	Worldwide

The following are deciduous flowering trees sold (in thousands):

Crape myrtle (*Lagerstroemia*)	1,911
Crabapple (*Malus*)	1,776
Callery pear (*Pyrus*)	1,553
Dogwood (*Cornus*)	1,475
Flowering cherry (*Prunus*)	1,342
Magnolia	790
Flowering plum (*Prunus*)	947
Redbud (*Cercis*)	624
Hawthorn (*Crataegus*)	521
Amelanchier	196
Goldenrain (*Koelreuteria*)	174

The total of the trees listed is 11,309,000. The total of other flowering trees sold is 7,638,000.

The 11 most common varieties of flowering tree were 50% more of the volume sold than all the others together.

The original sources of deciduous flowering trees listed previously are as follows:

Amelanchier	North America, Europe
Callery pear (*Pyrus*)	Europe, western to eastern Asia, North Africa
Crabapple (*Malus*)	Asia, Europe
Crape myrtle (*Lagerstroemia*)	Asia to Australia
Dogwood (*Cornus*)	Asia, eastern North America
Flowering cherry (*Prunus*)	Southeast Asia, northern temperate regions
Flowering plum (*Prunus*)	Asia, northern temperate regions
Goldenrain (*Koelreuteria*)	Asia
Hawthorn (*Crataegus*)	Northern temperate regions
Magnolia	Asia, eastern North America, tropical South America
Redbud (*Cercis*)	Mediterranean, Asia, North America

The following are broadleaf evergreens sold (in thousands):

Azalea (*Rhododendron*)	22,123
Holly (*Ilex*)	14,408
Euonymus	7,715
Buxus	6,357
Privet (*Ligustrum*)	2,655
Pittosporum	2,137
Cotoneaster	1,821
Viburnum	1,163
Rhododendron	996
Pieris	967
Magnolia	592

The total of the listed broadleaf evergreens sold is 59,934,000. The total of other broadleaf evergreen plants sold is 29,932,000.

The 11 top varieties were more than twice the volume of the others together.

The original sources of the broadleaf evergreen plants listed previously are as follows:

Azalea (*Rhododendron*)	Asia, eastern North America
Buxus	Europe, Asia, Africa
Cotoneaster	Asia, Europe, North Africa
Euonymus	Asia
Holly (*Ilex*)	Worldwide
Magnolia	Asia, eastern North America
Pieris	Africa, Pacific islands
Privet (*Ligustrum*)	Asia
Rhododendron	Asia, North America
Viburnum	Northern temperate regions, Asia

The following are coniferous evergreens sold (in thousands):

Pine (*Pinus*)	30,068
Juniperus	18,411
Arborvitae (*Thuja*)	6,048
Yew (*Taxus*)	4,987
Spruce (*Picea*)	4,363
Cypress (*Cupressus*)	2,277
Fir (*Abies*)	1,194
Cedar (*Cedrus*)	711
Hemlock (*Tsuga*)	646

The total number of listed coniferous evergreens is 68,705,000. The total number of other coniferous evergreen plants sold is 10,161,000.

The previously listed nine varieties outnumbered the others by almost seven to one.

The original sources of the plants listed previously are as follows:

Arborvitae (*Thuja*)	Asia, North America
Cedar (*Cedrus*)	Asia, North Africa
Cypress (*Cupressus*)	Asia, eastern Mediterranean
Fir (*Abies*)	Europe and North Africa, Asia, North America
Hemlock (*Tsuga*)	Asia, North America
Juniperus	Northern Hemisphere
Pine (*Pinus*)	Northern Hemisphere
Spruce (*Picea*)	Europe, Asia, North America
Yew (*Taxus*)	Northern Hemisphere

The following are the deciduous shrubs sold (in thousands):

Rosa	39,185
Hydrangea	6,969
Spiraea	5,862
Hibiscus	3,967
Lilac (*Syringa*)	2,439
Viburnum	2,002
Buddleja	1,727
Weigela	1,348

The total of the previously listed deciduous shrubs sold is 63,499,000. The total of other deciduous shrubs sold is 23,238,000.

The top eight varieties accounted for more than three times the volume of the others together. This was largely the effect of the number of rose bushes sold.

The original sources of the plants previously listed are as follows:

Buddleja	Asia
Hibiscus	Africa, Asia, Western Pacific
Hydrangea	East Asia, North America
Lilac (*Syringa*)	Central and West Asia, southeastern Europe
Rosa	Asia, Europe
Spiraea	Europe, Asia, North America
Viburnum	Asia, North America
Weigela	Asia

All of these lists show the massive influence of plants of foreign descent on the nursery business and the flora of this country to a greater or lesser extent. The categories selected by the U.S. Department of Agriculture are fairly distinctive, without too much overlap, and are intuitively appropriate.

These categories do not take account of hybrids and cultivars, concentrating on species and genera. Even in cases where the original plants were endemic in more than one continent, commercial selection in the United States tends to favor foreign origin. The extensive work of hybridization has rendered all the findings much more complex but cannot mask the original sources.

APPENDIX 3

PRINCIPAL PLANTS IMPORTED INTO ENGLAND*

Plant Name/Common Name	Collector	Country of Origin
Abelia chinensis R. Br.	Abel, Clark	China
A. floribunda (M. Martens & Galeotti) Decne.		Mexico
A. schumannii (Graebn.) Rehder	Wilson, E. H.	China
A. uniflora R. Br.	Fortune, Robert	China
Actinidia chinensis Planch.	Fortune, Robert	China
A. kolomikta Maxim.	Maries, Charles	Japan
Aesculus parviflora Walter/bottlebrush tree	Bartram, William	North America
A. pavia L./Carolina buckeye (red buckeye)		North America
Aloysia triphylla Royle/lemon verbena	Commerson, Philibert	Argentina
Amelanchier laevis Wiegand/ choak berry	Argyll, Duke Archibald	North America
A. oblongifolia M. Roem./swamp sugar pear		North America
A. ovalis Medik./dwarf medlar		North America
Arbutus andrachne L.		
A. menziesii Pursh/American madrone	Menzies, Archibald	North America
A. unedo L./strawberry tree		
Aristolochia macrophylla Lam./ Dutchman's pipe; birthwort in U.K.	Bartram, John	North America
Aucuba japonica Thunb./variegated laurel		Japan
Berberidopsis corallina Hook. f./ coralberry	Pearce, Richard	Chile
Berberis aggregata C. K. Schneid./ barberry	Wilson, E. H.	China
B. candidula C. K. Schneid./ barberry	Farges, Père Paul	China

*The information in this appendix was adapted from Alice Coats, 1964, *Garden Shrubs and their Histories* and 1969, *The Plant Hunters*. New York: McGraw-Hill. This taxa list has been condensed from the original source material. Whenever possible, the nomenclature represents the current, accepted names for plant taxa.

Plant Name/Common Name	Collector	Country of Origin
B. darwinii Hook./barberry	Darwin, Charles	South America
B. gaignepainii/barberry		China
B. ilicifolia L. f./barberry	Sparmann, A.	South America
B. thunbergii DC./barberry	Thunberg, C. P.	Japan
B. verruculosa Hemsl. & E. H. Wilson/barberry		China
B. wilsonii Hemsl./barberry	Wilson, E. H.	China
B. ×lologensis/barberry	Comber, H. C.	Chile
Buddleja alternifolia Maxim.	Piasetski, Pavel Jakovlevich	China
B. auriculata Benth.		South Africa
B. colvillei Hook. f. & Thomson	Hooker, Joseph D.	India, North
B. davidii Franch./butterfly bush	Henry, Augustine	China
B. globosa Hope/buddleia		Chile
B. lindleyana Fortune	Fortune, Robert	China
B. magnifica	Wilson, E. H.	China
B. salviifolia Lam./South African sagewood		South Africa
B. wilsonii	Wilson, E. H.	China
Calycanthus accidentalis/ Florida allspice	Douglas, David	North America
C. floridus L./Carolina allspice or sweet bush	Catesby, Mark	North America
Camellia cuspidata (Kochs) C. H. Wright ex H. J. Veitch	Wilson, E. H.	China
C. japonica L./Japan rose	Cunningham, James	China, but native to Japan
C. oleifera C. Abel/oil-bearing tea		Japan/China
C. reticulata Lindl.	Rawes, Capt. James	Japan/China
C. saluenensis Stapf ex Bean	Forrest, George	China
C. sasanqua Thunb.	Cunningham, Joseph	Japan?
C. sinensis (L.) Kuntze/tea	Goff, Captain	China
Campsis chinensis (Lam.) Voss	Kaempfer, Engelbert	Japan
C. radicans (L.) Seem. ex Bureau/ trumpet creeper	Catesby, Mark	North America
Carpenteria Torr.	Frémont, John Charles	North America
Caryopteris mastacanthus Schauer/ blue spiraea	Fortune, Robert	China
C. mongholica Bunge	Von Bunge, Alexander	China
C. tangutica Maxim.	Przewlaski, Nikolai	China
Ceanothus americanus L./red twig; red root; New Jersey tea		North America
C. dentatus Torr. & A. Gray var. *floribundus*	Lobb, William	North America
C. thyrsiflorus Eschsch./ California lilac	Eschscholz, J. F.	North America

Plant Name/Common Name	Collector	Country of Origin
C. ×*lobbianus*	Lobb, William	North America
C. ×*veitchianus* Hook.	Lobb, William	North America
Ceratostigma plumbaginoides Bunge/moustache plant	Bunge, Alexander von/ Fortune, Robert	China
C. willmottianum Stapf	Wilson, E. H.	China
Chaenomeles japonica (Thunb.) Lindl./japonica	Thunberg, C. P.	Japan
Chamaecytisus austriacus (L.) Link/broom		Asia; Europe
C. hirsutus (L.) Link/broom		
C. purpureus (Scop.) Link	Pallas	
Chimonanthus praecox (L.) Link/ wintersweet	De Semedo, Alvarez	China
C. praecox f. *grandiflorus*	Ancel, R. P.	China
Choisya ternata Kunth/Mexican orange blossom	Humboldt, A. von/ Bonpland, A.	Mexico
Chusquea aristata Munro/bamboo		South America
Cistus albidus L./male cistus	Unclear	
C. ladanifer L./gum cistus	Unclear	
C. salviifolius L./female cistus; sage rose		China
Clematis alpina (L.) Mill.		Europe?; Italy?
C. armandii Franch./clematis	Wilson, E. H.	China
C. flammula L./biting clematis; purging periwinkle	Unclear	
C. florida Thunb.	Henry, Augustine	China
C. lanuginosa Lindl./clematis	Fortune, Robert	China
C. macropetala Ledeb./clematis	Purdom, William	China
C. montana Buch.-Ham. ex DC./ great Indian virgin's bower		India/China
C. orientalis L.	Tournefort, Joseph	Unclear
C. patens C. Morren & Decne.	Siebold, Philipp von	Japan
C. rehderiana Craib	Aubert, Père	China
C. tangutica (Maxim.) Korsh./ clematis		China
C. tangutica subsp. *obtusiuscula* (Rehder & E. H. Wilson) Grey-Wilson	Wilson, E. H.	China
C. texensis Buckley	Buckley	United States
C. viticella L.		Asia; Europe
Colutea arborescens L./ bladder senna		
C. istria Mill.	Pococke, Rev. Dr.	Turkey
C. orientalis Lam.	Tournefort, Joseph	Unclear
Cornus alba L./dogwood		Siberia

Plant Name/Common Name	Collector	Country of Origin
C. canadensis L./dogwood "baked apples and pears"		Canada
C. capitata Wall./dogwood		
C. florida L./dogwood	Catesby, Mark	North America
C. kousa F. Buerger ex Miq./ dogwood		Japan
C. kousa var. *chinensis* Osborn/ dogwood	Wilson, E. H.	China
C. mas L./male cornel; cornelian cherry		Asia; Europe
C. nuttallii Audubon/western dogwood	Douglas, David	North America
C. stolonifera Michx./red osier dogwood	Tradescant, John Jr.?	North America
Cotoneaster bullatus Bois		China
C. conspicuus Comber	Ward, F. K.	Tibet
C. divaricatus Rehder & E. H. Wilson	Wilson, E. H.	China
C. franchetii Bois		China
C. frigidus Wall. ex Lindl.	(Native collectors)	India
C. henryanus (C. K. Schneid.) Rehder & E. H. Wilson	Wilson, E. H.	China
C. horizontalis Decne./fishbone; herringbone cotoneaster	David, Père Armand	China
C. lacteus W. W. Sm.	Forrest, George	China
C. microphyllus Wall. ex Lindl.	(Native collectors)	India
C. rotundifolius Wall. ex Lindl.	(Native collectors)	India
C. salicifolius Franch.	David, Père Armand	China
C. serotinus Hutch.	Forrest, George	China
Cytisus battandieri Maire/broom	de Peyerimhoff, M.	Morocco
C. canariensis (L.) Steud./broom (Genista)	Masson, Francis	
C. monspessulanus L./broom		
C. multiflorus (Aiton) Sweet/sweet white Spanish or Portugal broom		Spain or Portugal?
C. nigricans L./broom		
C. oromediterraneus Rivas Mart. et al./ broom		
Daboecia azorica Tutin & E. F. Warb./ heath		Azores
Daphne cneorum L.		Austria
D. genkwa Siebold & Zucc.	Maries, Charles	China
D. mezereum L.		Poland?
D. odora Thunb.		China
D. pontica L.	Tournefort, Joseph	Asia, West
D. retusa Hemsl.	Wilson, E. H.	China

Plant Name/Common Name	Collector	Country of Origin
Deutzia ×candida (Lemoine) Rehder/pride of Rochester		
D. discolor Hemsl. var. *major*	Wilson, E. H.	China
D. gracilis Siebold & Zucc./ snowflower		Japan
D. longifolia var. *farreri*	Farrer, Reginald	China
D. longifolia var. *veitchii*	Wilson, E. H.	China
D. monbeigii W. W. Sm.	Forrest, George	China
D. pulchra S.Vidal	Wilson, E. H.	China
D. purpurascens (Franch. ex L. Henry) Rehder	Delavay, Père	China
D. rubens Rehder	Wilson, E. H.	China
D. scabra Thunb.	Kaempfer, Engelbert	Japan
D. vilmorinae Lemoine & Bois	Farges, Père	China
Diervilla coraeensis (Thunb.) DC.		Japan
D. floribunda Siebold & Zucc.		Japan
D. florida (Bunge) Siebold & Zucc.	Fortune, Robert	China
D. hortensis Siebold & Zucc.		Japan
D. hortensis var. *nivea*		Japan
D. japonica (Thunb.) DC.		Korea
D. lonicera Mill.		Canada
D. middendorfiana Carrière		
D. praecox Lemoine		Korea
D. venusta Stapf		Korea
Drimys aromatica (R. Br. ex DC.) F. Muell.		Tasmania
D. winteri J. R. Forst. & G. Forst./ Winter's bark	Winter, Capt. (with Drake' s squadron)	South America
Elaeagnus angustifolia L./oleaster, Jerusalem willow, wild olive		
E. argentea Pursh/silverberry		North America
E. glabra Thunb.		Japan
E. inermis Mill.		
E. macrophylla Thunb.	Thunberg, C. P.	Japan
E. multiflora Thunb.		Japan
E. orientalis L./zungeed; zinzerd (Farsi)		Persia?
E. pungens Thunb.		Japan
Eleagnus aculeatus		
Embothrium coccineum J. R. Forst. & G. Forst./Chilean fire bush	Lobb, William	Chile
E. lanceolatum Ruiz & Pav.		Chile
E. longifolium		Chile
Enkianthus campanulatus (Miq.) G. Nicholson	Maries, Charles	Japan

Plant Name/Common Name	Collector	Country of Origin
E. cornuus		Japan
E. quinqueflorus Lour.	Jenkinson, Robert	China
Erica arborea L./tree heath		
E. australis L./Spanish heath		Spain?
E. carnea L./Austrian heath		
E. ciliaris L./heath		
E. concinna/heath		South Africa
E. lusitanica Rudolphi/Portugal heath		Portugal?
E. mediterranea L./heath		
E. terminalis Salisb./Corsican heath		
E. tubiflora/heath		South Africa
Escallonia illinita C. Presl	Lobb, William	
E. macrantha Hook. & Arn.	Lobb, William	Island of Chiloé
E. myrtilloides L. f.		
E. organensis Gardner	Lobb, William	Brazil
E. pterocladon Hook.	Lobb, William	Patagonia
E. rubra (Ruiz & Pav.) Pers.		Chile
E. virgata (Ruiz & Pav.) Pers.		Chile
Eucryphia cordifolia Cav.		Chile
E. glutinosa (Poepp. & Endl.) Baill.	Pearce, Richard	Chile
Euonymus americanus L./strawberry bush; burning bush		North America
E. atropurpureus Jacq./burning bush		
E. fortunei (Turcz.) Hand.-Mazz.	Fortune, Robert	Japan
E. japonicus Thunb./Chinese box		Japan
E. latifolius (L.) Mill.		
Fallopia baldschuanica (Regel) Holub/ Russian vine	Regel, Albert von	Russia
Fargesia murielae (Gamble) T. P. Yi	Wilson, E. H.	China
F. nitida (Mitford) Keng f. ex T. P. Yi	Potanin, G.	China
Fatsia japonica (Thunb.) Decne. & Planch./fig-leaf palm	Thunberg, C. P.	Japan
Forsythia europaea Degen & Bald.		Albania
F. giraldiana Lingelsh.		China
F. ovata Nakai	Wilson, E. H.	Korea
F. suspensa (Thunb.) Vahl	Thunberg, C. P.	Japan
F. viridissima Lindl.	Fortune, Robert	China
Fuchsia coccinea Aiton	Feuillee, Père R. P. L.	Brazil
F. fulgens DC.	Hartweg, Theodore	Mexico
F. triphylla L.	Plumier, Charles	Caribbean islands
Garrya elliptica Douglas ex Lindl./ quinine bush; silk tassel bush	Menzies, Archibald	North America
Gaultheria procumbens L./checker-berry; boxberry; creeping wintergreen		North America
G. shallon Pursh	Menzies, Archibald/ Douglas, David	North America

Plant Name/Common Name	Collector	Country of Origin
Genista monosperma (L.) Lam.	Bentinck, William	Spain
G. monspessulana (L.) L. A. S. Johnson		
G. sagittalis L.		
G. virgata Lam./madeira broom	Masson, Francis	Madeira
Halimium acymoides		
H. halimifolium		Spain?/Portugal?
H. lasianthum (Lam.) Spach.		Portugal
H. umbellatum Spach		Spain?/Portugal?
Hamamelis japonica Siebold & Zucc. var. *aborea*/witch hazel		China?/Japan?
H. mollis Oliv./witch hazel	Maries, Charles	China
H. virginiana L./witch hazel		North America
Hedera chrysocarpa Walsh/Italian ivy		
H. colchica (K. Koch) K. Koch/ivy	Kaempfer, E./Fortune, R./ Wallich, N.	Japan/China/Nepal
Helianthemum alpestre		
H. glaucum		
Hibiscus syriacus L./Syrian mallow		
Hydrangea arborescens L.		North America
H. hortensia Siebold	Kaempfer, Engelbert	Japan
H. paniculata Siebold		Japan
H. paniculata var. *floribunda*	Maximowicz, Carl	Japan
H. petiolaris Siebold & Zucc.	Leichtlin, Max	Japan?
H. quercifolia W. Bartram/swamp snowball; oak-leaved hydrangea	Bartram, William	North America
H. sargentiana Rehder	Wilson, E. H.	China
Hypericum calycinum L./ St. John's wort	Wheler, Sir George	Turkey
H. hookerianum Wight & Arn.	Hooker, Sir Joseph	Assam
H. hookerianum var. *rogersii*		Burma
H. leschenaultii Choisy	de la Tour, Leschenault	Java
H. patulum Thunb.	Oldham, Richard	Asia
H. patulum var. *forrestii*	Forrest, George	China
H. patulum var. *henryi*	Henry, Augustine	China
Jasminum grandiflorum L./ great Spanish jasmine		Spain?
J. humile L./Italian jasmine		Italy
J. nudiflorum Lindl./winter jasmine	Bunge, Alexander von	China
J. officinale L./jasmine		Asia
J. officinale var. *affine*	Royle, John	India
J. polyanthum Franch.	Delavay, Père	China
J. primulinum Hemsl.	Wilson, E. H.	China
J. revolutum Sims		Asia
J. sambac (L.) Aiton/Arabian jasmine		Arabia?
J. wallichianum Lindl.		Asia
Kalmia angustifolia L./laurel		North America

Plant Name/Common Name	Collector	Country of Origin
K. latifolia L.	Catesby, Mark	North America
Kerria japonica (L.) DC.	Kaempfer, Engelbert	Japan
Kolkwitzia amabilis Graebn./ beauty bush	Giraldi, Père Giuseppe	China
Laurus nobilis L./bay tree		Germany?
Leycesteria crocothyrsos Airy Shaw	Ward, F. K.	Assam
L. formosa Wall./flowering nutmeg; Elisha's tears		Nepal
Ligustrum japonicum Thunb. var. *rotundifolia*		Japan
L. lucidum	Aiton, W. T.	China
L. quihoui Carrière	Fortune, Robert	China
L. sinense Lour.	Fortune, Robert	China
Lonicera alpigena L./honeysuckle		
L. caerulea L./honeysuckle		
L. caprifolia L./honeysuckle		
L. etrusca Santi/honeysuckle		Italy?
L. fragrantissima Lindl. & Paxton/ honeysuckle	Fortune, Robert	China
L. japonica Thunb. ex Murray/ honeysuckle	Kerr, William	China
L. japonica var. *aureo-reticulata*/ honeysuckle	Fortune, Robert	Japan
L. nigra L./honeysuckle		
L. nitida E. H. Wilson/ honeysuckle	Wilson, E. H.	China
L. sempervirens L./North American trumpet honeysuckle		North America
L. standishii Jacques/honeysuckle	Fortune, Robert	China
L. syringantha Maxim./lilac honeysuckle	Pzrewalski, Nikolai	China
L. tatarica L./honeysuckle		Siberia
L. tragophylla Hemsl./honeysuckle	Wilson, E. H.	China
L. xylosteum L./honeysuckle		Europe
Lycium chinense Mill./Duke of Argyll's tea tree		China
L. halimifolium Mill.		China
Magnolia dealbata Zucc.	Hernández, Francisco	Mexico
M. denudata Desr.	Oldham, Richard	China
M. grandiflora L.	Catesby, Mark	North America
M. liliflora Desr.	Thunberg, C. P.	Japan
M. salicifolia (Siebold & Zucc.) Maxim.		Japan
M. sinensis (Rehder & E. H. Wilson) Stapf	Wilson, E. H.	China
M. stellata (Siebold & Zucc.) Maxim.	Oldham, Richard	Japan

Plant Name/Common Name	Collector	Country of Origin
M. virginiana L./North American swamp or sweet bay	Banister, John	North America
M. wilsonii (Finet & Gagnep.) Rehder	Wilson, E. H.	China
Mahonia aquifolium (Pursh) Nutt./ Oregon grape	Menzies, Archibald	North America
M. bealei (Fortune) Carrière	Fortune, Robert	China
M. japonica (Thunb.) DC.	Thunberg, C. P.	Japan
M. lomariifolia Takeda	Johnston, Major Laurence	China
M. napaulensis DC.		Nepal
Myricaria germanica (L.) Desv./ tamarisk	Grindal, Bishop	Germany
Myrtus communis L./myrtle		
Nandina domestica Thunb./ heavenly bamboo	Kaempfer, Engelbert	Japan
Nerium oleander L./oleander (rose bay, rose laurel)		
Olearia gunniana (DC.) Hook. f./ daisy bush		Australia
O. haastii Hook. f./daisy bush		
O. macrodonta Baker		New Zealand
O. semidentata Decne. ex Hook. f./ daisy bush		New Zealand
Osmanthus delavayi Franch.	Delavay, Père Jean-Marie	China
O. fragrans Lour.		China
O. ilicifolius Standish		Japan
Paeonia delavayi Franch./peony	Delavay, Père Jean-Marie	China
P. delavayi var. *acutiloba*/peony	Wilson, E. H.	China
P. lutea Delavay ex Franch./peony	Delavay, Père Jean-Marie	China
P. moutan Sims/tree peony		China
Paliurus spina-christi Mill./ Christ's thorn		
Parthenocissus henryana (Hemsl.) Diels & Gilg	Henry, Augustine	China
P. quinquefolia (L.) Planch./ Virginia creeper		North America
P. tricuspidata (Siebold & Zucc.) Planch.	Veitch, John Gould	Japan
P. vitacea (Knerr) Hitchc.		North America
Passiflora caerulea L./blue passion flower		South America
P. incarnata L.		North America
Pernettya mucronata (L. f.) Gaudich. ex A. Spreng.	Pernetty, Dom Antoine Joseph	South America
Philadelphus coronarius L./syringa	Busbecq, Ogier Ghiselin	Turkey
P. coulteri S. Watson	Coulter, Thomas	Mexico
P. grandiflorus Willd.		North America

Plant Name/Common Name	Collector	Country of Origin
P. inodorus L.	Catesby, Mark	North America
P. microphyllus A. Gray	Sargent, Charles Sprague	North America
P. pubescens Loisel.		North America
Phillyrea angustifolia L.		Africa; Europe
P. latifolia L.		Africa; Europe
Phlomis fruticosa L./Jerusalem sage		Asia
Phyllostachys bambusoides Siebold & Zucc./timber bamboo		Japan
P. henonis Mitford/bamboo		Japan
P. mitis auct./edible bamboo		Japan
P. nigra (Lodd. ex Lindl.) Munro/ bamboo		Japan?
Pieris floribunda (Pursh) Benth. & Hook. f.	Lyons, John	North America
P. formosa (Wall.) D. Don		India?/Burma?/ Assam?
P. japonica (Thunb.) D. Don ex G. Don	Thunberg, C. P.	Japan
P. taiwanensis Hayata	Wilson, E. H.	China
Poncirus trifoliata (L.) Raf./ Japanese bitter orange	Kaempfer, Engelbert	Japan
Potentilla davurica Nestl.	Busch, Joseph	Russia
Prunus incisa Thunb./Japanese Fuji cherry	Thunberg, C. P.	Japan
P. japonica Thunb.		China
P. laurocerasus L./cherry-laurel		
P. lusitanica L./Portuguese laurel	Fairchild, Thomas	Portugal
P. pumila L./American sand cherry		Canada
P. tenella Batsch/Russian dwarf almond	Sutherland, James	Russia
P. triloba Lindl./Chinese flowering plum	Fortune, Robert	China
Pseudosasa japonica (Siebold & Zucc. ex Steud.) Makino ex Nakai var. *japonica*	Siebold, Philipp von	Japan
Punica granatum L./pomegranate		Persia/Afghanistan
Pyracantha angustifolia (Franch.) C. K. Schneid./firethorn	Soulié, Père	China
P. atalantioides (Hance) Stapf/ firethorn	Wilson, E. H.	China
P. coccinea M. Roem./firethorn		
P. crenulata var. *rogersiana* A. B. Jacks./firethorn	Forrest, George	China
Rhamnus alaternus L./buckthorn		
Rhododendron arboreum Sm.	Hardwicke, Capt.	Nepal

Plant Name/Common Name	Collector	Country of Origin
R. calendulaceum (Michx.) Torr./ fiery azalea	Bartram, William	North America
R. campylocarpum Hook. f.	Hooker, Joseph D.	Sikkim
R. catawbiense Michx.	Fraser, John	North America
R. caucasicum Pall.	Mussin-Puschkin, Count Apollos	Russia
R. chrysanthum Pall.		Russia
R. ciliatum Hook. f.	Hooker, Joseph D.	Sikkim
R. cinnabarinum Hook. f.	Hooker, Joseph D.	Sikkim
R. dauricum L.	Chamier, Anthony	Siberia
R. falconeri Hook. f.	Hooker, Joseph D.	Sikkim
R. ferrugineum L.		Europe
R. fortunei Lindl.	Fortune, Robert	China
R. griersonianum Balf. f. & Forrest	Forrest, George	China
R. griffithianum Wight	Hooker, Joseph D.	Sikkim
R. hirsutum L./alpine rose		Europe
R. indicum (L.) Sweet/sweet azalea; dwarf evergreen		Japan
R. japonicum (A. Gray) J. V. Suringar/ azalea	Siebold, Philipp von	Japan
R. kaempferi Planch./azalea		Japan
R. kiusianum Makino/azalea; dwarf "Kurume"		Japan
R. luteum Sweet/azalea	Tournefort, Joseph	Turkey
R. malvaticum/azalea		Japan
R. maximum L./rock rose of Pennsylvania; rose bay of Carolina	Bartram, John	North America
R. molle (Blume) G. Don/azalea		Japan
R. nudiflorum Torr./azalea pinxter-flower		North America
R. occidentale (Torr. & Gray) A. Gray/ azalea	Lobb, William	North America
R. oldhamii Maxim./azalea		Japan
R. ponticum L./rhododendron	Tournefort, Joseph	Turkey
R. simsii Planch./azalea		Japan?
R. speciosum/azalea		North America
R. thomsonii Hook. f.	Hooker, Joseph D.	Sikkim
R. viscosum (L.) Torr./azalea "swamp honeysuckle"	Banister, John	North America
Rhus coriaria L./Currier's shrub		
R. cotinoides Nutt. ex Torr. & A. Gray	Nuttall, Thomas	North America
R. cotinus L./Venice or silken sumach		
R. glabra L.		North America
R. radicans L./poison ivy		North America

Plant Name/Common Name	Collector	Country of Origin
R. toxicodendron L./poison oak		North America
R. typhina L./stag's-horn sumac	Banister, John	North America
Ribes aureum Pursh/golden, Missouri, or Buffalo currant	Lewis, Meriwether	North America
R. laurifolium Jancz.	Wilson, E. H.	China
R. sanguineum Pursh/flowering currant	Menzies, Archibald	North America
R. speciosum Pursh	Menzies, Archibald	North America
Robinia hispida L./rose acacia	Catesby, Mark	North America
Romneya coulteri Harv./Matilija poppy; Californian bush poppy	Coulter, Dr. Thomas	Mexico
R. trichocalyx Eastw./California poppy		North America
Rosa alba L./white rose		Europe
R. banksiae W. T. Aiton/Banksia rose	Kerr, William	China
R. bracteata J. C. Wendl./Macartney's rose; Chickasaw rose	Staunton, Sir George	China
R. centifolia L./moss or cabbage rose		Europe
R. chinensis Jacq./China rose (double)		China
R. damescena Mill./autumn damask rose		West Asia
R. foetida Herrm./Austrian rose		Europe
R. gallica L./red rose		Europe
R. gigantea Collett ex Crep.		China
R. hemisphaerica Herrm.	Clusius, C.	Turkey
R. indica auct./'Parson's Pink,' common blush		China, but sent from India
R. indica/'Slater's Crimson China'		
R. indica minima/miniature rose		Mauritius
R. laevigata Michx./Cherokee rose		China
R. moschata/musk rose		West Asia
R. moyesii Hemsl. & E. H. Wilson	Pratt, Antwerp	China/Tibet
R. multiflora var. *platyphylla* Thory/seven sisters rose	Greville, Charles	China
R. odorata (Andrews) Sweet/'Hume's Blush Tea-scented'		China
R. rouletii Correvon/miniature rose	Roulet, Major	Switzerland
R. rubrifolia Villars		
R. rugosa Thunb./Japanese ramanas rose		Japan
R. sericea f. *pteracantha* (Franch.) Mast	Wilson, E. H.	China
R. virginiana Mill./St. Mark's rose		North America
R. wichuraiana Crép.	Wichura, Max	Japan
Rubus biflorus Buch.-Ham. ex Sm.	Hamilton, Dr. Buchanan	Nepal
R. deliciosus Torr./bramble	James, Dr. Edwin	North America
R. odoratus L./bramble		North America

Plant Name/Common Name	Collector	Country of Origin
R. parviflorus Nutt./bramble	Douglas, David	North America
R. phoenicolasius Maxim./Japanese wineberry	Maximowicz, Carl	Japan?
R. spectabilis Pursh/bramble	Douglas, David	North America
R. trilobus Moc. & Sessé ex Ser./bramble		Mexico
Ruscus hypoglossum L./horse tongue		Asia
R. hypophyllum L.		
Sambucus canadensis L./elder		North America
S. racemosa L./elder		
Sarcococca confusa Sealy	Wilson, E. H.	China
S. hookeriana Baill.		
S. hookeriana var. *digyna* Franch.	Wilson, E. H.	China
S. ruscifolia Stapf	Wilson, E. H.	China
S. saligna (D. Don) Müll. Arg.		
Sasa tessellata (Munro) Makino & Shibata/dwarf bamboo		China?/Japan?
Semiarundinaria fastuosa (Mitford) Makino/bamboo		Japan
Skimmia fragrans	Fortune, Robert	China
S. japonica Thunb.	Kaempfer, Engelbert	Japan
Spartium junceum L./Spanish broom		Spain
Spiraea bullata Maxim.	Maximowicz, Carl	Japan
S. chamaedryfolia L.		
S. crenata L.		
S. douglasii Hook.	Douglas, David	North America
S. henryi Hemsl.	Wilson, E. H.	China
S. hypericifolia L.		
S. japonica L. f.		Japan
S. menziesii Hook.		North America
S. prunifolia Siebold & Zucc.	Fortune, Robert	China
S. salicifolia L.	Clusius, C.	Germany
S. thunbergii Siebold ex Blume		Japan
S. tomentosa L./steeplebush		North America
S. veitchii Hemsl.	Wilson, E. H.	China
S. wilsonii Duthie	Wilson, E. H.	China
Staphylea colchica Steven		Russia
S. holocarpa Hemsl.	Wilson, E. H.	China
S. pinnata L./bladder nut		
Symphoricarpos orbiculatus Moench/ coral berry; Indian currant		North America
S. rivularis Suksd./common snowberry	Lewis, Meriwether	North America
Syringa emodi Wall. ex G. Don/lilac	Royle, John	
S. josikaea J. Jacq./lilac	Josika, Baroness	Hungary
S. julianae C. K. Schneid./lilac	Wilson, E. H.	China
S. microphylla Diels/lilac	Purdom, William	China
S. oblata Lindl./lilac	Fortune, Robert	China

Plant Name/Common Name	Collector	Country of Origin
S. persica L. var. *lacinia*/Persian lilac	Meyer, Frank	China
S. pubescens Turcz./lilac	Bretschneider, Dr. Emil	China
S. reflexa C. K. Schneid./lilac	Wilson, E. H.	China
S. sweginzowii Koehne & Lingelsh./lilac	Wilson, E. H.	China
S. tomentella Bureau & Franch./lilac	Wilson, E. H.	China
S. velutina Kom./lilac	Wilson, E. H.	China
S. villosa Vahl/lilac	Bretschneider, Dr. Emil	China
S. vulgaris L./lilac	De Busbecq, Ogier	Turkey
S. wolfii C. K. Schneid./lilac	Wilson, E. H.	China
Tamarix gallica L./tamarisk		Italy, Germany
T. pentandra Pall./tamarisk		
T. tetrandra/tamarisk		
Tricuspidaria lanceolata Miq./ lantern tree	Lobb, William	Chile
Veronica anomala J. B. Armstr./hebe	Armstrong, J. B.	New Zealand
V. cupressoides Hook. f.	Sinclair, Dr. Andrew	New Zealand
V. elliptica G. Forst.		Argentina
V. gigantea		New Zealand
V. hulkeana F. Muell.	Mueller, Sir Ferdinand	New Zealand
V. salicifolia G. Forst.		New Zealand
V. speciosa R. Cunn. ex A. Cunn.	Cunningham, Richard	New Zealand
V. traversii Hook. f.		New Zealand
Viburnum betulifolium Batalin	Wilson, E. H.	China
V. carlesii Hemsl.	Carles, William Richard	Korea
V. fragrans Bunge	Purdom, William	China
V. macrocephalum Fortune/ Chinese snowball tree	Fortune, Robert	China
V. tinus L./laurestinus; wild bay		
V. tomentosum var. *plicatum* (Thunb.) Maxim./Japanese snowball tree	Fortune, Robert	China
Vitex agnus-castus L./chaste-tree		Sicily
V. negundo L.		China
Vitis aestivalis Michx./grapevine		North America
V. amurensis Rupr./grapevine		Manchuria
V. coignetiae Pulliat ex Planch./ grapevine	Coignet, Mme.	Japan
V. davidii (Rom. Caill.) Foëx/ grapevine		China
V. flexuosa Thunb./grapevine		China
V. labrusca L./grapevine		North America
V. thunbergii Siebold & Zucc./ grapevine		Japan
V. vulpina L./riverside grape		North America
Wisteria floribunda (Willd.) DC./ wisteria	Siebold, Philipp von	Japan
W. fruticosa/Carolina kidney bean	Catesby, Mark	North America

Plant Name/Common Name	Collector	Country of Origin
W. sinensis Sweet/wisteria	Reeves, John	China
Yucca aloifolia L.		North America?
Y. filamentosa L./silk grass		North America?
Y. flaccida Haw.	Lyon, John	North America
Y. glauca Nutt.		North America?
Y. gloriosa L./Adam's needle		North America
Y. recurvifolia Salisb.	Lecomte, M.	North America

APPENDIX 4

PLANTS GROWN IN EARLY AMERICA, BOTH NATIVE AND IMPORTED*

TREES

Name	Date Imported
Native	
Abies balsamea (L.) Mill.	
Acer negundo L.	
A. saccharinum L.	
A. saccharum Marshall	
Castanea dentata Borkh.	
Catalpa Scop. sp.	
Cercis canadensis L.	
Cornus florida L.	
Fraxinus americana L.	
Gleditsia triacanthus Mill.	
Ilex opaca Sol.	
Juglans nigra L.	
Juniperus virginiana L.	
Liquidambar styraciflua L.	
Liriodendron tulipifera L.	
Maclura pomifera (Raf.) C. K. Schneid.[1]	
Magnolia acuminata L.	
M. grandiflora L.	
Malus ioensis (Alph. Wood) Britton f. *plena* Rehder	
Picea glauca (Moench) Voss	
P. pungens Engelm.	
Pinus strobus L.	
Platanus occidentalis L.	
Populus deltoides Bartram ex Marshall	
Pseudotsuga menziesii (Mirb.) Franco	
Quercus alba L.[2]	
Q. palustris Münchh.	
Q. rubra L.	

(continued)

*All tables in this appendix were adapted from Denise Wiles Adams, 2004, *Restoring American Gardens: An Encyclopedia of Heirloom Ornamental Plants 1640–1940*, Portland, Oreg.: Timber Press. Whenever possible, the nomenclature represents the current, accepted names for plant taxa.

TREES *(continued)*

Name	Date Imported
Native (continued)	
Robinia pseudoacacia L.	
Thuja occidentalis L.	
Tilia americana L.	
Tsuga canadensis (L.) Carrière	
Ulmus americana L.	
Imported: Sixteenth Century	
Melia azedarach L.	16th century
Aesculus hippocastanum L.	1576
Imported: Seventeenth Century	
Prunus persica (L.) Batsch	1636 (pre)
Imported: Eighteenth Century	
Salix babylonica L.	1730
Platycladus orientalis (L.) Franco	1737
Lagerstroemia indica L.	1747
Populus nigra var. *italica* Du Roi	1750 (pre)
Pinus nigra Aiton	1759
Koelreuteria paniculata Laxm.	1763
Ailanthus altissima (Mill.) Swingle	1784
Populus alba L.	1784
Sorbus aucuparia L.	1792
Imported: Nineteenth Century	
Eucalyptus globulus Labill.	1804
Magnolia ×*soulangeana* Hort.	1820
Cedrus deodara (Roxb.) G. Don	1831
Paulownia imperialis Siebold & Zucc.	1834
Chamaecyparis pisifera (Siebold & Zucc.) Endl.	1861
Imported: No Dates	
Acer platanoides L.	
Betula pendula Roth	
Crataegus laevigata DC.	
Fagus sylvatica L.	
Juniperus communis L. 'Hibernica'	
Larix decidua Mill.	
Palmae Juss.	
Picea abies (L.) H. Karst	
Pinus sylvestris L.	
Tilia platyphyllos Scop.	

1. Lewis and Clark brought *Maclura pomifera* back from their expedition.
2. *Quercus alba* is a favorite of the trade.

SHRUBS

Name	Date Imported	Date First Sold	Nursery
Native			
Calycanthus floridus L.			
Chionanthus virginicus L.			
Clethra alnifolia L.			
Cornus stolonifera Michx.			
Euonymus americanus L.			
Kalmia latifolia L.			
Magnolia virginiana L.			
Mahonia aquifolium (Pursh) Nutt.			
Rhododendron catawbiense Michx.			
Robinia hispida L.			
Symphoricarpos albus (L.) S. F. Blake			
S. orbiculatus Moench			
Imported: Sixteenth Century			
Laburnum anagyroides Medik.	1560		
Philadelphus coronarius L.	1560	1771	Prince
Daphne mezereum L.	1561	1792	Bartram
Syringa vulgaris L.	1563	1652	U.S.[1]
Nerium oleander L.	1596	ca. 1760	
Imported: Seventeenth Century			
Hibiscus syriacus L.	1600	1736	John Custis
Syringa ×persica L.	1614	1734	John Custis
Buxus sempervirens L.	1652[2]		
Cotinus coggygria Scop.[3]	1656	1786	U.S.[1]
Rhododendron indicum (L.) Sweet	1680	1815	SC Nursery
Prunus glandulosa Thunb.	1683	1790	Prince
Lantana camara L.	1692	ca. 1800	
Imported: Eighteenth Century			
Buxus sempervirens L. 'Suffruticosa'	1737		
Camellia japonica L.	1740		
Lonicera tatarica L.	1752		
Gardenia augusta Merr.	1754		
Viburnum opulus var. *roseum* L.	1771		Prince
Eriobotrya japonica Lindl.	1784		
Chaenomeles Lindl. spp.	1784	1832	U.S.[1]

(continued)

SHRUBS *(continued)*

Name	Date Imported	Date First Sold	Nursery
Imported: Eighteenth Century (continued)			
Hydrangea macrophylla (Thunb.) Ser.	1790		
Colutea arborescens L.	1790		U.S.[1]
Imported: Nineteenth Century			
Ligustrum vulgare L.[4]	1800		
Kerria japonica (L.) DC.	1800?	1805	
Rhododendron calendulaceum (Michx.) Torr.	1800	1832	Kendrick
Pittosporum tobira (Thunb.) W. T. Aiton	1804	1846	Paulsen
Deutzia gracilis Siebold & Zucc.	1840		
Spiraea prunifolia Siebold & Zucc.	1843		
Weigela florida A. DC.	1843		
Forsythia viridissima Lindl.	1844		
Ligustrum ovalifolium Hassk.	1847		
Berberis vulgaris f. *atropururea* Regel[4]	1853		
Osmanthus fragrans Lour.	1856		
Hydrangea paniculata var. *grandiflora* Siebold	1862		
Berberis thunbergii DC.	1864		
Spiraea ×*vanhouttei* (Briot) Carrière	1866		
Spiraea japonica L. 'Anthony Waterer'	ca. 1895		
Imported: No Date			
Amorpha fruticosa L.			

1. Available in the United States, but actual nursery not known.
2. The first reference to this plant was on this date.
3. George Washington was the first to own this plant.
4. This plant received a U.S. citation as a useful plant.

HARDY ANNUALS

Name	Date Imported
Native	
Clarkia unguiculata Lindl.	
Eschscholzia californica Cham.	
Phlox drummondii Hook.	
Imported: Seventeenth Century	
Calendula officinalis L.	1631
Matthiola incana var. *annua* (Sweet) Voss	1631
Amaranthus caudatus L.	1665
Dianthus caryophyllus L.	1672
Imported: Eighteenth Century	
Viola tricolor L.	1718
Callistephus chinensis Nees	1735
Ricinus communis L.	1737
Celosia argentea var. *cristata* (L.) Kuntze	1738
Heliotropium arborescens L.	1757
Tropaeolum majus L.	1759
Centaurea cyanus L.	1760
Consolida ajacis (L.) Schur	1760
Gomphrena globosa L.	1760
Impatiens balsamina L.	1760
Papaver somniferum L.	1760
Scabiosa atropurpurea L.	1760
Tagetes erecta L.	1760
Mirabilis jalapa L.[1]	1767
Dianthus chinensis L.	1769
Amaranthus tricolor L.	1786
Reseda odorata L.	1792
Mesembryanthemum crystallinum L.	1793
Pelargonium ×*hortorum* L. H. Bailey	1793
Tagetes patula L.	1793
Zinnia elegans Jacq. (single)	1793
Imported: Nineteenth Century	
Lobularia maritima (L.) Desv.	1806
Gypsophila elegans M. Bieb.	1828
Salvia splendens Ker Gawl.	1833
Schizanthus pinnatus Ruiz & Pav.	1836
Verbena ×*hybrida*	1838
Helichrysum bracteatum (Vent.) Haw.	1839
Portulaca grandiflora Hook.	1839

(continued)

HARDY ANNUALS *(continued)*

Name	Date Imported
Imported: Nineteenth Century (continued)	
Salpiglossis sinuata Ruiz & Pav.	1840
Ageratum houstonianum Mill.	1846
Petunia ×hybrida	1847
Zinnia elegans Jacq. (double)	1858
Lobelia erinus L.	1867
Solenostemon scutellarioides (L.) Codd	1868
Cosmos pinnatus Jacq. ex Steud.	1889

1. Thomas Jefferson was the first to own this plant.

HARDY HERBACEOUS

Name	Date Imported	Date First Sold	Nursery
Native			
Aquilegia canadensis L.			
Asclepias tuberosa L.			
Aster novae-angliae L.			
Baptisia australis (L.) R. Br.			
Coreopsis lanceolata L.			
Dodecatheon meadia L.			
Echinacea purpurea (L.) Moench			
Hibiscus moscheutos L.			
Liatris spicata Willd.			
Lobelia cardinalis L.			
Lupinus polyphyllus Lindl.			
Monarda didyma L.			
Oenothera macrocarpa Pursh			
Penstemon barbatus (Cav.) Roth			
Phlox L.			
P. subulata L.			
Physostegia virginiana Benth.			
Polemonium caeruleum L.			
Rudbeckia laciniata L.			
Senna marilandica (L.) Link			
Tradescantia virginiana L.			
Yucca filamentosa L.			
Imported: Seventeenth Century			
Aconitum napellus L.	1631		
Alcea rosea L.	1631		

(continued)

HARDY HERBACEOUS *(continued)*

Name	Date Imported	Date First Sold	Nursery
Imported: Seventeenth Century (continued)			
Aquilegia vulgaris L.	1631		
Erysimum cheiri (L.) Crantz	1631		
Iris germanica L.	1672		
Tanacetum parthenium Sch. Bip.	1672		
Chrysanthemum L.	1688	1735	Bartram
Imported: Eighteenth Century			
Viola odorata L.	1718		
Digitalis purpurea L.	1737		
Paeonia L. spp.	1737		
Convallaria majalis L.	1738		
Iberis sempervirens L.	1739		
Dictamnus albus L.	1741		
Papaver orientale L.	1741		
Primula veris L.	1747		
Campanula medium L.	1760		
Dianthus barbatus L.	1760		
D. plumarius L.	1760		
Bellis perennis L.	1761		
Antirrhinum majus L.	1767		
Lychnis coronaria (L.) Desr.	1767		
Hosta plantaginea Aschers.	1784		
H. ventricosa (Salisb.) Stearn	1790		
Hemerocallis lilioasphodelus L.	1793		
Hesperis matronalis L.	1793		
Aurinia saxatilis (L.) Desv.	1796		
Linum perenne L.	1796		
Lychnis chalcedonica L.	1796		
Imported: Nineteenth Century			
Achillea ptarmica L.	1806		
Armeria maritima Willd.	1806		
Campanula carpatica Jacq.	1806		
C. pyramidalis L.	1806		
Centranthus ruber DC.	1806		
Delphinium elatum L.	1806		
D. grandiflorum L.	1806		
Filipendula vulgaris Moench	1806		
Lychnis flos-cuculi L.	1806		
Platycodon grandiflorus A. DC.	1806		

(continued)

HARDY HERBACEOUS *(continued)*

Name	Date Imported	Date First Sold	Nursery
Imported: Nineteenth Century (continued)			
Cortaderia selloana (Schult. & Schult. f.) Asch. & Graebn.	1848		
Myosotis L. spp.	1851		
Anemone L. spp.	1851		
Iris ensata Thunb.	1857		
Gypsophila paniculata L.	1862		
Leucanthemum ×*superbum*	1890		

BULBS AND TUBEROUS PLANTS

Name	Date Imported
Native	
Lilium canadense L.	
L. superbum L.	
Imported: Seventeenth Century	
Anemone coronaria L.	1635
Fritillaria imperialis L.	1635
Lilium candidum L.	1654
Tulipa L.	1682
Crocus L.	1684
Imported: Eighteenth Century	
Canna indica L.	1735
Gladiolus L.	1735
Polyanthes tuberosa L.	1735
Galanthus nivalis L.	1737
Narcissus L.	1737
Lilium martagon L.	1739
Hyacinthus orientalis L.	1765
Imported: Nineteenth Century	
Sprekelia formosissima Herb.	1802
Agapanthus africanus Hoffmanns.	1806
Colocasia esculenta (L.) Schott	1806
Iris xiphium L.	1806
Ornithogalum umbellatum L.	1806
Lilium pumilum Delile in Redouté	1816

(continued)

BULBS AND TUBEROUS PLANTS *(continued)*

Name	Date Imported
Imported: Nineteenth Century (continued)	
Dahlia Cav.	1819
Crinum bulbispermum (Burm. f.) Milne-Redh. & Schweick.	1820
Zantedeschia aethiopica (L.) Spreng.	1822
Lilium lancifolium Thunb.	1827
L. speciosum Thunb.	1830
Begonia L.	1839
Hippeastrum ×*johnsonii*	1839
Lilium ×*testaceum* (Lindl.) Turrill	1860
L. auratum Lindl.	1861

VINES AND CLIMBERS

Name	Date Imported	Date First Sold	Nursery
Native			
Apios americana Medik.			
Aristolochia macrophylla Lam.			
Celastrus scandens L.			
Clematis virginiana L.			
Gelsemium sempervirens (L.) J. St.-Hil.			
Lonicera flava Sims			
L. sempervirens L.			
Parthenocissus quinquefolia (L.) Planch.			
Wisteria frutescens (L.) Poir.			
Imported: Sixteenth Century			
Momordica balsamina L.	1568	1793	
Imported: Seventeenth Century			
Ipomoea purpurea (L.) Roth	1621	1793	
I. quamoclit L.	1629	1791	
Imported: Eighteenth Century			
Hedera helix L.[1]	1748		
Lathyrus latifolius L.	1760		
Lonicera periclymenum L.	1790		

(continued)

VINES AND CLIMBERS *(continued)*

Name	Date Imported	Date First Sold	Nursery
Imported: Eighteenth Century (continued)			
Lablab purpureus (L.) Sweet	1790	1806	M'Mahon
Periploca graeca L.	1792		
Cobaea scandens Cav.	1792	1827	Thorburn
Imported: Nineteenth Century			
Jasminum officinale L.	ca. 1800		
Lathyrus odoratus L.	1800		
Passiflora caerulea L.	1806		
Vinca major L.	1806		
V. minor L.	1807		
Tropaeolum peregrinum L.	1810	1839	
Thunbergia alata Bojer ex Sims	1823	1833	
Campsis grandiflora K. Schum.	1827		
Cardiospermum halicacabum L.	1827		
Clematis flammula L.	1827		
Dioscorea batatas Decne.	1827		Prince
Wisteria sinensis Sweet	1832		
Akebia quinata Decne.	1845		
Anredera cordifolia (Ten.) Steenis	1846		
Clematis ×jackmanii T. Moore	1858	1876	
Lonicera japonica Thunb. ex Murray	1862		
Parthenocissus tricuspidata (Siebold & Zucc.) Planch.	1862		
Clematis terniflora DC.	1864		
Lonicera japonica 'Aureareticulata'	1868		Denmark Nursery, IA
Pueraria montana var. *lobata* DC. (kudzu)	1885		
Gourds (five types)			

1. Pehr Kalm collected this plant.

APPENDIX 5

REPRESENTATIVE PLANTS IMPORTED INTO AMERICA (BY DATE)*

TREES

Name	Date
Sixteenth Century	
Aesculus hippocastanum L.	1576
Melia azedarach L.	16th century
Seventeenth Century	
Prunus persica (L.) Batsch	1636
Tilia platyphyllos Scop.	1646
Eighteenth Century	
Salix babylonica L.	1730
Larix decidua Mill.	1736
Platycladus orientalis (L.) Franco	1737
Lagerstroemia indica L.	1747
Pinus nigra Aiton	1759
P. sylvestris L.	1760
Quercus alba L.	1760
Koelreuteria paniculata Laxm.	1763
Picea abies (L.) H. Karst.	1771
Ailanthus altissima (Mill.) Swingle	1784
Populus alba L.	1784
Crataegus laevigata DC.	ca. 1786
Sorbus aucuparia L.	1792
Nineteenth Century	
Eucalyptus globulus Labill.	1804
Betula pendula Roth	1806
Fagus sylvatica L.	1806
Cedrus deodara (Roxb.) G. Don	1831

(continued)

*All tables in this appendix were adapted from Denise Wiles Adams, 2004, *Restoring American Gardens: An Encyclopedia of Heirloom Ornamental Plants 1640–1940*, Portland, Oreg.: Timber Press. Whenever possible, the nomenclature represents the current, accepted names for plant taxa.

TREES *(continued)*

Name	Date
Nineteenth Century (continued)	
Paulownia imperialis Siebold & Zucc.	1834
Juniperus communis 'Hibernica'	ca. 1836
Malus ioensis (Alph. Wood) Britton f. *plena* Rehder	ca. 1840
Chamaecyparis pisifera (Siebold & Zucc.) Endl.	1861
Palmae Juss. (numerous species)	various dates

Total number of trees in original list 60: 34 native, 26 imported.

The numbers are very small and do not reflect every tree imported into the United States. These are only the ones with reliable provenance.

Thirteen species of tree were traced in sufficient detail to be certain about their arrival before 1800. Most were from Europe or the Middle East. Only *Koelreuteria paniculata*, *Lagerstroemia indica*, and *Ailanthus altissima* were from East Asia. The basis of the dates was inclusion in a nursery catalog such as M'Mahon's in 1806.

After 1800, another nine species of tree were noted. A few had no recorded specific date of arrival but were known not to have been available before. *Eucalyptus globulus* came from Australia, *Cedrus deodara* from the Himalayas, *Chamaecyparis pisifera* from Japan, and *Paulownia imperialis* from China. Others were still coming in from Europe, partly because diplomats and wealthy travelers were requesting them specifically.

SHRUBS

Name	Date Imported	Date First Sold	Nursery
Sixteenth Century			
Laburnum anagyroides Medik.	1560		
Seventeenth Century			
Syringa vulgaris L.	1652		
Buxus sempervirens L.	1652 (1st ref.)		
Eighteenth Century			
Syringa ×persica L.	1734		
Hibiscus syriacus L.	1736		
Buxus sempervirens 'Suffruticosa'	1737		
Camellia japonica L.	1740		
Lonicera tatarica L.	1752		
Gardenia augusta Merr.	1754		
Nerium oleander L.	ca. 1760		

(continued)

SHRUBS *(continued)*

Name	Date Imported	Date First Sold	Nursery
Eighteenth Century (continued)			
Philadelphus coronarius L.	1771		
Viburnum opulus var. *roseum* L.	1771		Prince
Eriobotrya japonica Lindl.	1784		
Cotinus coggygria Scop.[1]	1786		
Colutea arborescens L.	1790		
Hydrangea macrophylla (Thunb.) Ser.	1790		
Prunus glandulosa Thunb.	1790		Prince
Daphne mezereum L.	1792		
Nineteenth Century			
Lantana camara L.	ca. 1800		
Ligustrum vulgare L.	1800		
Kerria japonica var. *pleniflora* Witte	1800?	1805	
Rhododendron calendulaceum (Michx.) Torr.	1800	1832	Kendrick
R. indicum (L.) Sweet	1815		
Chaenomeles Lindl. spp.		1832	
Deutzia gracilis Siebold & Zucc.	1840		
Spiraea prunifolia Siebold & Zucc.	1843		
Weigela florida A. DC.		1843	
Forsythia viridissima Lindl.		1844	
Pittosporum tobira (Thunb.) W. T. Aiton	1846		
Ligustrum ovalifolium Hassk.	1847		
Berberis vulgaris f. *atropurpurea* Regel[2]	1853		
Osmanthus fragrans Lour.		1856	
Hydrangea paniculata var. *grandiflora* Siebold	1862		
Berberis thunbergii DC.		1864	
Spiraea ×*vanhouttei* (Briot) Carrière	1866		
Spiraea japonica L. f.	ca. 1895		
Amorpha fruticosa L.	n.d.		

Total number of shrubs in original list 49: 12 native, 37 imported.

n.d., no date known.

1. George Washington was the first to own this plant.
2. This plant received a U.S. citation as a useful plant.

The trend in the shrubs is more clearly defined. Those that can be reliably traced in the nineteenth century from Asia are in the preponderance. Once again, a few had been quietly arriving in the eighteenth century, such as *Hydrangea* and *Gardenia.*

HARDY ANNUALS

Plant	Date
Seventeenth Century	
Calendula officinalis L.	1631
Matthiola incana var. *annua* (Sweet) Voss	1631
Amaranthus caudatus L.[1]	1665
Dianthus caryophyllus L.	1672
Eighteenth Century	
Viola tricolor L.	1718
Callistephus chinensis Nees	1735
Ricinus communis L.	1737
Celosia argentea var. *cristata* (L.) Kuntze	1738
Heliotropium arborescens L.	1757
Tropaeolum majus L.	1759
Centaurea cyanus L.	1760
Consolida ajacis (L.) Schur	1760
Gomphrena globosa L.	1760
Impatiens balsamina L.	1760
Papaver somniferum L.	1760
Scabiosa atropurpurea L.	1760
Tagetes erecta	1760
Mirabilis jalapa L.[2]	1767
Dianthus chinensis L.	1769
Amaranthus tricolor L.	1786
Reseda odorata L.	1792
Mesembryanthemum crystallinum L.	1793
Pelargonium ×*hortorum* L. H. Bailey	1793
Tagetes patula L.	1793
Zinnia elegans Jacq. (single)	1793
Z. elegans Jacq. (double)	1858
Nineteenth Century	
Lobularia maritima (L.) Desv.	1806
Gypsophila elegans M. Bieb.	1828
Salvia splendens Ker Gawl.	1833
Schizanthus pinnatus Ruiz & Pav.	1836
Verbena L. sp.	1838
Helichrysum bracteatum (Vent.) Haw.	1839
Portulaca grandiflora Hook.	1839
Salpiglossis sinuata Ruiz & Pav.	1840

(continued)

HARDY ANNUALS *(continued)*

Plant	Date
Nineteenth Century (continued)	
Ageratum houstonianum Mill.	1846
Petunia Juss. sp.	1847
Lobelia erinus L.	1867
Solenostemon scutellarioides (L.) Codd	1868
Cosmos pinnatus Jacq. ex Steud.	1889

Total number of hardy annuals in original list 41: 3 native, 38 imported.
1. In England since Norman times.
2. Imported by Thomas Jefferson.

The hardy annuals reflect a very broad set of origins, involving the Southern Hemisphere just as much as Asia. The South African *Mesembryanthemum crystallinum* arrived in 1793, about 15 years after Masson collected it. *Pelargonium* ×*hortorum* was a hybrid of *P. zonale* L'Hér. and *P. inquinans* (L.) L'Hér. ex Aiton, both species found in South Africa. *Zinnia*, *Cosmos*, and *Portulaca* came from South America.

HARDY HERBACEOUS PLANTS

Plant	Date
Seventeenth Century	
Aconitum napellus L.	1631
Alcea rosea	1631
Aquilegia vulgaris L.	1631
Erysimum cheiri (L.) Crantz	1631
Iris germanica L.	1672
Tanacetum parthenium Sch. Bip.	1672
Eighteenth Century	
Viola odorata L.	1718
Chrysanthemum L.[1]	1735
Digitalis purpurea L.	1737
Paeonia L. spp.	1737
Convallaria majalis L.	1738
Iberis sempervirens L.	1739
Dictamnus albus L.	1741
Papaver orientale L.	1741
Primula veris L.	1747
Campanula medium L.	1760
Dianthus barbatus L.	1760
D. plumarius L.	1760
Bellis perennis L.	1761
Antirrhinum majus L.	1767

(continued)

HARDY HERBACEOUS PLANTS *(continued)*

Plant	Date
Eighteenth Century (continued)	
Lychnis coronaria (L.) Desr.	1767
Hosta plantaginea Aschers.	1784
H. ventricosa (Salisb.) Stearn	1790
Hemerocallis lilioasphodelus L.	1793
Hesperis matronalis L.	1793
Aurinia saxatilis (L.) Desv.	1796
Linum perenne L.	1796
Lychnis chalcedonica L.	1796
Nineteenth Century	
Achillea ptarmica L.	1806
Armeria maritima Willd.	1806
Campanula carpatica Jacq.	1806
C. pyramidalis L.	1806
Centranthus ruber DC.	1806
Delphinium elatum L.	1806
D. grandiflorum L.	1806
Filipendula vulgaris Moench	1806
Lychnis flos-cuculi L.	1806
Platycodon grandiflorus A. DC.	1806
Dicentra spectabilis (L.) Lem.	1846
Cortaderia selloana (Schult. & Schult. f.) Asch. & Graebn.	1848
Anemone L. spp.	1851
Myosotis L. spp.	1851
Iris ensata Thunb.	1857
Gypsophila paniculata L.	1862
Leucanthemum ×*superbum* (Bergmans ex J. W. Ingram) D. H. Kent	1890

Total number of hardy herbaceous plants in original list 67: 22 native, 45 imported.
1. John Bartram was one of the first to grow this flower.

Eastern and Southern Hemisphere plants continued to be broadly introduced. Chinese and Japanese peonies and chrysanthemums appeared surprisingly early, also the daylily *Hemerocallis*.

BULBS AND TUBEROUS PLANTS

Plant	Date
Seventeenth Century	
Anemone coronaria L.	1635
Fritillaria imperialis L.	1635

(continued)

BULBS AND TUBEROUS PLANTS *(continued)*

Plant	Date
Seventeenth Century (continued)	
Lilium candidum L.	1654
Tulipa L.	1682
Crocus L.	1684
Eighteenth Century	
Canna indica L.	1735
Gladiolus L.	1735
Polyanthes tuberosa L.	1735
Galanthus nivalis L.	1737
Narcissus L.	1737
Lilium martagon L.	1739
Hyacinthus orientalis L.	1765
Nineteenth Century	
Sprekelia formosissima Herb.	1802
Agapanthus africanus Hoffmanns.	1806
Colocasia esculenta (L.) Schott	1806
Iris xiphium L.	1806
Ornithogalum umbellatum L.	1806
Lilium pumilum Delile in Redouté	1816
Dahlia Cav.	1819
Crinum bulbispermum (Burm. f.) Milne-Redh. & Schweick.	1820
Zantedeschia aethiopica (L.) Spreng.	1822
Lilium lancifolium Thunb.	1827
Lilium speciosum Thunb.	1830
Begonia L.	1839
Hippeastrum ×*johnsonii*	1839
Lilium ×*testaceum* (Lindl.) Turrill	1860
Lilium auratum Lindl.	1861

Total number of bulbs and tuberous plants in original list 29: 2 native, 27 imported.

The pattern of imports is similar in this category of plants. South Africa is heavily represented in the nineteenth century, but Mexico and South America were also significant. This reflects the fact that nursery firms were beginning to send collectors to those countries. William Lobb is one of the best known.

VINES AND CLIMBERS

Plant	Date	Nursery
Eighteenth Century		
Hedera helix L.[1]	1748	
Lathyrus latifolius L.	1760	

(continued)

VINES AND CLIMBERS *(continued)*

Plant	Date	Nursery
Eighteenth Century (continued)		
Lonicera periclymenum L.	1790	
Ipomoea quamoclit L.	1791	
Periploca graeca L.	1792	
Ipomoea purpurea (L.) Roth	1793	
Momordica balsamina L.	1793	
Nineteenth Century		
Jasminum officinale L.	ca. 1800	
Lathyrus odoratus L.	1800	
Lablab purpureus (L.) Sweet	1806	M'Mahon
Passiflora caerulea L.	1806	
Vinca major L.	1806	
V. minor L.	1807	
Campsis grandiflora K. Schum.	1827	
Cardiospermum halicacabum L.	1827	
Clematis flammula L.	1827	
Cobaea scandens Cav.	1827	Thorburn
Dioscorea batatas Decne.	1827	Prince
Wisteria sinensis Sweet	1832	
Thunbergia alata Bojer ex Sims	1833	
Tropaeolum peregrinum L.	1839	
Akebia quinata Decne.	1845	
Anredera cordifolia (Ten.) Steenis	1846	
Lonicera japonica 'Halliana'	1862	
Parthenocissus tricuspidata (L.) Planch.	1862	
Clematis terniflora DC.	1864	
Lonicera japonica 'Aureareticulata'	1868	Denmark Nursery, IA
Clematis ×*jackmanii* T. Moore	1876	
Pueraria montana spp. (kudzu)	1885	
Gourds (five types)	Various dates	

Total number of vines and climbers in original list 39: 9 native, 30 imported.

1. Peter Kalm collected this plant.

It is not farfetched to say that the world truly was coming into American gardens by this time. The cumulative effect of these numbers, small as they are and far from statistically valid, is very striking. In the categories given above, there were 82 native species and 203 imported ones by the third quarter of the nineteenth century.

NOTES

1. Coats, Alice M. 1969. *The Plant Hunters.* New York: McGraw-Hill.
2. Preston, Diana, and Michael Preston. 2004. *A Pirate of Exquisite Mind.* London: Walker and Co.
3. Mann, Charles C. 2005. *1491: New Revelations of the Americas before Columbus.* New York: Alfred Knopf.
4. Hooker, Sir Joseph. May 1856. Letter to Darwin. Darwin Correspondence Project, Department of Manuscripts, University Library, Cambridge University.
5. Darwin, Charles. 1844. Manures and Steeping Seeds. *Gardeners' Chronicle,* June 8.
6. Darwin, Charles. 1875. *The Movements and Habits of Climbing Plants.* London: John Murray.
7. Whittle, Tyler. 1970. *The Plant Hunters.* Philadelphia: Chilton Book Company.
8. Ellis, Edward S., and Charles F. Horne. 1921. *The Story of the Greatest Nations and the World's Famous Events,* vol. 1. New York: Auxiliary Educational League.
9. Rackham, Oliver. 2001. *Trees and Woodland in the British Landscape.* London: Phoenix Press.
10. Goodburn, Damian. 1999. An Image of Ancient English Woodland. *British Archaelogy* 42 (March): 10–11.
11. Isaacson, Walter. 2003. *Benjamin Franklin, an American Life.* New York: Simon and Schuster.
12. Campbell-Culver, Maggie. 2001. *The Origin of Plants.* London: Headline.
13. Phillips, Roger, and Martyn Rix. 1993. *The Quest for the Rose.* New York: Random House.
14. Bede, the Venerable. ca. 731. *Historia Ecclesiatica Gentis Anglorum* (*The Ecclesiastical History of the English People*).
15. Arano, Luisa Cogliati, ed. 1976 (reissued 1996). *Medieval Health Handbook* (*Tacuinum Sanitatis*). New York: George Braziller.
16. Cantor, Norman F. 1969. *Medieval History: The Life and Death of a Civilization.* New York: Macmillan.
17. Hobhouse, Penelope. 1992. *Gardening through the Ages.* New York: Simon and Schuster.
18. Monardes, Nicholas. 1569. *Joyfull Newes out of the Newe Found World.* Translated by John Frampton (1574). London.
19. Gerard, John. 1597 (reissued 1636). *The Herball.* London: Adam and Richard Whittakers.
20. Gleason, Henry A., and Arthur Cronquist. 1964. *The Natural Geography of Plants.* New York: Columbia University Press.
21. Ewan, Joseph. 1969. *A Short History of Botany in the United States.* New York: Hafner Publishing Company.
22. Hariot, Thomas. 1588. *A Briefe and True Report of the New Found Land of Virginia.* London.
23. Key, Trevor. 2002. Hardy Yuccas. *Plant Heritage* 9(1).
24. Bacon, Sir Francis. 1629. *Essayes,* 3rd ed. London: R. Allot.
25. Gorer, Richard. 1978. *The Growth of Gardens.* London: Faber and Faber.
26. Champlain, Samuel de. 2004. *Voyages of Samuel de Champlain.* Whitefish, Mont.: Kessinger Publishing.
27. Bauhin, Gaspard. 1623. *Pinax Theatri Botanici.* Basel.
28. Pavord, Anna. 1999. *The Tulip: The Story of a Flower that Has Made Men Mad.* London: Bloomsbury.
29. da Orta, Garcia. 1563. *Colóquios dos simples e drogas he cousas medicinais da India.* Lisbon.
30. Sutherland, James. 1683. *Hortus Medicus Edinburgensis.* Edinburgh.
31. Parkinson, John. 1640. *Theatrum Botanicum.* London.
32. Harvey, John Hooper. 1976. Turkey as a Source of Garden Plants. *Garden History* IV(3): 21–24.

33. Dioscorides, Pedanius. 1655. *De Materia Medica* (*The Greek Herbal of Dioscorides*). John Goodyer, trans., and Robert T. Gunther, ed. (1933, reissued 1985). London: Oxford University Press.
34. Calmann, Gerta. 1977. *Ehret: Flower Painter Extraordinary.* New York: Graphic Society.
35. Clusius, Carolus. 1576. *Rariorum Aliquot Stirpium per Hispanias Observatorum Historia* (*History of Spanish Plants*). Antwerp: Christopher Plantin.
36. Arber, Agnes. 1970. *Herbals: Their Origin and Evolution. A Chapter in the History of Botany 1470–1670.* New York: Hafner.
37. Reed, Howard. 1942. *A Short History of the Plant Sciences.* Waltham, Mass.: Chronica Botanica Company.
38. Parkinson, John. 1629. *Paradiso in Sole.* London.
39. Ray, John. 1686–1704. *Historia Generalis Plantarum.* London.
40. Grew, Nehemiah. 1682. *Anatomy of Plants* (4 volumes). London.
41. Reinikka, Merle A. 1972 (reissued 1995). *A History of the Orchid.* Portland, Oreg.: Timber Press.
42. Linnaeus, Carolus. 1737. *Genera Plantarum.* Uppsala.
43. Allan, M. E. A. 1964. *The Tradescants.* London: M. Joseph.
44. Leith-Ross, Prudence. 1984. *The John Tradescants: Gardeners to the Rose and Lily Queen.* London: Peter Owen.
45. Tradescant, John. 1657. *Musaeum Tradescantianum, or a Collection of Rarities Preserved at South Lambeth near London.* London.
46. Hadfield, Miles. 1957. *British Trees: A Guide for Everyman.* London: J. M. Dent and Sons.
47. Hadfield, Miles. 1960. *History of British Gardening.* London: Penguin Books.
48. Wheler, Sir George. 1682. *Journey into Greece.* London.
49. Banks, R. E. R., B. Elliott, J. G. Hawkes, D. King-Hele, and G. L. Lucas, eds. 1994. *Sir Joseph Banks: A Global Perspective.* London: Royal Botanical Gardens at Kew.
50. Blunt, Wilfrid. 1978. *In for a Penny: A Prospect of Kew Gardens.* London: Hamish Hamilton.
51. Hooker, Joseph Dalton. 1847. *Flora Antarctica.* London: Reeves Brothers.
52. Hooker, Joseph Dalton. 1852–1855. *Flora Novae-Zelandiae.* London: Lovell Reeve.
53. Hooker, Joseph Dalton. 1855–1859. *Flora Tasmaniae.* London: Lovell Reeve.
54. Catesby, Mark. 1730. *The Natural History of Carolina, Florida and the Bahama Islands.* London.
55. Catesby, Mark. 1767. *Hortus Europae-Americanus.* London: J. Millan.
56. Miller, Philip. 1724. *The Gardeners and Florists Dictionary.* London.
57. Linnaeus, Carolus. 1753. *Species Plantarum.* Uppsala.
58. Berkeley, Edmund, and Dorothy Smith Berkeley. 1963. *John Clayton: Pioneer of American Botany.* Chapel Hill: University of North Carolina Press.
59. Gronovius, Johan Frederik. 1739–1743. *Flora Virginica.* Leiden.
60. Darlington, William. 1849 (reissued 1967). *Memorials of John Bartram and Humphrey Marshall.* Philadelphia: Lindsay and Blakiston.
61. Bartram, William. 1791. *Travels through North and South Carolina.* Philadelphia.
62. Spongberg, Stephen A. 1990. *A Reunion of Trees: The Discovery of Exotic Plants and Their Introduction into North American and European Landscapes.* Cambridge, Mass.: Harvard University Press.
63. Lincoln, Almira. 1836. *Familiar Lectures in Botany, Practical, Elementary and Physiological*, 5th ed. Hartford, Conn.: F. J. Huntington.
64. Baron, Robert C., ed. 1987. *The Garden and Farm Books of Thomas Jefferson.* Golden, Colo.: Fulcrum.
65. Fletcher, Harold R. 1969. *The Story of the Royal Horticultural Society, 1804–1968.* London: Oxford University Press.
66. Redouté, Pierre-Joseph. 1817–1824. *Les Roses.* Paris: Didot.
67. Redouté, Pierre-Joseph. 1802–1816. *Les Liliacées.* Paris: Didot.
68. Minter, Sue. 2000. *The Apothecaries' Garden.* London: Sutton Publications.
69. William Grant, personal communication.
70. Adams, Denise Wiles. 2004. *Restoring American Gardens: An Encyclopedia of Heirloom Ornamental Plants 1640–1940.* Portland, Oreg.: Timber Press.

71. Bailey, Liberty Hyde. 1914–1917. *The Standard Cyclopedia of Horticulture.* New York: MacMillan.
72. Marshall, Humphrey. 1788 (reissued 1967). *Arbustum Americanum.* New York: Hafner Publishing.
73. M'Mahon, Bernard. 1806. *The American Gardener's Calendar.* Philadelphia.
74. Eaton, Amos. 1817. *A Manual of Botany for the Northern and Middle States of America.* New York: Horticultural Society of New York.
75. Gray, Asa. 1848. *A Manual of the Botany of the Northern United States.* New York: James Monroe and Company.
76. Torrey, John, and Asa Gray. 1838–1843. *Flora of North America.* New York: Wiley and Putnam.
77. Vick, James. 1875. *Vick's Floral Guide* (first installment). Rochester, New York.
78. Siebold, Philipp von, and Joseph Gerhard Zuccarini. 1835. *Flora Japonica*, vol. 1. Leiden, published by the authors.
79. Dallman, Peter R. 1998. *Plant Life in the World's Mediterranean Climates: California, Chile, South Africa, Australia, and the Mediterranean Basin.* Published for the California Native Plant Society. Berkeley: University of California Press.
80. Darwin, Charles. 1856. Letter to Asa Gray. Darwin Correspondence Project, letter 1863, sent 2 May 1856. Department of Manuscripts, University Library, Cambridge University.
81. Gray, Asa. 1859. Diagnostic Characters of New Species of Phaenogamous Plants, Collected in Japan by Charles Wright, Botanist of the U.S. North Pacific Exploring Expedition, with Observations upon the Relations of the Japanese Flora to that of North America and other Parts of the Northern Temperate Zone. *Memoirs of the American Academy of Arts and Sciences* II(6): 377–452.
82. McKelvey, Susan Delano. 1955 (reissued 1991). *Botanical Exploration of the Trans-Mississippi West, 1790–1850.* Corvallis, Oreg.: Oregon State University Press.
83. Justice, Clive L. 2000. *Mr. Menzies' Garden Legacy: Plant Collecting on the Northwest Coast.* Vancouver: Cavendish Books.
84. Hooker, Sir William Jackson. 1829–1840. *Flora Boreali-Americana or the Botany of Northern Parts of British America.* London: Treuttel and Würtz.
85. Morwood, William. 1973. *Traveler in a Vanished Landscape: The Life and Times of David Douglas.* New York: Clarkson Potter Inc.
86. Short, Philip. 2004. *In Pursuit of Plants.* Portland, Oreg.: Timber Press.
87. Graustein, Jeannette E. 1967. *Thomas Nuttall, Naturalist: Explorations in America 1808–1841.* Cambridge, Mass.: Harvard University Press.
88. Nuttall, Thomas. 1818. *The Genera of North American Plants and a Catalogue of the Species to 1817.* Philadelphia.
89. Pursh, Frederick. 1814. *Flora Americae Septentrionalis.* London: White, Cochrane, and Co.
90. Dana, Richard Henry. 1840 (reissued 1909). *Two Years before the Mast.* New York: Macmillan.
91. Philbrick, Nathaniel. 2004. *Sea of Glory: America's Voyage of Discovery, the U.S. Exploring Expedition 1838–1841.* New York: Viking Penguin.
92. Shephard, Sue. 2003. *Seeds of Fortune: A Gardening Dynasty.* London: Bloomsbury Publishing.
93. Christopher, Thomas. 1989. *In Search of Lost Roses.* New York: Summit.
94. Keays, Ethelyn. 1935 (reissued 1978). *Old Roses.* New York: Specialty Books.
95. Moxham, Roy. 2001. *The Great Hedge of India.* New York: Carroll and Graf.
96. Blanchard, Peter. 1991. *Markham in Peru.* Austin: University of Texas Press.
97. MacKenzie, John M., ed. 1990. *Imperialism and the Natural World.* Manchester, U.K.: Manchester University Press.
98. Hooker, Joseph Dalton, and Thomas Thomson. 1855. *Introduction to the Flora Indica.* London: W. Pamplin.
99. Hooker, Joseph Dalton. 1872–1897. *Flora of British India.* London: Reeves.
100. McCracken, Donal P. 1997. *Gardens of Empire: Botanical Institutions of the Victorian British Empire.* London: Leicester University Press.
101. Kamat, K. L. 2004. *Tribute to Sir William Jones.* Available at: http:///www.kamat.com.
102. Roxburgh, William. 1832. *Flora Indica.* Serampore, India.
103. Michael Sterll, Botanical Museum, Copenhagen, personal communication.

104. De Candolle, Roger, and Alan Radcliffe-Smith. 1981. Nathaniel Wallich, MD, PhD, FRS, FRGS (1786–1854) and the Herbarium of the Honourable East India Company, and Their Relation to the de Candolles of Geneva and the Great Prodromus. *British Journal of the Linnean Society* 83: 248–325.
105. Wallich, Nathaniel. 1824–1826. *Tentamen Flora Nepalensis Illustratae.* London.
106. Wallich, Nathaniel. 1830–1832. *Plantae Asiaticae Rariores.* London: Treuttel and Würtz.
107. Desmond, Ray. 1992. *European Discovery of the Indian Flora.* Oxford: Oxford University Press.
108. Voigt, Johann Otto. 1845 (reissued 1984). *Hortus Suburbanus Calcuttensis or a Catalogue of the Plants. . . Lindley's Natural System of Botany.* New Delhi: International Book Distributors.
109. Noltie, H. J. 2002. *The Dapuri Drawings: Alexander Gibson and the Bombay Botanic Gardens.* Wappingers Falls, N.Y.: Antique Collectors' Club.
110. Manandhar, Narayan P. 2002. *Plants and People of Nepal.* Portland, Oreg.: Timber Press.
111. Stearn, W. T., ed. 1999. *John Lindley 1799–1865: Gardener-Botanist and Pioneer Orchidologist. Bi-centenary Celebration Volume.* Woodbridge, U.K.: Antique Collectors' Club for the Royal Horticultural Society.
112. Meyer, Karl E., and Shareen Blair Brysac. 1999. *Tournament of Shadows: The Great Game and the Race for Empire in Central Asia.* Washington, D.C.: Counterpoint.
113. Alder, Garry. 1985. *Beyond Bokhara.* London: Century Publishing.
114. Gorer, Richard. 1975. *The Flower Garden in England.* London: B. T. Batsford Ltd.
115. Fletcher, Harold. 1975. *A Quest of Flowers.* Edinburgh: Edinburgh University Press.
116. John Roy, personal communication.
117. Bretschneider, Emil V. 1898 (reissued 1981). *History of European Botanical Discoveries in China.* Leipzig, DDR: Leipzig-Antiquariat.
118. Stafleu, Frans A. 1971. *Linnaeus and the Linnaeans: The Spreading of their Ideas in Systematic Botany 1735–1789.* Utrecht, Netherlands: A. Oostehoek's Uitgeversmaatschappij NV for the International Association for Plant Taxonomy.
119. Petiver, James. 1702–1709. *Gazophylacium Naturae et Artis.* London.
120. Aiton, William Townsend. 1810–1813. *Hortus Kewensis,* 2nd ed. London.
121. Le Rougetel, Hazel. 1982. The Fa Tee Nurseries of South China. *Garden History* X(1) (Spring): 70–73.
122. Fortune, Robert. 1847 (reissued 1987). *Three Years' Wanderings in China.* London: Mildmay Books.
123. Sargent, Charles Sprague. 1913 (reissued 1988). *Plantae Wilsonianae.* Portland, Oreg.: Dioscorides Press.
124. Wilson, Ernest H. 1927. *Plant Hunting.* Boston: The Stratford Company.
125. Wilson, Ernest H. 1929. *China: Mother of Gardens.* Boston: The Stratford Company.
126. Cowan, J. McQueen, ed. 1952. *George Forrest: Journeys and Plant Introductions.* London: Oxford University Press for the Royal Horticultural Society.
127. Forrest, George, Jr. 1973. George Forrest "the Man" by His Eldest Son. *Journal of the Scottish Rock Garden Club* XIII(3): 169–175.
128. Sitwell, Sir Osbert. 1946. *Left Hand, Right Hand: An Autobiography.* London: Macmillan.
129. Farrer, Reginald. 1926. *On the Eaves of the World.* London: E. Arnold.
130. Cox, Kenneth, ed. 2001. *Frank Kingdon Ward's Riddle of the Tsangpo Gorges.* Woodbridge, Suffolk: Antique Collectors' Club.
131. Ward, F. K. 1913. *The Land of the Blue Poppy: Travels of a Naturalist in Eastern Tibet.* Cambridge: Cambridge University Press.
132. Aiton, William. 1789. *Hortus Kewensis,* 1st ed. London.
133. Gorer, Richard. 1970. *The Development of Garden Flowers.* London: Eyre and Spottiswoode.
134. Stearn, William T. 1999. Engelbert Kaempfer 1651–1716: Pioneer Investigator of Japanese Plants. *Curtis's Botanical Magazine* 16(2): 103-115.
135. Thunberg, Carl. 1774. *Flora Japonica.* Leipzig.
136. Michel, Wolfgang. 1993. Engelbert Kaempfer und die Medizin in Japan. Pp. 248–293 in Detlef Haberland, ed., Engelbert Kaempfer: Werk und Wirkung. Stuttgart: Boethius-Verlag. Available at:

http://www.flc.kyushu-u.ac.jp/~michel/publ/books/04/04outline.html, accessed 4 August 2008.
137. Pim, Sheila. 1966. *The Wood and the Trees: A Biography of Augustine Henry.* London: Macdonald.
138. Serena Marner, Curator, Fielding Druce Herbarium, Oxford University Department of Plant Sciences, personal communication.
139. George, Alex S. 1999. *William Dampier in New Holland: Australia's First Natural Historian.* Hawthorn, Victoria: Bloomings' Books.
140. Erickson, Rica. 1969. *The Drummonds of Hawthornden.* Osborne Park, Australia: Lamb Paterson Pty. Ltd.
141. Lines, William J. 1994. *An All Consuming Passion: Origins, Modernity, and the Australian Life of Georgiana Molloy.* Los Angeles: University of California Press.
142. Lee, Ida. 1925. *Early Explorers in Australia.* London: Methuen.
143. Curry, S., B. R. Maslin, and J. A. Maslin. 2002. *Allan Cunningham: Australian Collecting Localities.* Canberra, Australia: Australian Biological Resources Study.
144. Bentham, George, and Sir Ferdinand von Mueller. 1863–1878. *Flora Australiensis.* London: Lovell Reeve.
145. Bentham, George. 1861. *Flora Hongkongensis.* London: Lovell Reeve.
146. Aiken, Richard, and Michael Looker. 2002. *The Oxford Companion to Australian Gardens.* Oxford: Oxford University Press.
147. Mabberley, David. 2002. Bidwill of the Bunya-Bunya. *Curtis's Botanical Magazine* 19(4): 252–264.
148. Santos, Robert Leroy. 1997. *The Eucalyptus of California.* Denair, Calif.: Alley-Cass Publications.
149. Glenn, Rewa. 1950. *Botanical Explorers of New Zealand.* Wellington, New Zealand: A. H. and A. W. Reed.
150. Le Lievre, Audrey. 1990. William Colenso, New Zealand Botanist: Something of His Life and Work. *Kew Magazine* 7: 186–200.
151. Dunlop, Eric. 2002. *The Story of the Dunedin Botanic Garden—New Zealand's First.* Dunedin, New Zealand: Friends of the Dunedin Botanic Garden.
152. Saltmarsh, Anna C. 2003. Francis Masson: Collecting Plants for King and Country. *Curtis's Botanical Magazine* 20(4): 225–244.
153. Bentham, George. 1839–1857. *Plantae Hartwegianae.* London: George Pamplin.
154. Matschat, Cecile Hulse. 1935. *Mexican Plants for American Gardens.* New York: Houghton Mifflin Co.
155. de Sahagún, Fray Bernadino. 1830. *History of Mexico.* Madrid.
156. Seler, Eduard, ed. 1927. *Fray Bernadino de Sahagún's History of Mexico.* Stuttgart: Strecker and Schröder.
157. Hernández, Francisco. 1580 (reissued 1923). *Historia de las Plantas de Nueva España.* Mexico City: Imprenta Universitaria.
158. Varey, Simon, ed. 2000. *The Mexican Treasury: The Writings of Dr. Francisco Hernández.* Stanford, Calif.: Stanford University Press.
159. Cutter, Donald C. 1960. *Malaspina in California.* San Francisco: John Howell.
160. Sessé, Martin, and Jose Mariano Mociño. 1891–1896. *Flora Mexicana* (published as appendixes to *La Naturaleza*). Mexico City: Sociedad de Historia Natural.
161. Von Hagen, Victor Wolfgang. 1945. *South America Called Them.* New York: Alfred A. Knopf.
162. Wallace, Alfred Russel, ed. 1908 (reissued 1970). *Notes of a Botanist on the Amazon and Andes.* New York: Johnson Reprint Company.
163. Hyams, Edward. 1969. *Great Botanical Gardens of the World.* London: MacMillan.
164. Humboldt, Alexander von. 1808. *Plantae Equinoctiales.* Paris.
165. Bentham, George. 1862–1883. *Genera Plantarum.* London.
166. Davis, Wade. 1996. *One River.* New York: Simon and Schuster.
167. Mee, Margaret. 1988. *In Search of Flowers of the Amazon Forests.* Woodbridge, U.K.: Nonesuch Editions.
168. The Spruce Project. Joint Project of the Natural History Museum (British Museum), London and the Royal Botanic Gardens, Kew.

169. Bateman, James. 1843. *The Orchidaceae of Mexico and Guatemala.* London: J. Ridgway and Sons.
170. Bateman, James. 1874. *A Monograph of Odontoglossum.* London: Lovell Reeve and Co.
171. McLean, Brenda. 1997. *A Pioneering Plantsman: A. K. Bulley and the Great Plant Hunters.* London: Her Majesty's Stationery Office.
172. Cribb, Philip, and Michael Tibbs. 2004. *A Very Victorian Passion: The Orchid Paintings of Thomas Day 1863–1888.* London: Blacker Publishing and the Royal Botanic Gardens at Kew in association with Thames and Hudson.
173. Taylor, Judith M., and Harry M. Butterfield. 2003. *Tangible Memories: Californians and Their Gardens 1800–1950.* Philadelphia: Xlibris.
174. Darwin, Charles. 1862. *The Various Contrivances by which British and Foreign Orchids Are Fertilized by Insects.* London: John Murray.
175. Fairchild, David. 1938. *The World Was My Garden.* New York: Scribner's.
176. U.S. Department of Agriculture, Bureau of Plant Industry. Office of Foreign Seed and Plant Introduction Bulletins. 1901–1953. Washington, D.C.
177. Cunningham, Isabel Shipley. 1984. *Frank N. Meyer: Plant Hunter in Asia.* Ames: Iowa State University Press.
178. Cox, E. H. M. 1944. *Plant Hunting in China: A History of Botanical Exploration in China and the Tibetan Marches.* London: Scientific Book Guild.
179. McLewin, William, and Dezhong Chen. 2006. *Peony rockii and Gansu Mudan.* Wellesley, Mass.: Wellesley-Cambridge Press.
180. Hovey, Dana, ed. 1838–1868. *Magazine of Horticulture.* Boston: Massachusetts Horticultural Association.
181. *The American Gardener.* 1922–present. Alexandria, Va.: American Horticultural Society.
182. Linnaeus, Carl. 1751. *Philosophia Botanica.* Stockholm.
183. Mabberley, D. J. 1985. *Jupiter Botanicus: Robert Brown of the British Museum.* London: Lubrecht and Cramer Ltd.
184. Montero, Marta Iris. 2001. *Roberto Burle Marx: The Lyrical Landscape.* Los Angeles: University of California Press.
185. Karsten, Mia. 1951. *The Old Company's Garden at the Cape and Its Superintendents, Involving a Historical Account of Early Cape Botany.* Cape Town, South Africa: Maskew Miller.
186. Curtis, William. 1787. *Botanical Magazine; or Flower-Garden Displayed.* London.
187. Loudon, J. C. 1822. *Encyclopaedia of Gardening.* London: Longman.
188. Loudon, J. C. 1829. *Encyclopaedia of Plants.* London: Longman.
189. Desmond, Ray. 1994. *Dictionary of British and Irish Botanists and Horticulturists.* London: Taylor and Francis.
190. David Ellis, American Horticultural Society, personal communication.
191. Jane Wallace Alling, public services librarian at the McLean Library, Pennsylvania Horticultural Society, personal communication.
192. Chadwick, A. A. 2005. The Great Cattleya Species of the Cut-Flower Era: Eight Cattleyas That Provided a Year of Spectacular Colors. *Orchids* 74(12) (December): 910–917.
193. Brickell, Christopher, and Judith D. Zuk, eds. 1996. *The American Horticultural Society A–Z Encyclopedia of Garden Plants.* New York: DK Publishing, Inc.
194. Fossum, M. Truman. 1955. *Trade in Horticultural Specialties: A Statistical Compendium 1890–1950.* Marketing Research Report April (Report 33). Washington, D.C.: U.S. Department of Agriculture, Bureau of Agricultural Economics.
195. U.S. Department of Agriculture Census of Horticultural Specialties. 2003.

BIBLIOGRAPHY

Aaronsohn, Aaron. 1910. *Agricultural and Botanical Explorations in Palestine.* Government Printing Office Bulletin. Washington, D.C.: Bureau of Plant Industry.

Adams, Denise Wiles. 2004. *Restoring American Gardens: An Encyclopedia of Heirloom Ornamental Plants 1640–1940.* Portland, Oreg.: Timber Press.

Aiken, Richard, and Michael Looker. 2002. *The Oxford Companion to Australian Gardens.* Oxford: Oxford University Press.

Aiton, William. 1789. *Hortus Kewensis.* London.

Aiton, Wiliam Townsend. 1810–1813. *Hortus Kewensis*, 2nd ed. London.

Alder, Garry. 1985. *Beyond Bokhara.* London: Century Publishing.

Allan, M. E. A. 1964. *The Tradescants.* London: M. Joseph.

Arber, Agnes. 1970. *Herbals: Their Origin and Evolution. A Chapter in the History of Botany 1470–1670.* New York: Hafner.

Bacon, Sir Francis. 1629. *Essayes*, 3rd ed. London: R. Allot.

Bailey, Liberty Hyde. 1914–1917. *The Standard Cyclopedia of Horticulture.* New York: MacMillan.

Banks, R. E. R., B. Elliott, J. G. Hawkes, D. King-Hele, and G. Lloyd Lucas, eds. 1994. *Sir Joseph Banks: A Global Perspective.* London: Royal Botanical Gardens at Kew.

Bartram, William. 1791. *Travels through North and South Carolina.* Philadelphia.

Bateman, James. 1874. *A Monograph of Odontoglossum.* London: L. Reeve and Co.

———. 1843. *The Orchidaceae of Mexico and Guatemala.* London: J. Ridgway and Sons.

Bauhin, Gaspard. 1623. *Pinax Theatri Botanici.* Basel.

Bede, the Venerable. ca. 731. *Historia Ecclesiatica Gentis Anglorum* (*The Ecclesiastical History of the English People*).

Bentham, George. 1862–1883. *Genera Plantarum.* London.

Bharucha, F. R. 1983. *Plant Geography of India.* Bombay: Oxford University Press.

Biographical Dictionary of Botanists Represented in the Hunt Institute Portrait Collection, Pittsburgh, Pa. 1972. Pittsburgh: Hunt Botanical Library, Carnegie Mellon University.

Blunt, Wilfrid. 1978. *In for a Penny: A Prospect of Kew Gardens.* London: Hamish Hamilton.

Bretschneider, Emil Vassilievich. 1898 (reissued 1981). *History of European Botanical Discoveries in China.* Leipzig, DDR: Leipzig-Antiquariat.

Brickell, Christopher, and Judith D. Zuk, eds. 1997. *The American Horticultural Society A–Z Encyclopedia of Garden Plants.* New York: DK Publishing, Inc.

Calmann, Gerta. 1977. *Ehret: Flower Painter Extraordinary.* New York: Graphic Society.

Cambridge University. Darwin Correspondence Project, Department of Manuscripts, University Library.

Camp, Wendell H., Victor R. Boswell, and John R. Magness. 1975. *The World in Your Garden.* Washington, D.C.: National Geographic Society.

Campbell-Culver, Maggie. 2001. *The Origin of Plants.* London: Headline.

Cantor, Norman F. 1969. *Medieval History: The Life and Death of a Civilization.* New York: Macmillan.

Champlain, Samuel de. 2004. *Voyages of Samuel de Champlain.* Whitefish, Mont.: Kessinger Publishing.

Christopher, Thomas. 1989. *In Search of Lost Roses.* New York: Summit.

Clusius, Carolus. 1576. *Rariorum Aliquot Stirpium per Hispanias Observatorum Historia* (*History of Spanish Plants*). Antwerp: Christopher Plantin.

———. 1601. *Rariorum Plantarum Historia.* Antwerp: Christopher Plantin.

Coats, Alice M. 1964. *Garden Shrubs and Their Histories.* New York: E. P. Dutton.

———. 1969. *The Plant Hunters.* New York: McGraw-Hill.

Cox, Euan H. M. 1945. *Plant Hunting in China.* London: Scientific Book Guild.

Cox, Kenneth, ed. 2001. *Frank Kingdon Ward's Riddle of the Tsangpo Gorges.* Woodbridge, Suffolk: Antique Collectors' Club.

Cribb, Philip, and Michael Tibbs. 2004. *A Very Victorian Passion: The Orchid Paintings of Thomas Day 1863–1888.* London: Blacker Publishing and the Royal Botanic Gardens at Kew in association with Thames and Hudson.

Crotz, Keith, compiler. 1989. *Ewaniana: The Writings of Joe and Nesta Ewan.* Chilcothe, Ill.: The American Botanist, American Horticultural Series No. 2.

Cunningham, Isabel Shipley. 1984. *Frank N. Meyer: Plant Hunter in Asia.* Ames: Iowa State University Press.

Curry, S., B. R. Maslin, and J. A. Maslin. 2002. *Allan Cunningham: Australian Collecting Localities.* Canberra, Australia: Australian Biological Resources Study.

Dallman, Peter R. 1998. *Plant Life in the World's Mediterranean Climates: California, Chile, South Africa, Australia, and the Mediterranean Basin.* Published for the California Native Plant Society. Berkeley: University of California Press.

da Orta, Garcia. 1563. *Colóquios dos simples e drogas he cousas medicinais da India.* Lisbon.

Davis, Elisabeth B., and Diane Schmidt. 1996. *Guide to Information Sources in the Botanical Sciences.* Englewood, Colo.: Libraries Unlimited, Inc.

Davis, Wade. 1996. *One River.* New York: Simon and Schuster.

Desmond, Ray. 1992. *European Discovery of the Indian Flora.* Oxford: Oxford University Press.

———. 1994. *Dictionary of British and Irish Botanists and Horticulturists.* London: Taylor and Francis.

———. 1999. *Sir Joseph Dalton Hooker, Traveller and Plant Collector.* Woodbridge, U.K.: Antique Collectors' Club and Royal Botanic Gardens, Kew.

Dioscorides, Pedanius. 1655. *De Materia Medica (The Greek Herbal of Dioscorides).* John Goodyer, trans., and Robert T. Gunther, ed. (1933, reissued 1985). London: Oxford University Press.

Edwards, Mike. 1997. Our Man in China. *National Geographic Magazine* (January).

Erickson, Rica. 1969. *The Drummonds of Hawthornden.* Osborne Park, Australia: Lamb Paterson Pty. Ltd.

Ewan, Joseph. 1969. *A Short History of Botany in the United States.* New York: Hafner Publishing Company.

———. 1973. *William Lobb, Plant Hunter for Veitch and Messenger of the Big Tree.* Pamphlet. Berkeley: University of California Publications in Botany.

Fairchild, David. 1938. *The World Was My Garden.* New York: Scribner's.

———. 1947. *The World Grew Round My Door.* New York: Scribner's.

Filipiuk, Marion, ed. 1997. *George Bentham: Autobiography 1800–1834.* Toronto: University of Toronto Press.

Fletcher, Harold R. 1969. *The Story of the Royal Horticultural Society, 1804–1968.* London: Oxford University Press.

Fletcher, Harold R., and William H. Brown. 1970. *The Royal Botanic Garden Edinburgh 1670–1970.* Edinburgh: Her Majesty's Stationery Office.

Forrest, George, Jr. 1973. George Forrest "the Man" by His Eldest Son. *Journal of the Scottish Rock Garden Club* XIII(3): 169–175.

Fortune, Robert. 1847 (reissued 1987). *Three Years' Wanderings in China.* London: Mildmay Books.

———. 1852 (reissued 1987). *A Journey to the Tea Countries.* London: Mildmay Books.

Frodin, D. G. 1984. *Guide to Standard Floras of the World.* Cambridge: Cambridge University Press.

Gerard, John. 1597 (reissued 1636). *The Herball.* London: Adam and Richard Whittakers.

Gleason, Henry A., and Arthur Cronquist. 1964. *The Natural Geography of Plants.* New York: Columbia University Press.

Goodburn, Damian. 1999. An Image of Ancient English Woodland. *British Archaeology* 42 (March).

Goodspeed, T. Harper. 1961. *Plant Hunters in the Andes.* Berkeley: University of California Press.

Gorer, Richard. 1970. *The Development of Garden Flowers.* London: Eyre and Spottiswoode.

———. 1975. *The Flower Garden in England.* London: B. T. Batsford Ltd.

———. 1978. *The Growth of Gardens.* London: Faber and Faber.

Graustein, Jeannette E. 1967. *Thomas Nuttall, Naturalist: Explorations in America 1808–1841.* Cambridge, Mass.: Harvard University Press.

Gray, Asa. 1876. *Darwinia.* New York: D. Appleton.

———. 1848. *Manual of Botany.* New York: James Monroe and Company.

Greene, Edward Lee, and Frank N. Egerton, eds. 1983. *Landmarks of Botanical History.* Stanford, Calif.: Stanford University Press.

Grew, Nehemiah. 1682. *Anatomy of Plants* (4 volumes). London.

Hadfield, Miles. 1957. *British Trees: A Guide for Everyman.* London: J. M. Dent and Sons.

———. 1960. *History of British Gardening.* London: Penguin Books.

Handel-Mazzetti, Heinrich. 1927 (reissued 1996). *A Botanical Pioneer in South West China.* Vienna: Osterreichischer Bundesverlag.

Harshberger, John William. 1911 (reissued 1958). *Phytogeographic Survey of North America.* Monticello, N.Y.: Lubrecht and Cramer.

Harvey, John Hooper. 1988. *The Availability of Hardy Plants of the Late Eighteenth Century.* London: Garden History Society.

———. 1972. *Early Gardening Catalogues.* London: Phillimore.

———. 1974. *Early Nurserymen.* London: Philimore.

———. 1981. *Early Horticultural Catalogues.* Bath: University of Bath Library.

———. 1983. *Hardy Plants Introduced to Britain by 1799.* Cranborne, U.K.: Cranborne Garden Centre.

———. 1989. *Restoring Period Gardens.* Oxford: Shire Publications.

———. 1976. Turkey as a Source of Garden Plants. *Garden History* IV(3): 21–24

———, ed. 1983. *The Georgian Garden (Galpine's of Blandford 1782 Catalogue: Facsimile Reprint).* Wimborne, U.K.: Dovecote Press.

Hedrick, U. P., ed. 1919 (reissued 1972). *Sturtevant's Edible Plants of the World.* New York: Dover Publications.

Hemsley, W. Botting, and Francis Blackwell Forbes. 1911. *Index Florae Sinnensis* [sic]. London: Longmans Green for the Linnean Society.

Henrey, Blanche. 1975. *A History and Bibliography of British Botanical and Horticultural Books up to 1800.* London: Oxford University Press.

———. 1986. *No Ordinary Gardener: John Knowlton 1691–1781.* London: British Museum.

Heriz-Smith, Shirley. 1988. The Veitch Nurseries of Killerton and Exeter c. 1780–1863, part 1, *Garden History* 16(1): 41–58.

———. 1988. The Veitch Nurseries of Killerton and Exeter c. 1780–1863, part 2, *Garden History* 16(2): 174–189.

———. 1989. James Veitch and Sons, Chelsea 1853–1870, *Garden History* 17(2): 135–153.

———. 1992. James Veitch and Sons, Chelsea: Harry Veitch's Reign 1870–1890, *Garden History* 20(1): 57–71.

Hernández, Francisco. 1580 (reissued 1942). *Historia de las Plantas de Nueva España.* Mexico City: Imprenta Universitaria.

Hirtz, François. 1993. *Victor Lemoine, un grand nom de l'horticulture mondiale.* Thesis, Faculte de Sciences Pharmaceutiques et Biologiques, University of Nancy, France.

Hobhouse, Penelope. 1992. *Gardening through the Ages.* New York: Simon and Schuster.

Hook, Brian, and Dennis Twitchet, eds. 1991. *The Cambridge Encyclopaedia of China.* Cambridge: Cambridge University Press.

Hooker, Sir William Jackson. 1829–1840 (reissued 1960). *Flora Boreali-Americana.* London.

Hyams, Edward. 1969. *Great Botanical Gardens of the World.* London: Macmillan.

Ingalhalikar, Shrikant. 2001. *Flowers of Sahyadri: Field Guide to 500 Flowers of North Western-Ghats of India.* Pune, India: Self-published.

Jermyn, Jim. 2001. *The Himalayan Garden.* Portland, Oreg.: Timber Press.

Justice, Clive L. 2000. *Mr. Menzies' Garden Legacy: Plant Hunting on the Northwest Coast.* Vancouver: Cavendish Books.

Karsten, Mia. 1951. *The Old Company's Garden at the Cape and Its Superintendents, Involving a Historical Account of Early Cape Botany.* Cape Town, South Africa: Maskew Miller.

Keays, Ethelyn. 1935 (reissued 1978). *Old Roses.* New York: Specialty Books.

Knobloch, Irving W. 1983. *A Preliminary Verified List of Plant Collectors in Mexico.* Plainfield, N.J.: Moldenke.

Kruckeberg, Arthur. 2002. *Geology and Plant Life: The Effects of Landforms and Rock Type on Plants.* Seattle: University of Washington Press.

Lanjouw, J., and F. A. Stafleu. 1957. *Index Herbariorum II: Index of Collectors.* Utrecht, The Netherlands.

Lauener, Lucien A., and David K. Ferguson, eds. 1996. *The Introduction of Chinese Plants into Europe.* Amsterdam: SPB Academic Publishing.

Lawrence, George H. M. 1970. *Plant Explorers and Their Introductions.* Pamphlet. Pittsburgh: Carnegie Mellon University.

Lemmon, Kenneth. 1968. *The Golden Age of Plant Hunters.* London: Phoenix House.

Le Rougetel, Hazel. 1982. The Fa Tee nurseries of South China. *Garden History* X(1) (Spring): 70–73.

———. 1988. *A Heritage of Roses.* London: Unwin Hyman.

Lighton, Conrad. 1960. *Cape Floral Kingdom.* Cape Town: Juta and Co.

Lindsay, Ann. 2005. *Seeds of Blood and Beauty: Scottish Plant Explorers.* Edinburgh: Birlinn.

Lines, William J. 1994. *An All Consuming Passion: Origins, Modernity, and the Australian Life of Georgiana Molloy.* Los Angeles: University of California Press.

Linnaeus, Carolus. 1753. *Species Plantarum.* Uppsala.

Mabberley, David. 2002. Bidwill of the Bunya-Bunya. *Curtis's Botanical Magazine* 19(4): 252–264.

MacKenzie, John M., ed. 1990. *Imperialism and the Natural World.* Manchester, U.K.: Manchester University Press.

Madeira: A Garden. Funchal, Madeira: Francisco Ribeiro and Filhos LDA.

Mallet, Corinne. 1992. *Hydrangea: Species and Cultivars.* Self-published.

Manandhar, Narayan P. 2002. *Plants and People of Nepal.* Portland, Oreg.: Timber Press.

Mann, Charles C. 2005. *1491: New Revelations of the Americas before Columbus.* New York: Alfred Knopf.

Marshall, Humphrey. 1788 (reissued 1967). *Arbustum Americanum.* New York: Hafner Publishing.

Martius, Carl Philipp Friedrich von. 1868. *Flora Brasiliensis.*

Matschat, Cecile Hulse. 1935. *Mexican Plants for American Gardens.* Boston: Houghton Mifflin Co.

McCracken, Donal P. 1997. *Gardens of Empire: Botanical Institutions of the Victorian British Empire.* London: Leicester University Press.

McKelvey, Susan Delano. 1955. *Botanical Exploration of the Trans-Mississippi West, 1790–1850.* Jamaica Plain, Mass.: Arnold Arboretum of Harvard University.

McLean, Brenda. 1997. *A Pioneering Plantsman: A. K. Bulley and the Great Plant Hunters.* London: Her Majesty's Stationery Office.

McLewin, William, and Dezhong Chen. 2006. *Peony rockii and Gansu Mudan.* Wellesley, Mass.: Wellesley-Cambridge Press.

McVaugh, Rogers. 1956. *Edward Palmer: Plant Explorer of the American West.* Norman, Okla.: University of Oklahoma Press.

Mee, Margaret. 1988. *In Search of Flowers of the Amazon Forests.* Woodbridge, U.K.: Nonesuch Editions.

Menzies, Gavin. 2002. *1421: The Year China Discovered America.* New York: HarperCollins.

Merrill, Elmer Drew, and Egbert Walker. 1938. *A Bibliography of Eastern Asiatic Botany.* Forestburgh, N.Y.: Lubrecht and Cramer.

Meyer, Frank N. 1915. *USDA Yearbook: China A Fruitful Field for Plant Exploration.* Washington, D.C.: U.S. Printing Office.

Meyer, Karl E., and Shareen Blair Brysac. 1999. *Tournament of Shadows: The Great Game and the Race for Empire in Central Asia.* Washington, D.C.: Counterpoint.

Miller, David Philip, and Peter Hanns Reill, eds. 1996. *Visions of Empire: Voyages, Botany and Representations of Nature.* Cambridge: Cambridge University Press.

Miller, Philip. 1724. *The Gardeners and Florists Dictionary.* London.

Minter, Sue. 2000. *The Apothecaries' Garden.* London: Sutton Publications.

Monardes, Nicholas. 1569. *Joyfull Newes Out of the Newe Found World.* Translated by John Frampton (1574). London.

Morison, Robert. 1669. *Praeludia Botanica.* Edinburgh.

Morris, Edwin T. 1983. *The Gardens of China.* New York: Charles Scribner's Sons.

Munz, Philip A., and David C. Keck. 1968. *A California Flora.* Berkeley: University of California Press.

Musgrave, Toby, Chris Musgrave, and Will Gardner. 1998. *The Plant Hunters.* London: Ward-Lock.

Needham, Joseph. 1954 (reissued 1988). *Science and Civilization in China.* Cambridge: Cambridge University Press.

Noltie, H. J. 2002. *The Dapuri Drawings: Alexander Gibson and the Bombay Botanic Gardens.* Wappingers Falls, N.Y.: Antique Collectors' Club.

Parkinson, John. 1629. *Paradiso in Sole.* London.

———. 1640. *Theatrum Botanicum.* London.

Patterson, Henry. 2001. In the Footsteps of Fortune. *Gourmet Magazine* (December).

Pavord, Anna. 1999. *The Tulip: The Story of a Flower that Has Made Men Mad.* London: Bloomsbury.

Philbrick, Nathaniel. 2004. *Sea of Glory: America's Voyage of Discovery, the U.S. Exploring Expedition 1838–1841.* New York: Viking Penguin.

Phillips, C. E. Lucas, and Peter N. Barber. 1967 (reissued 1979). *The Rothschild Rhododendrons: A Record of the Gardens at Exbury.* London: Cassell.

Phillips, Roger, and Martyn Rix. 1993. *The Quest for the Rose.* New York: Random House.

Pim, Sheila. 1966. *The Wood and the Trees: A Biography of Augustine Henry.* London: Macdonald.

Plimpton, Pauline Ames. 1979. *Oakes Ames: Jottings of a Harvard Botanist.* Cambridge, Mass.: Botanical Museum of Harvard University.

Rackham, Oliver. 2001. *Trees and Woodland in the British Landscape.* London: Phoenix Press.

Ray, John. 1686–1704. *Historia Generalis Plantarum.* London.

Reed, Howard. 1942. *A Short History of the Plant Sciences.* Waltham, Mass.: Chronica Botanica Co.

Reinikka, Merle A. 1972 (reissued 1995). *A History of the Orchid.* Portland, Oreg.: Timber Press.

Report on the G. Weidman Groff Collection (in typescript). University Park, Pa.: Pennsylvania State University.

Reveal, James L. 1992. *Gentle Conquest: The Botanical Discovery of North America.* Washington, D.C.: Starwood Publishing.

Robertson, Forbes W. 2000. *Early Scottish Gardeners and Their Plants.* Edinburgh: Tuckwell Press.

Rowntree, Lester. 1936. *Hardy Californians.* London: Humphrey Milford/Oxford University Press.

———. 1939. *Flowering Shrubs of California and Their Value to the Gardener.* London: Humphrey Milford/Oxford University Press.

Saltmarsh, Anna C. 2003. Francis Masson: Collecting Plants for King and Country. *Curtis's Botanical Magazine* 20(4): 225–244.

Sargent, Charles Sprague. 1905 (reissued 1960). *Manual of Trees of North America.* New York: Dover Press.

———. 1913. *Plantae Wilsonianae.* Jamaica Plain, Mass.: Arnold Arboretum of Harvard University.

Schwanitz, Franz. 1966. *The Origin of Cultivated Plants.* Cambridge, Mass.: Harvard University Press.

Seaward, M. R. D., and S. M. D. FitzGerald, eds. 1996. *Richard Spruce, 1817–1893: Botanist and Explorer.* London: The Royal Botanic Gardens, Kew.

Seler, Eduard, ed. 1927. *Fray Bernadino de Sahagún's History of Mexico.* Stuttgart: Strecker and Schröder.

Sharman, Fay. 1990. A Pioneer in Japan. *The Garden* (Journal of the Royal Horticultural Society) 115(7) (July): 350–354.

Shephard, Sue. 2003. *Seeds of Fortune: A Gardening Dynasty.* London: Bloomsbury Publishing.

Sherwood, Shirley. 1996. *Contemporary Botanical Artists: The Shirley Sherwood Collection*. London: Weidenfeld and Nicholson.

———. 2001. *A Passion for Plants*. London: Cassell.

Short, Philip. 2004. *In Pursuit of Plants*. Portland, Oreg.: Timber Press.

Sitwell, Sir Osbert. 1946. *Left Hand, Right Hand: An Autobiography*. London: Macmillan.

Slosson, Elvenia, compiler. 1951. *Pioneer American Gardening*. New York: Coward-McCann.

Smith, Sir James. 1804. *Exotic Botany*. London: R. Taylor.

Spongberg, Stephen A. 1990. *A Reunion of Trees*. Cambridge, Mass.: Harvard University Press.

Stafleu, Frans A. 1971. *Linnaeus and the Linnaeans: The Spreading of their Ideas in Systematic Botany 1735–1789*. Utrecht, Netherlands: A. Oostehoek's Uitgeversmaatschappij NV for the International Association for Plant Taxonomy.

Stearn, W. T., ed. 1999. *John Lindley 1799–1865: Gardener-Botanist and Pioneer Orchidologist. Bi-centenary Celebration Volume*. Woodbridge, U.K.: Antique Collectors' Club for the Royal Horticultural Society.

Stuckey, Ronald L., and William R. Burk, eds. 2000. *Emanuel Rudolph's Studies in the History of North American Botany*. Fort Worth: Botanical Research Institute of Texas Botanical Miscellany Series.

Sutherland, James. 1683. *Hortus Medicus Edinburgensis*. Edinburgh.

Sutton, Ann, and Myron Sutton. 1963. *Exploring with the Bartrams*. Chicago: Rand McNally.

Takhtajan, Armin. 1986. *Floristic Regions of the World*. Berkeley: University of California Press.

Taylor, Sir George. 1979. British Gardens and North American Flora. *Pacific Horticulture* 40(3): 3–11.

Thompson, Peter. 2001. *The Looking-Glass Garden: Plants and Gardens of the Southern Hemisphere*. Portland, Oreg.: Timber Press.

Tyrell, Ian. 1999. *True Gardens of the Gods: California-Australian Environmental Reform, 1860–1930*. Berkeley: University of California Press.

U.S. Department of Agriculture, Bureau of Plant Industry Bulletins. 1901–present.

Valder, Peter. 1999. *The Garden Plants of China*. Portland, Oreg.: Timber Press.

Varey, Simon, ed. 2003. *The Mexican Treasury: The Writings of Dr. Francisco Hernández*. Stanford, Calif.: Stanford University Press.

Von Hagen, Victor Wolfgang. 1945. *South America Called Them*. New York: Alfred A. Knopf.

Wallace, Alfred Russel, ed. 1908 (reissued 1970). *Notes of a Botanist on the Amazon and Andes*. New York; Johnson Reprint Company.

Wallich, Nathaniel. 1824–1826. *Tentamen Flora Nepalensis Illustratae*. London.

———. 1830–1832. *Plantae Asiaticae Rariores*. London: Treuttel and Würtz.

Ward, Francis Kingdon. 1913. *The Land of the Blue Poppy: Travels of a Naturalist in Eastern Tibet*. Cambridge: Cambridge University Press.

———. 1935. *The Romance of Gardening*. London: Jonathan Cape.

———. 1960. *Pilgrimage for Plants*. London: George G. Harrap and Co. Ltd.

Webber, Ronald. 1968. *The Early Horticulturists*. Newton Abbey, U.K.: David and Charles.

Wheler, Sir George. 1682. *Journey into Greece*. London.

Whittle, Tyler. 1970. *The Plant Hunters*. Philadelphia: Chilton Book Co.

Willson, Eleanor J. 1961. *Jamie Lee and the Vineyard Nursery, Hammersmith*. London: Hammersmith Local History Group.

Wilson, Ernest H. 1927. *Plant Hunting*. Boston: The Stratford Co.

———. 1929. *China: Mother of Gardens*. Boston: The Stratford Co.

Wu, Zhengyi, and Peter H. Raven. 1994. *Flora of China*. St. Louis: Missouri Botanical Garden; Beijing: Science Press.

Yoshida, Toshio. 2002. *Portraits of Himalayan Flowers*. Portland, Oreg.: Timber Press.

INDEX

C

D

E

F

G

S

T